Ecological Studies

Analysis and Synthesis

Edited by

W. D. Billings, Durham (USA) F. Golley, Athens (USA)

O. L. Lange, Würzburg (FRG) J. S. Olson, Oak Ridge (USA)

Volume 16

Fennoscandian Tundra Ecosystems

Part 1 Plants and Microorganisms

Edited by F. E. Wielgolaski

Editorial Board

P. Kallio and T. Rosswall

With 90 Figures and 96 Tables

Springer-Verlag Berlin Heidelberg New York 1975

The picture on the cover is a simplified version of Fig. 1 on p. 4.

ISBN 3-540-07218-7 Springer-Verlag Berlin · Heidelberg · New York
ISBN 0-387-07218-7 Springer-Verlag New York · Heidelberg · Berlin

Distributed in the British Commonwealth Market by Chapman & Hall Limited, London.

Library of Congress Cataloging in Publication Data. Main entry under title: Fennoscandian tundra ecosystems. (Ecological studies; v. 16). CONTENTS: pt. 1. Plants and microorganisms. 1. Tundra ecology – Scandinavia. 2. Tundra ecology – Finland. 1. Wielgolaski, F. E., 1931– . II. Kallio, Paavo. III. Rosswall, Thomas. IV. Series.
QH164.F45 574.5'264 75–4809

Typesetting, printing, and binding: Brühlsche Universitätsdruckerei, Gießen.

Preface

The term Fennoscandia is used to denote Finland and the Scandinavian peninsula, which comprises Norway and Sweden and sometimes also neighbouring districts of the USSR. The western part of the Scandinavian peninsula is mountainous with peaks mostly within the range 1000–2000 m above sea level. Permafrost is not usually found in the southern districts of these alpine zones, but occurs in the northeastern part of Fennoscandia in mire, even at a low elevation. If tundra is defined as areas of permafrost, only very limited regions of Fennoscandia could be included. However, in the present volumes, we have used the word tundra in a broader sense as approximating areas with mean annual air temperature below 0° C at meteorological standard height (1.5–2 m); it is often used in this sense in North America. This usage allows all alpine zones to be included, along with the subalpine birch zone and certain open subarctic woodland ("forest tundra" in Russian usage).

Floral and faunal structures in Fennoscandia have been studied for many years. However, earlier studies were more limited on such aspects as productivity and other functional relationships of ecosystems. The functional relationships within an ecosystem have to be studied on a large scale requiring integrated and interdisciplinary research. This is possible only through joint planning by scientists and must be carried out in teamwork by meteorologists, soil scientists, botanists (taxonomists and physiologists), zoologists (both invertebrate and vertebrate), microbiologists, and more applied biologists (e.g. specialists in husbandry and agronomy). Ecosystem research is one of the goals of the International Biological Program (IBP), which is being implemented in different parts of the world. When similar projects are carried out in several countries over the same period, it is possible to compare the results on a wide geographical scale. In Fennoscandia it was early realized that much would be gained by strong cooperation between the countries within IBP, because of the limited resources available in funding and in scientifically qualified personnel in the various disciplines in each of the countries involved. This cooperation, it was felt, would make it possible to achieve results which could be of interest not only to scientific specialists within IBP, but also for Fennoscandian ecology as a whole, as well as providing a regional overview for an audience outside Fennoscandia.

Very little research had been carried out on the biological productivity of tundra areas (as defined above) in Finland, Norway and Sweden. It was therefore decided when the IBP projects were planned to make these ecosystem types one of the main areas of interest. In Norway, where tundra covers more of the total area than in Finland and Sweden, all the IBP terrestrial and freshwater productivity studies were located in southern mountain districts. In Finland and Sweden, areas north of the Arctic circle, close to the existing research stations at Kevo and Abisko, were chosen for the IBP tundra investigations. As IBP projects were also being carried out in these countries in other biomes, the tundra projects were somewhat smaller than in Norway. The project in Sweden started later than in the other two countries. In Finland and Norway the IBP tundra projects were planned in 1967–1968 and were operational from 1968–1969 through 1973, while in Sweden the project has been in operation only since 1970. The main results from the Fennoscandian IBP Tundra Biome studies are now available, and most of the papers in the present volumes report this recent knowledge. The editor and the Board wish to express their thanks to the IBP National Committees in Finland, Norway and Sweden for their permission to combine the IBP results in these volumes. The IBP tundra studies in Norway were supported by grants from the Norwegian Parliaments to IBP through the Norwegian Research Council; in Sweden by grants from the Bank of Sweden Tercentenary Fund, the Ecological Research Committee of the Swedish Natural Research Council, and the Swedish Royal Academy of Sciences; in Finland by the Finnish IBP Committee, and SITRA; their support is gratefully acknowledged.

In all three Fennoscandian countries the IBP tundra projects were planned as more or less total ecosystem studies. However, partly because of lack of personnel and partly due to lack of funds several details had to be omitted. In Finland the strongest emphasis was put on ecophysiological plant studies, especially of cryptogams (lichens and bryophytes), and on the relationship between the function of plants and their environment; in Sweden most effort was directed to the important plant-decomposer cycle; in Norway several aspects were treated of both the structure and function of an alpine tundra ecosystem. Each study related stand types to the surrounding landscape complex.

From the beginning, there has been close cooperation not only between the Fennoscandian IBP Tundra Biome projects, but also with the other IBP Tundra projects throughout the world. Differences in specialization in the three countries enabled us to use the limited resources in an attempt to obtain a synthesis of common Fennoscandian tundra problems. Conversely, the fact that the central problems were studied in all three countries made it possible to compare the results for similar and different communities in the areas chosen.

Many details from the projects have already been presented in various scientific journals, and more will be published within the next few years. This information, however, will be scattered and may not present a synthesis of the projects. These volumes integrate the various studies and relate the structure and function of the Fennoscandian tundra ecosystems. The editor and the Board are grateful to the publishing house of Springer for bringing the Fennoscandian IBP tundra projects and other projects into the Ecological Studies series. The present volume, and a volume on animals and systems analysis, would have been impossible without the great interest shown by several tundra scientists in Finland, Norway and Sweden, who worked together on the chapter sections. All manuscripts have been read by Mrs. M. Espeland for English style and we thank her for this valuable help. Because the Swedish project was smaller, there are relatively fewer pages from this country as compared with Finland and particularly Norway. We hope, however, that it will be possible for the reader to fit many facts and principles into an overall impression of the structure and function of the Fennoscandian tundra ecosystem.

Readers with a more general interest in these topics are referred to the description of sites and climate in the present volume and to introductory parts of the chapters Primary Producers, Decomposers, and Nutrient Cycling. The general concepts and conclusions are explained in the more detailed sections of each chapter. Very interesting results are reported in many of these detailed articles. For the first time, really functional details are given for Fennoscandian tundra areas, for instance on photosynthesis of the ecosystems, on mineral content and distribution in various species, or on decomposition rates of different biological groups in the mountainous region. The occurrence of everlasting days in the far north makes it possible to compare functional biological aspects in this area with similar aspects in mountain areas in southern Fennoscandia with shorter days. Other interesting influences on biological functional relationships in southern and northern areas, respectively, are due to the considerable differences in pollution in the Fennoscandian region. Global pollution is very high in the south, because cyclones often reach the area from industrial and urban areas in central, eastern and western Europe. The amount of water released in southern Fennoscandia is high, partly because of the topography of the area. The biological effect of this highly polluted, acid precipitation is strong in this generally poor area with low buffer capacity, and may be compared with the less polluted region in the north. In the treeless tundra this difference is most pronounced in its effect on the productivity of cryptogams. Other practical implications of the Fennoscandian tundra studies concern the possible utilization of the tundra by man, e.g. studies on increasing primary production to permit increased sheep husbandry, which may be important for the future food supply for man in the region. It is also important to know what is the optimal population of reindeer in Fennoscandian tundra areas, and to keep the area in good shape for recreational use by man, while some areas should be set aside as reserves. This multiple use of Fennoscandian tundra ecosystems presupposes a good ecological knowledge, particularly of how ecosystems function.

Oslo, November 1974 F. E. Wielgolaski

Contents

1. Site Description

2. Abiotic Variables

3. Primary Producers

3.1. Plant Structure

3.2. Biomass and Production

3.3. Plant Processes

4. Decomposers

5. Nutrient Cycling

Contents of Part 2

Contributors Part 1

Baker, J. H.	Freshwater Biological Association, River Laboratory, East Stoke, Wareham, Dorset BH20 6BB, United Kingdom
Berg, A.	Botanical Laboratory, University of Oslo. Present address: 2930 Bagn, Norway
Berg, B.	Department of Microbiology, Agricultural College, 75007 Uppsala 7, Sweden
Clarholm, M.	Department of Microbiology, Agricultural College, 75007 Uppsala 7, Sweden
Dahl, E.	Botanical Institute, Norwegian Agricultural College, 1432 Ås–NLH, Norway
Elven, R.	Botanical Garden, University of Oslo, Tøyen, Oslo 5, Norway
Flower-Ellis, J. G. K.	Royal College of Forestry, Section in Umeå, Umeå University, 90187 Umeå, Sweden
Goksøyr, J.	Institute of General Microbiology, University of Bergen, 5000 Bergen, Norway
Granhall, U.	Department of Microbiology, Agricultural College, 75007 Uppsala 7, Sweden
Hanssen, J. F.	Institute of General Microbiology, University of Bergen. Present address: Institute of Microbiology, Norwegian Agricultural College, 1432 Ås–NLH, Norway
Hayes, A. J.	Department of Forestry and Natural Resources, University of Edinburgh, Edinburgh EH9 3 JU, United Kingdom
Heinonen, S.	Department of Botany, University of Turku, 20500 Turku 50, Finland
Hesjedal, O.	Department of Agriculture, Division of Land Registration (DLR). Present address: Telemark distriktshøgskole, 3800 Bø i Telemark, Norway
Hinneri, S.	Department of Botany, University of Turku, 20500 Turku 50, Finland
Hurme, H.	Department of Botany, University of Turku, 20500 Turku 50, Finland
Jahren, A.	Botanical Laboratory, University of Oslo, Blindern, Oslo 3, Norway
Johansson, L.-G.	Department of Plant Physiology, Umeå University, 90187 Umeå, Sweden
Kallio, P.	Department of Botany, University of Turku, 20500 Turku 50, Finland
Kallio, S.	Department of Botany, University of Turku, 20500 Turku 50, Finland

KÄRENLAMPI, L.	Department of Botany, University of Turku. Present address: Department of Ecology, University of Kuopio, Box 140, 70101 Kuopio 10, Finland
KJELVIK, S.	Botanical Laboratory, University of Oslo. Present address: Botanical Institute, Norwegian Agricultural College, 1432 Ås–NLH, Norway
LARSSON, P.	Zoological Museum, University of Oslo, Sarsgt. 1, Oslo 5, Norway
LID-TORSVIK, V.	Institute of General Microbiology, University of Bergen, 5000 Bergen, Norway
LINDER, S.	Department of Reforestation, Royal College of Forestry, 10405 Stockholm 50, Sweden
LYE, K.A.	Botanical Institute, Norwegian Agricultural College, 1432 ÅS—NLH, Norway
NYGAARD, R.T.	Department of Botany, University of Trondheim. Present address: Molde gymnas, 6400 Molde, Norway
RHEINBERG, P.	Department of Forestry and Natural Resources, University of Edinburgh, Edinburgh EH9 3 JU, United Kingdom
ROSSWALL, T.	Department of Microbiology, Agricultural College, 75007 Uppsala 7, Sweden
RYDÉN, B.E.	Department of Hydrology, University of Uppsala, Box 554, 75122 Uppsala 1, Sweden
RYVARDEN, L.	Botanical Museum, University of Oslo. Present address: Botanical Laboratory, University of Oslo, Blindern, Oslo 3, Norway
SKARTVEIT, A.	Department of Geophysics, Meteorological Division, University of Bergen, 5014 Bergen, Norway
SKRE, O.	Botanical Laboratory, University of Oslo. Present address: Botanical Institute, Norwegian Agricultural College, 1432 Ås–NLH, Norway
SONESSON, M.	Department of Ecological Botany, University of Lund, Östra Vallgatan 14, 22361 Lund, Sweden
SVENSSON, B.	Department of Microbiology, Agricultural College, 75007 Uppsala 7, Sweden
TAMMISOLA, J.	Central Statistical Office of Finland, Box 504, 00101 Helsinki 10, Finland
TANGEN, K.	Zoological Museum, University of Oslo. Present address: Department of Marine Biology and Limnology, Section Marine Botany, University of Oslo, Blindern, Oslo 3, Norway
VALANNE, N.	Department of Botany, University of Turku, 20500 Turku 50, Finland
VEUM, A.K.	Department of Pedology, Norwegian Agricultural College. Present address: 3880 Dalen i Telemark, Norway
WIELGOLASKI, F.E.	Botanical Laboratory, University of Oslo, Blindern, Oslo 3, Norway

1. Site Description

Description of Fennoscandian Tundra Ecosystems

M. SONESSON, F. E. WIELGOLASKI, and P. KALLIO

The tundra landscape of Fennoscandia is mostly alpine. The southernmost areas above the tree line, with annual mean air temperatures below 0° C, are found at about 59° N in southern Norway. A mountain range runs along the entire Scandinavian peninsula up to about 70° N and is geologically part of the Caledonian mountain range, intensely folded mostly in the Silurian period (HOLTEDAHL, 1960). The highest mountains are found in the western districts. An eastward tilt of an imaginary surface connecting the summits was caused by a stronger land upheaval during the Tertiary period in these areas than further east. Generally, therefore, higher mountains and more mountainous habitats for all life are found in Norway than in any of the other Fennoscandian countries (cf. SØMME, 1968). The drainage divide is close to the western coast and the mountain slopes are therefore steeper to the west than to the east. During parts of Pleistocene time all Fennoscandia was covered with ice except for a few refuges in extreme maritime areas, and possibly also on a few of the highest mountain peaks. Glaciers are found even today all along the length of the mountain range, and they strongly influence the ecosystems near them. There are several plateaus in the mountain areas, partly incised by water erosion and abraded by ice movements. The ice movement caused formation of several moraines, and the quaternary deposit is often transported far from its geological origin (HOLTEDAHL, 1960; LÅG, 1965; RANKAMA, 1965; SØMME, 1968). Above the uplands rise snowy, high peaks, particulary in the central region of southern Norway. In Sweden most of the mountains are found along the Swedish-Norwegian border (a belt up to 100 km wide).

Fragments of arctic tundra are only found in the northernmost part of Finland and Norway. In these regions (Finmark and Lappland) much of the landscape is rolling or even relatively flat; well outside the mountain range; most elevations are below 300 m. The bedrock is mostly composed of granitic gneiss from the Precambrian Baltic shield. These rock types predominate even in many of the mountain areas further south, but here, and in the western mountain areas, younger bedrock formations (especially Cambro-Silurian) are also common. Most abundant in these formations are more or less metamorphic shisty bedrocks (as phyllite and mica shists), but crystalline limestone, dolomite and other types are also found (HOLTEDAHL, 1960; SØMME, 1968). Both in the southwestern mountains and in western areas further north in Norway, igneous bedrocks (from the Ordovician period) are common in areas predominated by the shisty rocks. In

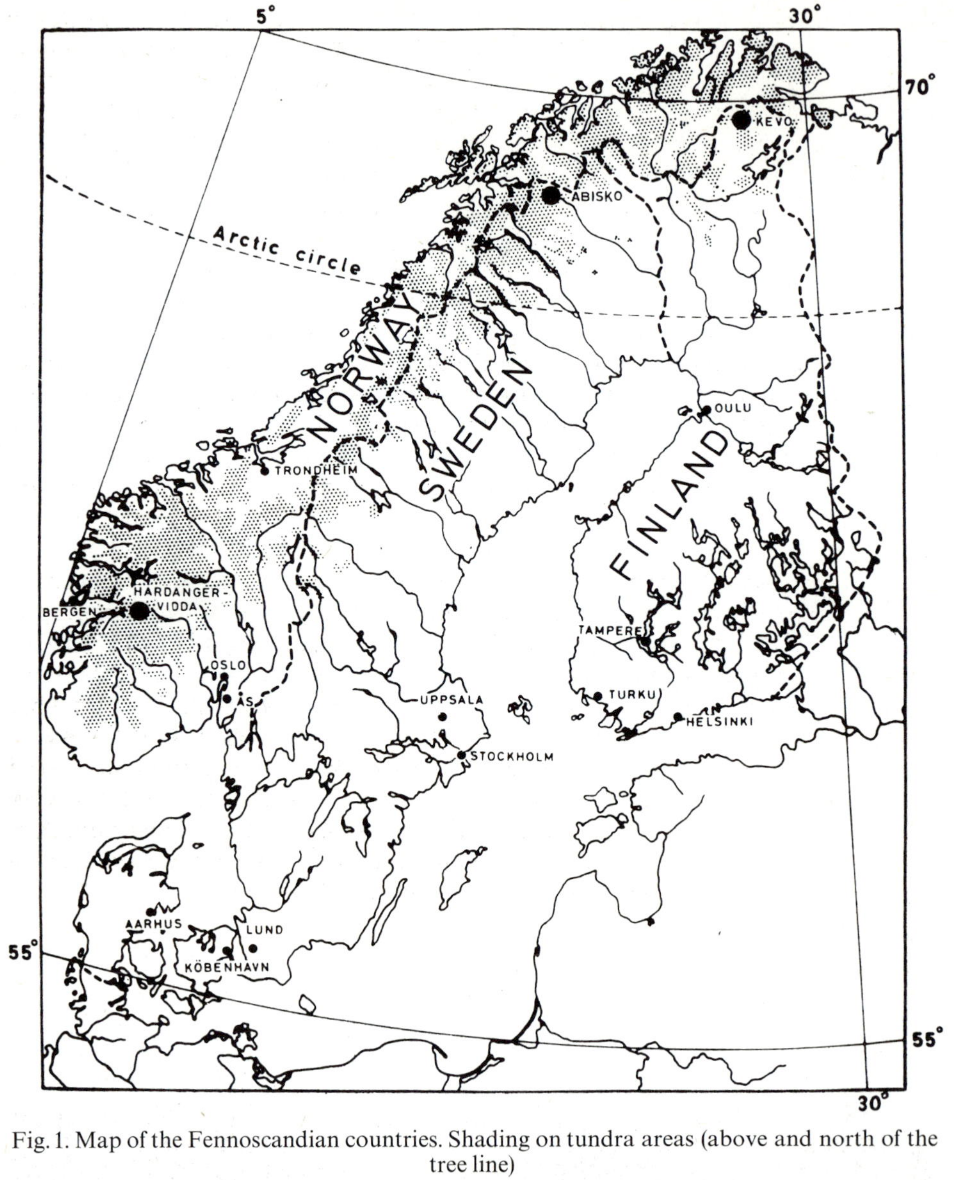

Fig. 1. Map of the Fennoscandian countries. Shading on tundra areas (above and north of the tree line)

usually small pockets, serpentine rocks are found, biologically indicated by specific plants (RUNE, 1953); other plant species as *Viscaria alpina* are found to be indicators on heavy metals in the bedrock, e.g. copper.

Man has used the Fennoscandian mountain areas for thousands of years for hunting and fishing purposes (e.g. from about 5000 B.C. at Hardangervidda, Norway). The possibilities have, however, been small for year round farming in most districts above the timber line, although it has been common to bring cattle and sheep from the lowlands up to the mountain areas for grazing during summer. In the northernmost Fennoscandian tundra, nomadic Lapps have usually taken their reindeer herds to different areas throughout the year. All these traditional types of tundra use by man are, however, decreasing at the present time, but may increase again if man's food supply situation is getting worse. The recreational use of Fennoscandian tundra has on the other hand strongly increased during the recent years, and several national parks and reserves are set aside or planned in these ecosystems in all three Fennoscandian countries.

The geological structure and the geography of Fennoscandia are very important for the climate of the region (SØMME, 1968; SKARTVEIT et al., 1975). The decrease in temperature with increasing altitude and latitude enormously influences the development of different types of tundra vegetation (DAHL, 1975).

Generally, observations from the Fennoscandian IBP tundra sites serve as examples of an array of typical and contrasting tundra types of the region, but they do not include the fragments of the arctic tundra found in Fennoscandia. When the IBP sites were chosen, several aspects were taken into account. The sites had to be homogenous to obtain comparable results from destructive sampling throughout several years. They should also be representative of larger areas and/or represent various points on gradients, e.g. on nutrients and/or soil moisture. As most tundra areas on a global scale have relatively continental climate, we decided to chose the more oceanic mountain areas in Norway for comparison (MOORE, 1975; HEAL and ROSSWALL, 1975). Similarly, an oligotrophic *Sphagnum* and dwarf shrub dominated mire in the rain shadow on the eastern slopes of the mountains of northern Sweden was chosen for comparison with results from bogs of more oceanic climate in Great Britain (HOLDING, 1974; HEAL and ROSSWALL, 1975). Our tundra mires and the maritime blanket bogs of England and Ireland can well be compared with bog mires and fens of the Boreal forest and cool temperate zones (MALMER, 1975). In all these regions, the evolutionary development of the bog as an ecosystem, can modify the hydrologic cycle, slow nutrient cycles and decay of peat, and transform the initial topography.

Fig. 2. Overview of the Hardangervidda plateau, Stigstuv tourist hut

The Fennoscandian IBP tundra sites at Hardangervidda, Norway, the Abisko area in Sweden, and Kevo in Finland (Fig. 1) are situated within a wide latitudinal and longitudinal range. Those lying furthest south and west are the Norwegian sites at approximately 60° N and 7° 30′ E, and the northernmost and easternmost are the Finnish sites at nearly 70° N and 27° E. The Swedish site (68° N and 19° E) has an intermediate location, although it is geographically closest to Kevo.

The climatic variation, especially that of temperature, depends to a large extent on the distance inland and the altitude. All the Fennoscandian IBP tundra sites are situated at approximately the same distance from the North Atlantic Ocean, viz. 50–100 km, but have very different altitudes. The Norwegian sites have altitudes of 780–1300 m, the Swedish 350 m, and the Finnish sites 90–330 m.

Hardangervidda is a mountain plateau (about 9000 km^2) above the tree line in the southern part of the Scandinavian mountain range, where gneissic bedrock of the Precambrian Baltic shield prevails. The topography is fairly level at the plateau itself (Fig. 2), with height differences of 100–400 m, although there are some high peaks (above 1800 m), especially in the northern part of the plateau (Fig. 3). The subalpine site at about 780 m at Maurset (Fig. 14) is situated on a river terrace on the western slope of Hardangervidda, where bedrock of Caledonian orogenesis also appears, and the topography is considerably more broken. The other sites are all in the low-mid alpine zones at 1200–1320 m (Fig. 3). The sites are described by WIELGOLASKI (1971, 1972), and the main characteristics are given in Table 1.

The Abisko area in the northern part of the mountain range has a highly accentuated relief, with relative heights between 300 and 2000 m (Fig. 4). The bedrock is mainly Caledonian, of varying ages and qualities. The Stordalen site is situated in the subalpine belt (Figs. 5 and 11) approximately 1 km from, and 10 m above, Lake Torneträsk—one of the large mountain lakes of Fennoscandia. It occupies a narrow level zone of the Precambrian granitic basement at the lake immediately northeast of the Caledonian overthrust nappes. The site is described by SONESSON et al. (1973), and some abiotic and biotic values are given in Table 2.

The three sites in the Kevo area are situated outside the Scandinavian mountain range on the Baltic shield. The Kevo area has a low relief with height differences usually between 50 and 200 m (Fig. 6). The bedrock is predominantly gneissic. Two of the sites, the pine, and birch forests in the subalpine belt, are situated at altitudes of 90 and 130 m respectively (Fig. 7), whereas the third, in the low alpine belt, is at 330 m. The area is more fully described by e.g. KALLIO (1964); KÄRENLAMPI and KAUHANEN (1972), and KÄRENLAMPI (1972). The main characteristics of the IBP sites are given in Table 2.

Altitude and topography contribute greatly to the variations in climatic conditions within, and between, the areas under consideration. This also applies to the different radiation conditions, but here latitude is especially important (SKARTVEIT et al., 1975). Although oceanic on a macro scale, the climate of the Fennoscandian sites shows wide variations. In comparison with the Norwegian sites, the local climate of the Swedish and Finnish sites is continental, judging both from meteorological data and from vegetational criteria (Fig. 8). The snow cover is of vital ecological significance. Of special importance are the date of the

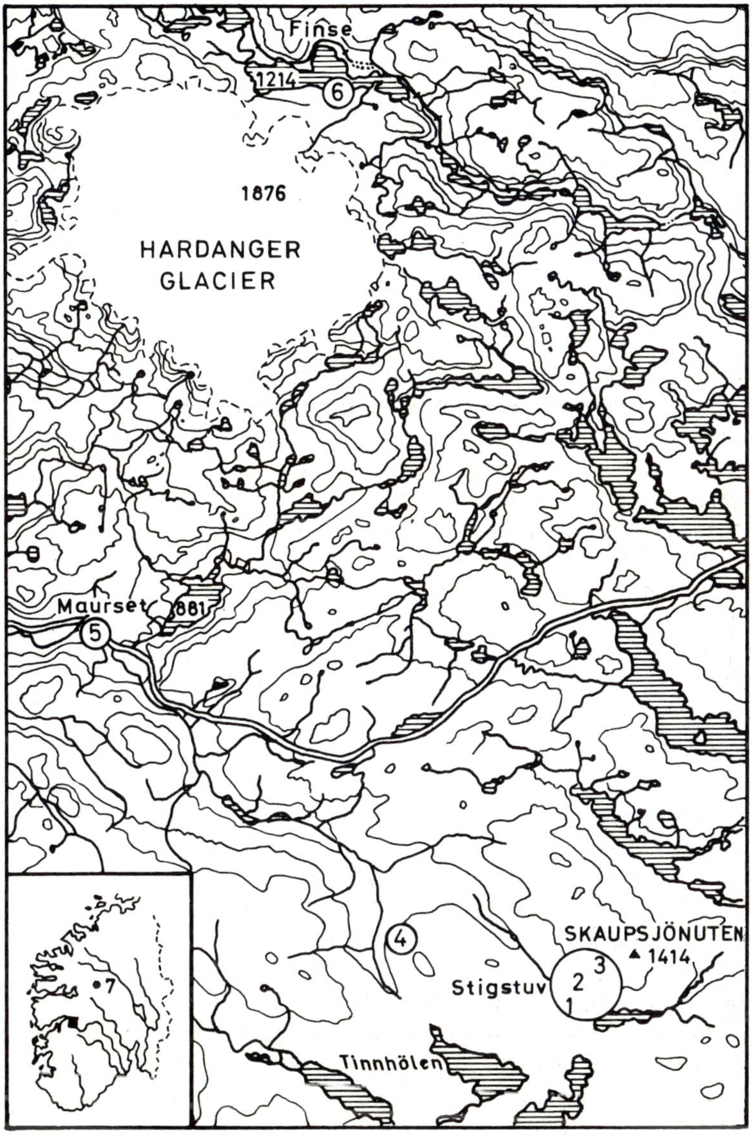

Fig. 3. Map of geographical position of six intensively studied IBP tundra sites at Hardangervidda, Norway. Contour intervals 100 m.

1. Lichen heath } Stigstuv area.
2. Dry meadow } Stigstuv area.
3. Wet meadow } Stigstuv area.
4. Willow thicket, Bjoreidalen.
5. Subalpine birch forest, Maurset.
6. Snow bed, Finse.
7. Lake Øvre Heimdalsvann, Jotunheimen. Tundra fresh-water studies (see LARSSON and TANGEN, 1975)

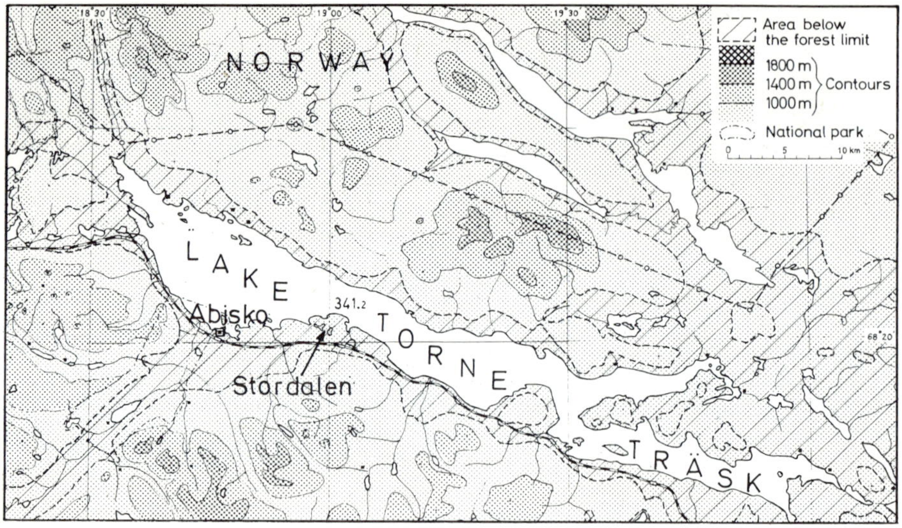

Fig. 4. Map of the central and eastern part of the Abisko area

Fig. 5. Stordalen, the IBP intensive site toward SE. Elevated, ombrotrophic areas with microrelief pattern of hummocks and depressions are shown as darker ridges running from left to right. (Photo S. Johansson, July 1973)

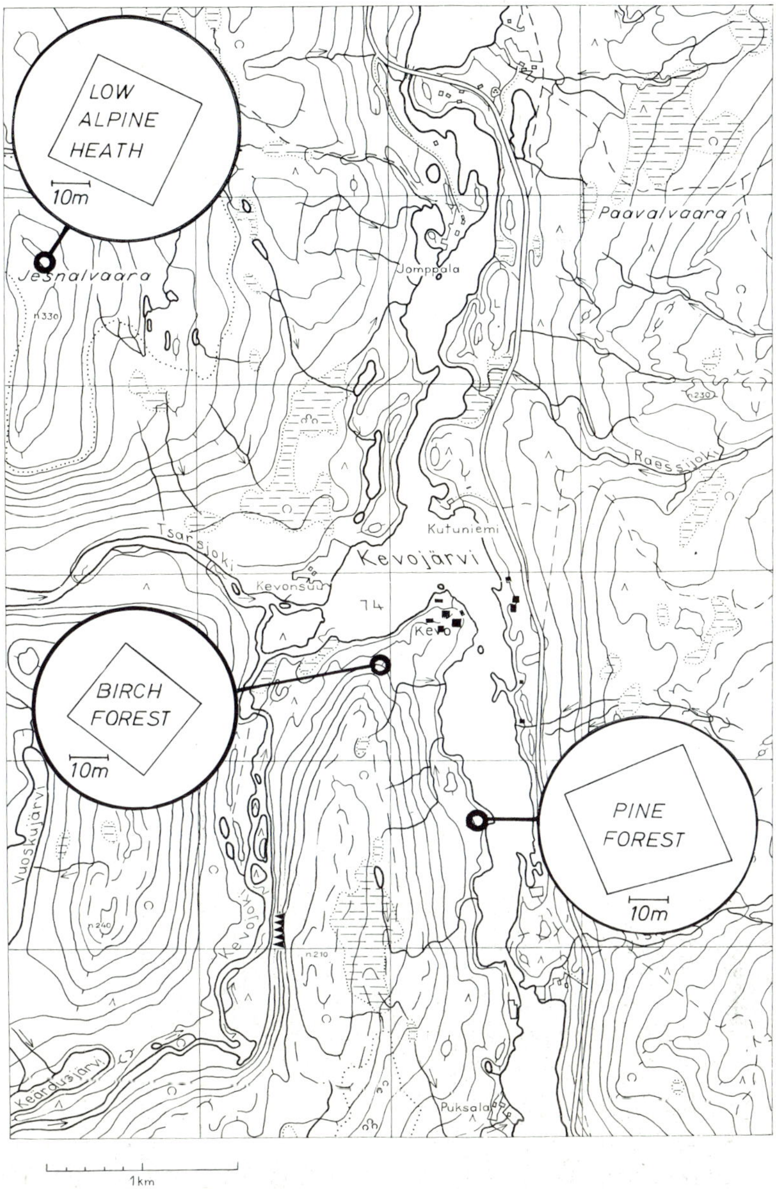

Fig. 6. Topographic map of the surroundings of Kevo and the location of the IBP-sites. Contour intervals 20 m

autumn snow falls and the depth and duration of the snow (SONESSON, 1969; SKARTVEIT et al., 1975).

The start of the growing season depends to a large extent on the duration of the snow cover (WIELGOLASKI and KÄRENLAMPI, 1975). For the Norwegian sites,

Fig. 7. The surroundings of Kevo, Finland. (A) Birch forest; (B) Low alpine heath

Fig. 8. Low alpine heath site, Kevo, Finland

the growing season accordingly varies between two months (the snow bed), to more than four months (the birch forest; potentially five months from the time of snow-free period). At the alpine lichen heath (Fig. 9), wind sweeps much of the snow away, leaving only about 50–60 cm at a maximum which can be penetrated

Fig. 9. Lichen heath site, Stigstuv, Hardangervidda, Norway

Fig. 10. *Salix herbacea* snow bed, Finse, Hardangervidda, Norway

by reindeer in winter, allowing the animals to graze the lichens. The growing season here is about three and a half months. For most of the vascular plants of the Swedish site the growing season is approximately four months, while the potential growing season for the hummock-top cryptogams is estimated at seven

Fig. 11a

Fig. 11b

months because of the sparseness or absence of snow during early spring, when there is sufficient insolation. For the Finnish sites the snow-free period is about five months on an average. The difference in duration of the period between the lower and upper sites at Kevo may be about two weeks or more. The mountain ridges are usually covered by snow of only a few centimeters depth, while a depth of 50–70 cm is normal at the Kevo Research Station (SKARTVEIT et al., 1975). As a result of the low precipitation and an annually repeated uneven distribution of snow, there is permafrost in many of the peatlands of the Abisko and Kevo areas, (but not at the Kevo IBP sites), and this is normally not found at Hardangervidda.

The lower parts of the Abisko area (SONESSON, 1974), and probably also most of the Hardangervidda ones were icefree at least at 7000 B.C., whereas the Kevo area was possibly deglaciated more than a thousand years earlier. A glacier (Hardangerjøkelen) is still situated north of Hardangervidda (Fig. 3); an adjacent site deglaciated during the last two-three hundred years constitutes the Norwegian snow bed site (Fig. 10). The Fennoscandian ecosystems are thus fairly young, and owing to glaciation their flora and fauna are impoverished (DAHL, 1975; ELVEN and RYVARDEN, 1975).

The mineral soils are mainly acid Quaternary deposits, poor in nutrients an derived from hard silicious bedrock. Such soils characterize all the Kevo sites and the lichen heath, snow bed and birch forest sites at Hardangervidda (HINNERI et al., 1975). The soils are podzols with a more or less well-developed leached A_2 horizon, depending on altitude and soil stability. The Stordalen site is a peatland (Fig. 11), probably developed on glaciofluvial sand. The plant cover at this site, as at the Norwegian lichen heath, mostly depends on nutrients from the atmosphere, as any contribution from the mineral substrate is very small due to the topography, and, at the Stordalen site, the permafrost in the peat as well. Although they occur sparsely, softer base-rich minerals also make up some of the soils, at Hardangervidda because of influence from phyllitic rocks at the elevation about 1250–1450 m. This is apparent in the wet and dry meadows (Fig. 12) at Hardangervidda from the occurrence of basicolous species and a higher primary production than at the oligotrophic sites. Especially favoured are sites such as the willow thicket (Fig. 13) at Hardangervidda with soil rich in available nutrients and kept sufficiently moist by mobile, oxygen-rich water.

Biogeographically the IBP sites are situated in the subalpine woodland belt or the alpine (non-wooded) belts of Fennoscandia. Woods of birch (*Betula pubescens* Ehrh. usually in a form called *B. tortuosa* Led.) prevail in the cool and oceanic areas of the Boreal coniferous zone (the Taiga), where they normally form the altitudinal limit towards the alpine areas (SJÖRS, 1963; DAHL, 1975). This limit runs at approximately 1050 m at Hardangervidda, at 500–800 m in the Abisko area and at about 350 m in the Kevo area. The structure of the birch forests (Fig. 14),

◁ Fig. 11 a and b. Stordalen, Sweden, IBP site. (a) Close-up of a tussocky elevated area with the dwarf shrubs *Rubus chamaemorus* and *Eriophorum vaginatum*, fruiting heads shown. In the background enclosure with the climatic station; the railway also shown (approximately 1 km distant). (b) Outside the elevated area in the foreground a dense, luxuriant *Eriophorum angustifolium* stands in a zone where a high rate of symbiotic nitrogen fixation has been measured. In the background the IBP field station. (Telephoto S. JOHANSSON, July 1973)

Fig. 12. Dry meadow site, Stigstuv, Hardangervidda, Norway with meteorological equipment for the micrometeorological studies

Fig. 13. Eutrophic willow thicket site, Bjoreidalen, Hardangervidda, Norway

Fig. 14. Birch forest site, Maurset, Hardangervidda, Norway

Fig. 15. Pine forest site, Kevo, Finland

Table 1. Site characteristics. Norwegian sites. Hardangervidda (for more details see different papers in the present volumes)

- 2.1.1 Birch forest, Maurset Lat. 60° 25′ N, Long. 7° 21′ E, subalpine belt (*Barbilophozio Pinetum* KIELLAND-LUND, 1967).
- 2.1.2 Lichen heath, Stigstuv, Lat. 60° 17′ N, Long. 7° 40′ E, low-mid alpine region (*Arcostaphyleto-Cetrarion nivalis* DAHL et al., 1971).
- 2.1.3 Wet meadow, Stigstuv, Lat. 60° 18′ N, Long. 7° 41′ E, mid alpine region (*Caricion canescentis fuscae* DAHL et al., 1971).
- 2.1.4 Dry meadow, Stigstuv, Lat. 60° 18′ N, Long. 7° 40′ E, low-mid alpine region (*Ranunculeto-Oxyrion digynae* + *Potentilleto-Polygonion vivipari* DAHL et al., 1971).
- 2.1.5 Willow thicket, Bjoreidalen, Lat. 60° 16′ N, Long. 7° 41′ E, low alpine region (*Lactucion alpine* DAHL et al., 1971).
- 2.1.6 Snow bed, Finse, Lat. 60° 35′ N, Long. 7° 31′ E, low-mid alpine region (*Nardeto-Caricion bigelowii/Cassiopeto-Salicion herbaceae* DAHL et al., 1971).

		Birch forest 2.1.1	Lichen heath 2.1.2	Wet meadow 2.1.3	Dry meadow 2.1.4	Willow thicket 2.1.5	Snow bed 2.1.6	Ref.
1. Altitude, m a.s.l.		780	1220	1320	1275	1200	1250	
2. Microtopography		Even	Even	Shallow depr.	Even	Shallow depr.	Shallow depr.	
3. Slope		Level	Level	5% SW	15% S	10% NW	15–25% N	
4. Soil		Podzol	Iron podzol	Peat	Weak iron podzol	Peat	Weak podzol	
Depth of horizon, cm/pH: pH: H_2O	A_{0-1}	4/4.0	2/4.0	35/5.3	9/5.3	n.d.	2/4.4	VEUM et al., (1975)
	A_2	13/4.2	6/4.3		6/5.3		A_2 + B 33/5.0	
	B	18/4.8	27/5.0		20/5.4			
C/N in A_{0-1}		29	23	21	16	n.d.	24	
5. Drainage and/or depth of water table		Well-drained	Well-drained	20 cm[a]	Well-drained	30 cm[a]	Well-drained	
Soil moisture max/min (1969–72) 0–10 cm depth (% wet weight)		35/14	19/5	90/77	74/30	n.d.	n.d.	KJELVIK and WIELGOLASKI (unpublished data)
6. Depth to permafrost in autumn		no permafrost						
7. Temperature °C (−5 cm −2 cm[c] −10 cm° −20 cm[b]):								
annual	(1970–71)	3.7[c] 3.4[b]	0.3[c] 0.2°	2.5	2.7	2.5[c]	2.0[c] 2.4[b]	SKARTVEIT (private communication)
January/February	(1970–71)	0.0[c] 0.2[b]	−5.8[c] −5.9°	0.0	−0.2	0.0[c]	−0.3[c] 0.0[b]	
July	(1970–71)	10.8[c] 9.9	8.9[c] 8.2°	8.2	9.7	8.5[c]	8.2[c] 8.3[b]	

Table 1 (continued)

	Birch forest 2.1.1	Lichen heath 2.1.2	Wet meadow 2.1.3	Dry meadow 2.1.4	Willow thicket 2.1.5	Snow bed 2.1.6	Ref.
8. Temperature, °C (+200 cm):							
annual (1970–71)	0.7 (1.3)	−3.0	−2.1 (−2.1[a])	−1.8	−1.8[a]	−1.9 (−1.4)	Skartveit et al. (1975)
January/February (1970–72)	−7.8 (−7.7)	−15.5	−10.7 (−10.6[a])	−10.8	−10.8[a]	−9.8 (−9.3)	
July (1969–72)	10.2 (11.7)	7.9	7.1 (8.0[a])	7.6	7.4[a]	7.4 (8.0)	
9. Precipitation, mm (1970–73):							
annual	1013	700[a]	750[a]	730[a]	750[a]	988	Skartveit (private communication)
lasting snow, date	beg. Nov.	end Oct.	mid Oct.	late Oct.	late Oct.	beg. Oct.	
snow melt, date	end May	end May	mid June	beg. June	beg. June	beg. July	
10. Time of sunshine during growing season, hours (1969–71)	500[a]	453	419	434	434[a]	300[a]	(Skartveit private communication)
11. Growing season:							
number days	125–130	110	100	105	105	60–70	Wielgolaski and Kärenlampi (1975)
start	end May	beg. June	mid June	early June	early June	early July	
end	beg. Oct.	late Sept.	late Sept.	late Sept.	late Sept.	mid Sept.	
12. Cover, %:							
Vascular plants	90	25	70	70	75[a]	51	Lye (1972)
trees	40	0	0	0		0	
shrubs	10	0	0	0		0	Lye (private communication)
dwarf shrubs	30	20	25	20		50	
herbs	5	0	10	30		0	
graminids	5	5	35	20		1	
Cryptogams	25	75	90–95	30	50[a]	95[a]	
bryophytes	20	5	90–95	25		80[a]	
lichens	5	70	0	5		15[a]	

Table 1 (continued)

	Birch forest 2.1.1	Lichen heath 2.1.2	Wet meadow 2.1.3	Dry meadow 2.1.4	Willow thicket 2.1.5	Snow bed 2.1.6	Ref.
13. Phytomass (live), g cm^{-2}							
Vascular plants							
total	4475	250	1430	675	2080	300	WIELGOLASKI (1975)
aboveground	2500	60	115	130	775	50	KJELVIK and KÄRENLAMPI (1975)
Cryptogams							
bryophytes	60	5	175	30	310	70	
lichens	10	370	0	20	0	60	
14. Production of plants g cm^{-2} yr^{-1} (net)	780	270	830	530	885	—	KJELVIK and KÄRENLAMPI (1975)
15. Vertebrate biomass and/or intake	Estimated fresh weight:						
	Passerines		4–15 kg km^{-2}				LIEN et al. (1974)
	Small rodents		0–550 kg km^{-2}				ØSTBYE et al. (unpublished)
	Reindeer		50–90 kg km^{-2}				GAARE (private communication)
16. Invertebrate biomass, max value g m^{-2}	n.d.	0.8	1.5	1.3	n.d.	n.d.	SOLHØY (private communication)
17. Gaseous carbon loss g C m^{-2} yr^{-1}	689	94	253	457	n.d.	177	SVENSSON et al. (1975)
18. Nitrogen fixation, g N m^{-2} yr^{-1}	0.17	0.06	0.10	0.11	n.d.	0.14	GRANHALL and LID-TORSVIK (1975)
19. Decomposition rate % yr^{-1} (1st yr)	33	12	33	38	35[a]	38	ROSSWALL et al. (1975)
20. Principal organisms							WIELGOLASKI and KJELVIK (unpublished data)
a) Plants							
Betula pubescens ssp *tortuosa*	×	—	—	—	—	—	
Salix lapponum and *S. glauca*	—	—	×	—	×	—	
Empetrum hermaphroditum	×	×	—	—	—	—	LYE (1972)
Salix herbacea	—	×	×	×	—	×	
S. reticulata	—	—	—	×	—	—	
Vaccinium myrtillus	×	—	—	—	—	—	

Table 1 (continued)

	Birch forest 2.1.1	Lichen heath 2.1.2	Wet meadow 2.1.3	Dry meadow 2.1.4	Willow thicket 2.1.5	Snow bed 2.1.6	Ref.
V. uliginosum	×	×	—	—	—	—	
V. vitis-idaea	×	×	—	×	—	—	
Alchemilla vulgaris	—	—	—	×	×	—	
Dryas octopetala	—	—	—	×	—	—	
Geranium silvaticum	—	—	—	×	×	—	
Polygonum viviparum	—	—	×	×	×	—	
Saussurea alpina	—	—	—	×	×	—	
Anthoxanthum odoratum	—	—	—	×	×	—	
Carex bigelowii and *C. vaginata*	—	×	—	×	×	—	
Carex nigra	—	—	×	—	—	—	
Deschampsia caespitosa	—	—	—	—	×	—	
D. flexuosa	×	—	—	×	×	—	
Festuca ovina	—	×	—	—	—	—	
Poa alpina	—	—	—	×	—	—	
Bryum pseudotriquetrum	—	—	×	—	×	—	
Mnium pseudopunctatum	—	—	×	×	×	—	
Philonotis fontana	—	—	×	—	—	—	
Pleurozium schreberi	×	—	—	×	×	—	
Cetraria spp.	—	×	—	×	—	×	
Cladonia spp.	×	—	×	×	×	×	
Hepaticae	×	—	×	—	×	×	
b) Vertebrates	n.d.	*Rangifer tarandus*	*Rangifer tarandus*	*Rangifer tarandus*	as dry meadow[a] + *Turdus pilaris*[a]	*Rangifer tarandus*	ØSTBYE et al. (private communication)
			Microtus oeconomus *Lemmus lemmus*	*Microtus oeconomus* *Lemmus lemmus*		*Lemmus lemmus*	

Table 1 (continued)

	Birch forest 2.1.1	Lichen heath 2.1.2	Wet meadow 2.1.3	Dry meadow 2.1.4	Willow thicket 2.1.5	Snow bed 2.1.6	Ref.
			Sorex sp.	*Clethrionomys rufocanus* *C. glareolus* *Lagopus lagopus* *L. mutus* *Pluvialis apricaria* *Calcarius lapponicus* *Anthus pratensis*			
c) Invertebrates	n.d.	*Acari* *Collembola* *Hemiptera* *Araneae* *Diptera* *Coleoptera* *Lepid-optera*	*Acari* *Collembola* *Enchytra.* *Araneae* *Coleopt.* *Diptera* *Thysanoptera*	*Acari* *Collembola* *Enchytra.* *Hemiptera* *Diptera* *Lepidoptera* *Coleoptera* *Araneae* *Thysanoptera* *Hymenoptera*	n.d.	n.d.	Solhøy (1972)
21. Microorganisms							
bacterial biomass $g\,m^{-2}$ (microscopic count)	730 0–20 cm	550 0–15 cm	940 0–30 cm	630 0–30 cm	n.d.	530 0–20 cm	Clarholm et al. (1975)
fungal biomass $g\,m^{-2}$ 0–10 cm	204	50	182	113		69	Hanssen and Goksøyr (1975)

[a] Estimated values.
Temperature values within brackets are 30 year normals (Bruun, 1967) for Maurset birch forest and Finse snow bed.

Table 2. Site characteristics Finnish sites (2.2.1–2.2.3) and Swedish site (2.3) (for more details see different papers in the present volume)

2.2.1 Pine forest, Kevo, Lat. 69°45′ N, Long. 27°01′ E. Subalpine belt/Subarctic subzone (*Cladonio-Pinetum/Vaccinio-Pinetum* KIELLAND-LUND, 1967).

2.2.2 Birch forest, Kevo, Lat. 69°45′ N, Long. 27°01′ E. Subalpine belt/Subarctic subzone (*Vaccinio-Pinetum/Eu-Piceetum myrtilletosum* KIELLAND-LUND, 1967).

2.2.3 Low alpine heath. Kevo, Lat. 69°45′ N, Long. 27°01′ E. Low alpine belt (*Arctostaphyleto Cetrarion nivalis* DAHL et al., 1971).

2.3 Swedish site

Oligotrophic mire, Stordalen, Lat. 68°22′ N, Long. 19°03′ E. Subalpine belt (*Oxycocco-Empetrion hermaphroditi/Leuco-Scheuchzerion* DAHL et al., 1971).

	Pine forest 2.2.1	Birch forest 2.2.2	Alpine heath 2.2.3	Oligotrophic mire, Sweden 2.3	Ref.
1. Altitude, m a.s.l.	90	130	330	351	
2. Microtopography	even	stony	stony	hummocks and depressions	KÄRENLAMPI (1972) SONESSON et al. (1973)
3. Slope	level	8° NNE	level	level	KÄRENLAMPI (1972) SONESSON et al. (1973)
4. Soil	podzol	podzol	podzol	peat	
Depth of horizon cm/pH_{H_2O} A_1	5/4.3	9/4.6	3/4.0		HINNERI et al. (1975) HINNERI (unpublished data)
A_2	3/4.2	4/4.8	4[a]/4.4	0–> 300	
B	24/5.0	22/5.2	17/5.3		SONESSON et al. (1973)
pH: H_2O/KCl: 0–10 cm				4.0/3.0	SONESSON (1973)
10–20 cm				4.1/3.0	
20–30 cm				4.2/3.2	
C/N 0–10 cm (alpine heath 0–3 cm)	48	36	37	48	HINNERI et al. (1975)
Water content % f.w. 0–10 cm	10	15	18	84	
5. Drainage and/or depth of water table	well-drained	slight seepage water effect	well-drained	groundwater frozen	HINNERI et al. (1975) SONESSON (1970b)
6. Depth to permafrost in autumn	no permafrost			hummocks: 30–50 cm depressions: 40–70 cm	KALLIO et al. (1969), SONESSON (1970a)

Table 2 (continued)

	Pine forest 2.2.1	Birch forest 2.2.2	Alpine heath 2.2.3	Oligotrophic mire, Sweden 2.3	Ref.
7. Soil temperature °C (-5^{b}cm or −10 cm)					
annual	n.d.	n.d.	n.d.	−0.8 −0.3[b]	SKARTVEIT et al. (1975), KÄRENLAMPI (1972), KÄRENLAMPI (unpublished data)
Jan./Feb.	−8	−5	−9	−4.2 −4.3[b]	
July	n.d.	n.d.	n.d.	5.3 9.3[b]	
8. Air temperature, °C (+200 cm)					
annual	2.4			−0.7 (−1.0)	KÄRENLAMPI (1972), RYDÉN (1974), ANDERSSON (unpublished data)
Jan./Feb.	−16.6			−11.7 (−10.8)	
July	11.8			11.5 (11.1)	
Thermal sum (1968–70)	595.0	569.0	436.6	—	
9. Precipitation, mm					
annual	420	420	420	300–350 (300)	KÄRENLAMPI (1972), RYDÉN (1974). SONESSON (1969)
lasting snow, date	mid-October	mid-October	mid-October	mid-October	
snow melt, date	mid-May	mid-May	mid-May	early May	
10. Sunshine duration during growing season, hours (1972, 1973)	887	887	887	(760); (~1200 for hummock top cryptogams)	RYDÉN (1974)
11. Growing season					
number days	115	115	100	120; 210 for hummock top cryptogams	KÄRENLAMPI (1972), JOHANSSON (1974), SONESSON (1969)
start	end-May	end-May	beginning-June	15 May; 15 March for hummock top cryptogams	
end	mid-September	mid-September	beginning September	15 Sept.; 15 Oct. for hummock top cryptogams	

Table 2 (continued)

	Pine forest 2.2.1	Birch forest 2.2.2	Alpine heath 2.2.3	Oligotrophic mire, Sweden 2.3	Ref.
12. Cover, %					KALLIO and KÄRENLAMPI (1971), KÄRENLAMPI (1972), SONESSON and JOHANSSON (1974)
Vascular plants					
trees	31	49	0	0	
shrubs	0	0	15	0	
dwarf shrubs	18	37	23	33	
herbs	0	< 1	0	10	
graminids	0	< 1	1	23	
Cryptogams					
bryophytes	17	43	(5)	87	
lichens	13	14	(40)	20	
13. Phytomass (live), $g\,m^{-2}$					KÄRENLAMPI (1973), SONESSON et al. (1973)
Vascular plants					
total	392	754	375	~580	
aboveground	112	243	121	153	
Cryptogams					
bryophytes	18	33	7	300	
lichens	85	19	77	12	
14. Production (net) of plants, $g\,m^{-2}\,yr^{-1}$ (the values do not include trees and shrubs)	75	125	70	~150	KÄRENLAMPI (1973), KÄRENLAMPI (unpublished data), SONESSON et al. (1973)
15. Vertebrate biomass and/or intake	n.d.	n.d.	n.d.	n.d.	
16. Invertebrate biomass, max. value $g\,m^{-2}$	n.d.	n.d.	n.d.	n.d.	
17. Gaseous carbon loss, $g\,C\,m^{-2}\,yr^{-1}$	n.d.	n.d.	n.d.	~80	SVENSSON et al. (1975)
18. Nitrogen fixation, $g\,N\,m^{-2}\,yr^{-1}$				0.18	GRANHALL and SELANDER (1973), KALLIO, S. and KALLIO, P. (1975)
Nephroma arcticum		0.10	0.05		
Stereocaulon paschale	0.38	0.05	0.17		
Free-living blue-green algae and bacteria	mean of the three sites 0.002				

Table 2 (continued)

	Pine forest 2.2.1	Birch forest 2.2.2	Alpine heath 2.2.3	Oligotrophic mire, Sweden 2.3	Ref.
19. Decomposition rate, % yr^{-1}		n.d.	n.d.	18 (6–37)	ROSSWALL et al. (1975)
bryophytes	n.d.	n.d.	n.d.	3–10	SONESSON (unpublished data)
lichen	6– 7				
birch leaves	24–29				
pine needles	14–22				
20. Principal organisms					KALLIO and KÄRENLAMPI
a) Plants					(1971), SONESSON et al. (1973)
Pinus silvestris	×	—	—	—	
Betula pubescens	—	×	—	—	
Andromeda polifolia	—	—	—	×	
Arctostaphylos alpina	—	—	×	—	
Empetrum hermaphroditum	×	×	×	×	
Ledum palustre	—	×	—	—	
Vaccinium myrtillus	—	×	—	—	
Vaccinium uliginosum	—	×	×	×	
Vaccinium vitis-idaea	×	×	×	×	
Rubus chamaemorus	—	—	—	×	
Eriophorum vaginatum	—	—	—	×	
Pleurozium schreberi	×	×	—	—	
Hylocomium splendens	—	×	—	—	
Dicranum elongatum	—	—	—	×	
Dicranum fuscescens	—	×	—	—	
Sphagnum balticum	—	—	—	×	
Sphagnum fuscum	—	—	—	×	
Ptilidium ciliare	×	—	—	—	
Cetraria nivalis	—	—	—	×	
Cladonia spp.	—	×	×	×	
Nephroma arcticum	—	×	—	—	
Stereocaulon sp.	×	—	—	—	
b) Vertebrates					KALLIO (unpublished data)
Alces alces	—	—	—	×	ANDERSSON (unpublished data)
Rangifer tarandus	×	×	×	×	

Table 2 (continued)

	Pine forest 2.2.1	Birch forest 2.2.2	Alpine heath 2.2.3	Oligothrophic mire, Sweden 2.3	Ref.
Clethrionomys spp.	×	×	×	×	
Lemmus lemmus	—	—	—	×	
Lepus timidus	×	×	×	×	
Anthus pratensis	×	×	×	×	
Lagopus lagopus	×	×	×	—	
Motacilla flava thunbergii	—	—	—	×	
Numenius phaeopus	—	—	—	×	
Pluvialis a. apricaria	—	—	—	×	
Turdus musicus	×	×	—	—	
c) Invertebrates					KOPONEN (1973), JONSSON (unpublished data)
Collembola	×	×	×	×	
Diptera	×	×	×	×	
Homoptera	—	—	—	×	
Acarina	×	×	×	×	
Enchytraeidae	—	—	—	×	
Trichoptera	—	—	—	×	
Ephemeroptera	—	—	—	×	
Araneae	×	×	×	×	
Coleoptera	×	×	×	×	
Hymenoptera	×	×	×	×	
Odonata	—	—	—	×	
21. Microorganisms					
bacterial biomass	n.d.	n.d.	n.d.	0–35 cm: 22 g m^{-2} (microscopical counting)	CLARHOLM et al. (1975)
fungal biomass	n.d.	n.d.	n.d.	0–10 cm: 58 g m^{-2}	HANSSEN and GOKSØYR (1975)
fruiting bodies, g ha^{-1}	4020	10790	650		KALLIO (1972)

2.2.1–3 Air temperature, precipitation and sunshine duration figures are averages 1962–1970 from the Kevo Meteorological Station. Soil temperatures are estimated from one year's measurements.

2.3 Values within brackets are 30-year averages from Abisko (10 km west of Stordalen). Other values are from the site at Stordalen (1971–1973). Soil temperatures at −10 cm depth are averages 1972–1973, at −5 cm values 1973.

[a] Only under shrub stands.

especially those at higher altitudes, is very similar to, or indistinguishable from, those of the subarctic areas of northernmost Fennoscandia. They are also usually regarded as being equivalent. Although birch woods predominate, pine forest stands (*Pinus silvestris* L.) also occur in the Kevo (Fig. 15), and Abisko areas as outliers or relict stands of the Boreal conifer forests.

In certain years, when herbivore populations (e.g. those of rodents, reindeer and caterpillars) are numerous, they exert a considerable influence on the sites. Thus the caterpillars of *Oporinia autumnata* defoliated large areas of birch forest at Kevo during 1965–1966 (KALLIO and LEHTONEN, 1975). These areas are now partly kept open by subsequent reindeer grazing which keeps the seedlings down. A similar herbivore impact, more restricted in area and not affecting the Stordalen site, was experienced in the Abisko area during 1955–1956. Extensive tracts are subjected to grazing and trampling, particularly by reindeer but also by moose. Although these animals are normal components of the ecosystem, the size of the populations and, in the case of reindeer, also their migrations, are determined to a large extent by man. This does not apply to Hardangervidda, where there is a wild reindeer population and moose are absent. Apart from a limited occurrence of sheep and cattle at Hardangervidda, other domesticated herbivores have no influence on the sites under consideration. In other respects, man's influence direct or indirectly is either practically non-existent or insignificant, except in the immediate vicinity of roads, railways and tourist tracks running through the areas. Global pollution is lowest over the northern sites (KALLIO and VEUM, 1975). Thus the concentrations of e.g. heavy metals in surface soil and plants in Northern Sweden and Northern Finland are about one fifth to one tenth of those recorded in Southern Norway (TYLER, 1972; SONESSON, 1973).

References

BRUUN, I.: Standard normals 1931–1960 of the air temperature in Norway. Oslo: Det Meteorologiske Institutt 1967.

CLARHOLM, M., LID-TORSVIK, V., BAKER, J. H.: Bacterial populations of some Fennoscandian tundra soils. In: WIELGOLASKI, F. E. (Ed.): Fennoscandian tundra ecosystems. Part 1: Plants and microorganisms, pp. 251–260. Berlin-Heidelberg-New York: Springer 1975.

DAHL, E.: Flora and plant sociology in Fennoscandian tundra areas. In: WIELGOLASKI, F. E. (Ed.): Fennoscandian tundra ecosystems. Part 1: Plants and microorganisms, pp. 62–67. Berlin-Heidelberg-New York: Springer 1975.

DAHL, E., KALLIOLA, R., MARKER, E., PERSSON, Å.: Nordisk vegetationsklassificering för kartläggning. IBP i Norden **7**, 3–12 (1971).

ELVEN, R., RYVARDEN, L.: Dispersal and primary establishment of vegetation. In: WIELGOLASKI, F. E. (Ed.): Fennoscandian tundra ecosystems. Part 1: Plants and microorganisms, pp. 82–85. Berlin-Heidelberg-New York: Springer 1975.

GRANHALL, U., LID-TORSVIK, V.: Nitrogen fixation by bacteria and free-living blue-green algae in tundra areas. In: WIELGOLASKI, F. E. (Ed.): Fennoscandian tundra ecosystems. Part 1: Plants and microorganisms, pp. 305–315. Berlin-Heidelberg-New York: Springer 1975.

GRANHALL, U., SELANDER, H.: Nitrogen fixation in a subarctic mire. Oikos **24**, 8–15 (1973).

HANSSEN, J. F., GOKSØYR, J.: Biomass and production of soil and litter fungi at Scandinavian tundra sites. In: WIELGOLASKI, F. E. (Ed.): Fennoscandian tundra ecosystems. Part 1: Plants and microorganisms, pp. 239–243. Berlin-Heidelberg-New York: Springer 1975.

HEAL, O. W., ROSSWALL, T. (Eds.): Structure and functioning of tundra ecosystems. Bull. Ecol. Res. Comm. (Stockholm). In press (1975).

HINNERI, S., SONESSON, M., VEUM, A.K.: Soils of Fennoscandian IBP tundra ecosystems. In: WIELGOLASKI, F.E. (Ed.): Fennoscandian tundra ecosystems. Part 1: Plants and microorganisms, pp. 31–40. Berlin-Heidelberg-New York: Springer 1975.

HOLDING, A.J., HEAL, O.W., MACLEAN, S.F., FLANAGAN, P.W. (Eds.): Soil organisms and decomposition in tundra. Stockholm: IBP Tundra Biome Steering Committee 1974.

HOLTEDAHL, O. (Ed.): Geology of Norway. Norges Geol. Undersøkelse **208**, 1–540 (1960).

JOHANSSON, L.-G.: Photosynthetic rates of some vascular plants on a subarctic mire at Stordalen. Progress Report 1973. Swed. IBP Tundra Project Tech. Rep. **16**, 145–155 (1974).

KALLIO, P.: The Kevo Subarctic Research Station of the University of Turku. Reports from the Kevo Subarctic Research Station. Ann. Univ. Turkuensis. **2**, 1–9 (1964).

KALLIO, P.: Svamparnas andel av tundraekosystemen i Kevo. IBP i Norden **8**, 67–68 (1972).

KALLIO, P., KÄRENLAMPI, L.: A review of the stage reached in the Kevo IBP in 1970. In: HEAL, O.W. (Ed.): Proceedings of the Tundra Biome working meeting on analysis of ecosystems, pp. 79–91. London: IBP Tundra Biome Steering Committee 1971.

KALLIO, P., LAINE, O., MÄKINEN, Y.: Vascular flora of Inari Lapland. Rep. Kevo Subarctic Res. Stat. **5**, 1–108 (1969).

KALLIO, P., LEHTONEN, J.: On the ecocatastrophe of birch forests by *Oporinia autumnata* (Bkh.) and the problem of reafforestation. In: WIELGOLASKI, F.E. (Ed.): Fennoscandian tundra ecosystems. Part 2: Animals and systems analysis. Berlin-Heidelberg-New York: Springer 1975 (in press).

KALLIO, P., VEUM, A.K.: Analysis of precipitation at Fennoscandian tundra sites. In: WIELGOLASKI, F.E. (Ed.): Fennoscandian tundra ecosystems. Part 1: Plants and microorganisms, pp. 333–338. Berlin-Heidelberg-New York: Springer 1975.

KALLIO, S., KALLIO, P.: Nitrogen fixation in lichens at Kevo, North Finland. In: WIELGOLASKI, F.E. (Ed.): Fennoscandian tundra ecosystems. Part 1: Plants and microorganisms, pp. 292–304. Berlin-Heidelberg-New York: Springer 1975.

KÄRENLAMPI, L.: Comparisons between the microclimates of the Kevo ecosystem study sites and the Kevo Meteorological Station. Rep. Kevo Subarctic Res. Stat. **9**, 50–65 (1972).

KÄRENLAMPI, L.: Biomass and estimated yearly net production of the ground vegetation at Kevo. In: BLISS, L.C., WIELGOLASKI, F.E. (Eds.): Primary production and production processes, Tundra Biome, pp. 111–114. Edmonton-Oslo: IBP Tundra Biome Steering Committee 1973.

KÄRENLAMPI, L., KAUHANEN, H.: A direct analysis of the vegetation of the surroundings of the Kevo subarctic station. Rep. Kevo Subarctic Res. Stat. **9**, 82–98 (1972).

KIELLAND-LUND, J.: Zur Systematik der Kiefernwälder Fennoscandians. Mitt. flor.-soz. Arb.Gemein. **11/12**, 127–141 (1967).

KJELVIK, S., KÄRENLAMPI, L.: Plant biomass and primary production of Fennoscandian subarctic and subalpine forests and of alpine willow and heath ecosystems. In: WIELGOLASKI, F.E. (Ed.): Fennoscandian tundra ecosystems. Part 1: Plants and microorganisms, pp. 111–120. Berlin-Heidelberg-New York: Springer 1975.

KOPONEN, S.: Herbivore invertebrates of the mountain birch forest at Kevo. Rep. Kevo Subarctic Res. Stat. **10**, 20–28 (1973).

LÅG, J.: Jordsmonnet vi lever av. Oslo: Aschehoug 1965.

LARSSON, P., TANGEN, K.: The input and significance of particulate terrestrial organic carbon in a subalpine freshwater ecosystem. In: WIELGOLASKI, F.E. (Ed.): Fennoscandian tundra ecosystems. Part 1: Plants and microorganisms, pp. 351–359. Berlin-Heidelberg-New York: Springer 1975.

LIEN, L., ØSTBYE, E., HOGSTAD, O., HAANDE, K.M., HAANDE, P.S., HAGEN, A., SKAR, H.-J., SKARTVEIT, A., SVALASTOG, D.: Bird surveys in the high mountain habitats of Finse and Stigstuv, Hardangervidda, South Norway, 1967–72. Norw. J. Zool. **22**, 1–14 (1974).

LYE, K.A.: Vegetation of selected localities for IBP investigation in Hardangervidda, Southern Norway. IBP i Norden **8**, 101–111 (1972).

MALMER, N.: To be published in a subsequent volume of the Series Ecological Studies. Berlin-Heidelberg-New York: Springer. In preparation.

MOORE, J.J. (Ed.): Tundra: Comparative analysis of ecosystems. Cambridge University Press 1975 (in press).

RANKAMA, K.: The geologic systems. The quaternary. New York-London-Sidney: John Wiley and Sons 1965.

ROSSWALL, T., VEUM, A.K., KÄRENLAMPI, L.: Plant litter decomposition at Fennoscandian tundra sites. In: WIELGOLASKI, F.E. (Ed.): Fennoscandian tundra ecosystems. Part 1: Plants and microorganisms, pp. 268–278. Berlin-Heidelberg-New York: Springer 1975.

RUNE, O.: Plant life on serpentines and related rocks in the north of Sweden. Acta Phytogeogr. Suecica **31**, 1–135 (1953).

RYDÉN, B.E.: Abiotic studies, Stordalen 1973. Progress Report 1973. Swedish IBP Tundra Project Tech. Rep. **16**, 185–202 (1974).

SJÖRS, H.: Amphi-atlantic zonation. Nemoral to Arctic. In: LØWE, A., LØWE, D. (Eds.): North Atlantic biota and their history, pp. 109–125. Oxford: Pergamon Press 1963.

SKARTVEIT, A., RYDÉN, B.E., KÄRENLAMPI, L.: Climate and hydrology of some Fennoscandian tundra ecosystems. In: WIELGOLASKI, F.E. (Ed.): Fennoscandian tundra ecosystems. Part 1: Plants and microorganisms, pp. 41–53. Berlin-Heidelberg-New York: Springer 1975.

SOLHØY, T.: Quantitative invertebrate studies in mountain communities at Hardangervidda, South Norway. I. Norsk ent. Tidsskr. **19**, 99–108 (1972).

SØMME, A. (Ed.): A Geography of Norden, Scandinavian University Books. Oslo: J. W. Cappelen 1968.

SONESSON, M.: Studies on mire vegetation in the Torneträsk area, northern Sweden. II. Winter conditions of the poor mires. Botan. Notiser **122**, 481–511 (1969).

SONESSON, M.: Studies on mire vegetation in the Torneträsk area, northern Sweden. III. Communities of the poor mires. Opera Botanica **26** (1970a).

SONESSON, M.: Studies on mire vegetation in the Torneträsk area, northern Sweden. IV. Some habitat conditions of the poor mires. Botan. Notiser **123**, 67–111 (1970b).

SONESSON, M.: Some chemical characteristics of the Stordalen site. Progress Report 1972. Swedish IBP Tundra Project Tech. Rep. **14**, 31–43 (1973).

SONESSON, M.: Late Quaternary forest development of the Torneträsk area, North Sweden. II. Pollen analytical evidence. Oikos **25**, 288–307 (1974).

SONESSON, M., FLOWER-ELLIS, J.G.K., JONSSON, S., ROSSWALL, T., RYDÉN, B.E.: Word model for Abisko (Stordalen), Sweden. Progress Report 1972. Swedish IBP Tundra Project Tech. Rep. **14**, 4–13 (1973).

SONESSON, M., JOHANSSON, S.: Bryophyte growth, Stordalen 1973. Progress Report 1973. Swedish IBP Tundra Project Tech. Rep. **16**, 17–27 (1974).

SVENSSON, B.H., VEUM, A.K., KJELVIK, S.: Carbon losses from tundra soils. In: WIELGOLASKI, F.E. (Ed.): Fennoscandian tundra ecosystems. Part 1: Plants and microorganisms, pp. 279–286. Berlin-Heidelberg-New York: Springer 1975.

TYLER, G.: Heavy metals pollute nature may reduce productivity. Ambio 1972, 52–59 (1972).

WIELGOLASKI, F.E.: IBP tundra studies, Hardangervidda, Norway. In: HEAL, O.W. (Ed.): Working meeting on analysis of ecosystems, Kevo, Finland, pp. 101–124. London: IBP Tundra Biome Steering Committee 1971.

WIELGOLASKI, F.E.: Production, energy flow and nutrient cycling through a terrestrial ecosystem at a high altitude area in Norway. In: WIELGOLASKI, F.E., ROSSWALL, T. (Eds.): Proceedings IBP IV International meeting on the biological productivity of tundra, Leningrad, pp. 283–291. Oslo-Stockholm: IBP Tundra Biome Steering Committee 1972.

WIELGOLASKI, F.E.: Primary productivity of alpine meadow communities. In: WIELGOLASKI, F.E. (Ed.): Fennoscandian tundra ecosystems. Part 1: Plants and microorganisms, pp. 121–128. Berlin-Heidelberg-New York: Springer 1975.

WIELGOLASKI, F.E., KÄRENLAMPI, L.: Plant phenology of Fennoscandian tundra areas. In: WIELGOLASKI, F.E. (Ed.): Fennoscandian tundra ecosystems. Part 1: Plants and microorganisms, pp. 94–102. Berlin-Heidelberg-New York: Springer 1975.

2. Abiotic Variables

Soils of Fennoscandian IBP Tundra Ecosystems

S. HINNERI[1], M. SONESSON[1], and A. K. VEUM[1]

The general outlines of soil formation in subarctic and subalpine areas of Fennoscandia are fairly well known (TEDROW, 1970), but few actual surveys are available (e.g. TAMM, 1930; POLYNTSEVA, 1958). This paper summarizes some of the overall results of the recent investigations on nutrient pools and cycling of bioelements carried out in Fennoscandian IBP tundra projects at Kevo, Finland, Hardangervidda, Norway, and Stordalen (Abisko), Sweden (SONESSON et al., 1975). Detailed comparisons are not always possible, since the investigations carried out and the methods used vary somewhat from country to country (general survey by BROWN, 1971). Soil profiles, physico-chemical soil analyses for the sites, and methods used for these analyses are presented in SONESSON (1973), VEUM (1973), BAKER (pers. comm., 1972) and HINNERI (1974) and also in BROWN and VEUM (1974). The bedrock and Quaternary geology of the areas have been described briefly by SONESSON et al. (1975). More details on the geology of the Hardangervidda area are found in REKSTAD (1903) and ROSENDAHL (1934).

Soil Types

The well-drained mineral soils of the subalpine ecosystems studied, the birch forests at Hardangervidda and Kevo and the pine forest at Kevo, and of the two alpine heaths represented by the lichen heath at Hardangervidda and the low alpine heath at Kevo, are distinctly podzolized (Table 1). The eluviated A_2 horizons are mostly somewhat thicker at the Norwegian than at the Finnish sites, and this may reflect differences in precipitation (SKARTVEIT et al., 1975). The content and distribution of Fe, Mn, major nutrients, C, pH and different soil water parameters are all typical of podzols (TAMM, 1930; RODE, 1937). The removal of Fe from the A horizon and its precipitation in the B horizon is marked in the Finnish podzols. Some of the extractable nutrients (Table 2) showed a reverse translocation and redistribution in all the soils examined, due partly to capillary transport and sorption of the elements by plants (RODE, 1937; MATTSON et al., 1944; ASSARSSON, 1961) and partly to frost processes. In soils of the Norwegian dry meadow and snow bed the podzol is weakly developed, especially in the eluviated

[1] The authors are listed in alphabetical order and all have the same responsibility for the paper. Each author is responsible for the data from his own country.

Table 1. Physical analyses of soils in Fennoscandian IBP Tundra Biome Projects. AWC = available water capacity; FC = field capacity at pF = 2; PWP = permanent wilting point at pF = 4.2; Air capacity = soil air content at pF = 2

	Horizon	Depth of horizon (cm)	Gravel > 2 mm %	Bulk density (kg/dm^3)	Loss on ig-nition	Water content (% wet weight) aver. (range)	AWC (%)	FC (%)	PWP (%)	Satura-tion (%)	Air ca-pacity (%)
KEVO:											
Pine forest	A0	0– 5		0.19							
	A2	5– 8	13			10 (2–18)					
	B21	8–23	21								
	B22	23–32	13			10 (4–18)					
	BC	32–60	0								
	C	60–	1			7 (2–21)					
Birch forest	A0–1	0– 9		0.23							
	A2	9–13				15 (8–18)					
	B	13–35	33			17 (7–24)					
	BC	35–55	46								
	C	55–	40			22 (12–17)					
Low alpine heath	A0	0– 3		0.14							
	A2	3– 7				18 (10–33)					
	B	7–24	49			16 (8–28)					
	C	24–70	29			18 (8–28)					
HARDANGERVIDDA:											
Birch forest	A0–1	0– 4		0.24	79.8	61 (32–74)[b]	36	48	12	78	39
	A2	4–17	14	1.29	1.4	23 (14–35)[c]	19	22	3	45	30
	B2	17–45	35	1.27	6.5		17	25	8	48	28
	BC	45–	64		1.8						
Lichen heath	A0–1	0– 2		0.6[a]	28.5	28 (15–50)[b]	23	29	6	54	32
	A2	2– 8	2	1.20	2.7	11 (5–19)[c]	13	17	4	46	38
	B21	8–18	8	1.35	1.9		7	10	3	37	40
	B22	18–36	7	1.47	0.8		2	3	1	36	43
	B23	36–41	10	1.47	1.0		3	4	1	37	43
	BC	41–49	12		0.4						
	C	49–	19		0.3						

Dry meadow	A0–1	0– 9		0.25	80.0	59 (45–70)[b]	44	65	21	83	21
	A2	9–15	12	1.37	4.0	54 (30–74)[c]	11	15	4	38	34
	B2	15–45	22	1.41	2.0		12	15	3	37	33
	BC	45–	29	1.30	1.5		11	15	4	38	39
Snow bed	A0–1	0– 2		0.4[a]	40.8	45 (15–68)[b]					
	A2	2–35	16	1.5[a]	1.8						
	B2		21	1.5[a]	3.0						
	BC	35–48	40		1.0						
Wet meadow	A	0–10		0.14	83.5	83 (77–92)[b]	47	58	11	85	34
	A	12–52		0.20	82.6	84 (77–90)[c]	56	67	11	82	19
	AC	52–		1.47	4.6		23	27	4	38	17
STORDALEN:											
Oligotrophic mire	A	0–10		0.084		84					
	A	10–20		0.103		85					
	A	20–30		0.102		80					

[a] Estimated from laboratory investigations (BONDORFF, 1950).
[b] 0–2 cm soil depth observations 1972.
[c] 0–10 cm soil depth observations 1969–1972 (KJELVIK and WIELGOLASKI, unpubl.).

Table 2. Chemical soil analyses in Fennoscandian IBP Tundra Biome Projects

	Horizon	pH H_2O	Tot. C (%)	C/N in humus	Tot. N (%)	NH_4 (ppm)	NO_3 (ppm)	Extractable nutrients (ppm) (Ammonium acetate: Kevo, Stordalen, Ammonium acetatelactate: Hardangervidda)				Trace elements extractable in 2 N HCl (ppm)			
								P	Ca	Mg	K	Fe	Mn	Cu	Zn
KEVO:															
Pine forest	A0	4.3	44.0	48.2	0.91			57.8	930	320	360	270	240	5.7	22
	A2	4.2	0.4					2.3	5.6	3.5	9.2	390	14	4.3	1.0
	B21	5.0	0.5					1.2	3.6	3.5	12	4570	15	5.0	5.9
	B22	5.0	0.3					1.1	4.1	1.6	11	3130	16	5.2	5.2
	BC	5.0	0.2					0.8	9.3	1.2	6.7	680	42	6.5	2.3
	C	5.2	0.2					0.2	37	2.3	7.9	620	10	5.8	1.8
Birch forest	A0–1	4.6	30.0	36.1	0.83			43.6	610	350	520	520	180	8.3	30
	A2	4.8	1.2					0.9	14	9.6	12	410	2	3.7	0.9
	B	5.2	1.9					1.6	7.6	6.0	10	5630	16	8.1	7.4
	BC	5.2	2.0					1.9	30	5.2	9.6	3300	17	8.5	5.0
	C	5.3	0.6					0.8	16	3.2	7.6	3120	19	10	15
Low alpine heath	A0	4.0	47.5	36.8	1.29			104.0	540	400	270	150	280	9.4	84
	A2	4.4	2.0					2.1	83	16	19	870	4	4.5	1.8
	B	5.3	2.6					1.5	63	7.5	15	5650	9	7.1	7.4
	C	5.4	0.6					1.7	33	4.8	11	1460	28	13	4.6

	Horizon	pH H_2O	Tot. C (%)	C/N in humus	Tot. N (%)	NH_4 (ppm)	NO_3 (ppm)	Extractable nutrients (ppm) (Ammonium acetate: Kevo, Stordalen; Ammonium acetate-lactate: Hardangervidda)				Exch. H^+ (m.e./100 g)	Cation exchange capacity (m.e./100 g)	Base saturation (%)
								P	Ca	Mg	K			
HARDANGERVIDDA:														
Birch forest	A0–1	4.0	42.4	28.6	1.48	113.4	1.1	194	1026	507	875	76.6	94.2	18.9
	A2	4.2	0.9		0.02	2.7	0.3	7	16	7	18			
	B2	4.8	2.3		0.44	3.2	0.4	15	9	4	21			
	BC	5.0	0.9		0.03	1.9	0.3	10	13	1	9			
Lichen heath	A0–1	4.0	17.3	22.8	0.76	29.2	1.0	25	430	75	142	42.0	49.0	14.2
	A2	4.3	1.2		0.06	2.2	0.4	4	16	5	12	5.9	6.1	3.0
	B21	4.8	1.2		0.05	1.3	0.4	10	11	2	5	5.4	5.5	1.3
	B22	5.2	0.4		0.01	0.9	0.4	10	11	2	3	2.5	2.6	2.4
	B23	5.3	0.4		0.01	3.6	0.3	7	12	2	4			
	BC	5.5	0.1		0.01	0.9	0.3	4	10	2	3	1.6	1.6	3.0
Dry meadow	A0–1	5.3	42.1	16.2	2.60	81.6	3.2	93	7906	227	544	37.8	116.2	67.2
	A2	5.3	1.7		0.14	2.2	0.6	4	1026	23	23	5.4	11.0	42.6
	B2	5.5	0.8		0.07	2.0	0.4	4	426	15	10	4.3	6.4	32.1
	BC	5.6	0.6		0.03	1.7	0.5	9	298	11	7	5.2	6.7	19.6
Snow bed	A0–1	4.4	20.9	24.0	0.87	7.8	0.7	45	249	129	612	38.9	44.8	12.1
	A2	4.9	1.7		0.07	1.5	0.6	17	23	3	15	11.9	12.0	1.0
	B2	4.8	1.1		0.04	1.4	0.4	32	47	3	13	11.1	11.2	0.9
	C	5.4	0.2					90	180	2	11	6.7	6.8	1.2
Wet meadow	cm													
	A													
	0–12	5.3	44.4	20.9	2.11	79.2	2.6	22	5387	159	570	46.4	102.0	53.9
	12–52	5.2	47.5	17.4	2.73	30.7	2.1	4	4260	34	38	50.6	119.2	55.7
	AC	4.9	1.7		0.04	2.2	0.6	2	896	41	10	7.1	11.3	37.6
STORDALEN:														
Oligotrophic mire	cm													
	A													
	0–10	4.0	44.6	48.0	0.93	—	—	—	221	83	60	—	80.0	—
	10–20	4.1	45.8	38.9	1.18	—	—	—	253	57	27	—	—	—
	20–30	4.2	43.8	31.1	1.41	—	—	—	265	49	14	—	120.0	—

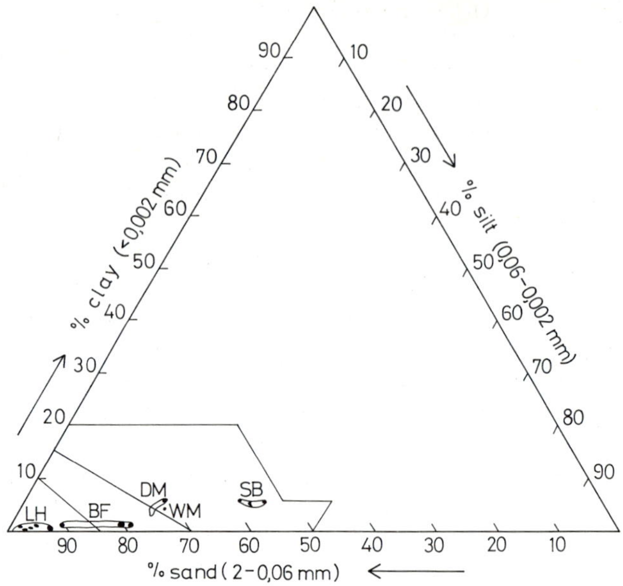

Fig. 1. Mechanical analyses of the profiles of the Norwegian IBP tundra sites, fractions <2 mm. LH = lichen heath, BF = birch forest, DM = dry meadow, WM = wet meadow and SB = snow bed

A_2 horizon. There is generally a great variation in the texture of the Finnish soils, both between and within separate profiles (HINNERI, 1974). If only soil fractions finer than 2 mm are considered, there is, however, a distinct order in the soils of the Norwegian sites (Fig. 1). The soil of the lichen heath is coarsest (sand) and that of the snow bed finest (sandy loam). The differences between the profile horizons at the Norwegian sites are small (cf. the narrow clusters in Fig. 1). Because of adequate drainage no B_1 horizon with accumulation of precipitated humus is discernible, which shows that only true iron podzols are involved (POLYNTSEVA, 1958).

The very poorly drained organic soils of the oligotrophic mire at Stordalen and of the wet eutrophic meadow at Hardangervidda are fairly undifferentiated peat, although the parent material and degree of humidification differ. The surface peat at Stordalen is mainly composed of slightly decomposed *Sphagnum* and *Dicranum*, whereas at Hardangervidda it is composed of more decomposed graminids and Amblystegiaceae. The distribution of major cations in peat profiles indicates leaching and reverse translocation processes in peat soils, similar to those in mineral soils (SONESSON, 1970, 1973). Although poorly drained organic soils are typical of subarctic areas (POLYNTSEVA, 1958), podzols are also common in well-drained sites beyond the subarctic-subalpine (forest) limit (TEDROW, 1970). Thus, from a pedological point of view there is nothing to indicate that the present IBP sites are not representative of tundra ecosystems.

Nutrient Pools—Soil Fertility

Base saturation in the humus layers especially of the Norwegian subalpine and alpine heath profiles is low, and pH usually below 4.5 both in Norway and Finland (Table 2). pH is lower in the humus at the Norwegian oligotrophic sites than for the Finnish sites, probably to some degree due to differences in the amount and chemical composition of the precipitation. Only in the two meadow sites at Hardangervidda does the surface humus pH exceed 5.0. The high pH of the humus of the Norwegian meadows is related to both the base-rich parent material (phyllite) and to seepage water, rich in electrolytes from the hillsides. The high pH of the subsoil in the Finnish low alpine heath is affected by local autochthonous base-rich parent material underlying glacial till, but adjacent to the bedrock leached and precipitated acid humus is found which decreases pH.

The exchange capacity of the humus layer of the Norwegian oligotrophic birch forest is similar to the two eutrophic meadow sites (Table 2), but, since the proportion of the exchangeable H^+ is high at the birch forest, the base saturation is of the same magnitude as, for example, seen at the lichen heath.

Coarse mineral soils, poor in fine material and colloids, are low in nutrients and exchange capacity (SJÖRS, 1961; HINNERI, 1974). Thus, at the Norwegian, oligotrophic sites the low nutrient values of the lichen heath mineral layers and the usually higher values of the snow bed may partly be due to differing contents of fine material (Fig. 1).

The poor availability of N to plants is indicated by often high C/N ratios in the present studies (Table 2). For the Finnish and Swedish sites they are above 30–35. The general deficiency of N in subarctic areas and the significance of some lichens for nitrogen fixation at forest sites at Kevo is reported by KALLIO et al. (1972). NH_4–N and NO_3–N analyses, which were carried out at the Norwegian sites, indicate that NO_3 is absent from mid-July to the beginning of September at most of the sites, while NH_4 values are almost constant throughout the growing season. At the Swedish site neither NO_2 nor NO_3 formation has been observed, but only NH_4 (SJÖQVIST and ROSSWALL, 1973).

Apart from N, P seems to be the most important limiting factor for the productivity of the ecosystems in this study. In acid soils most P is present as very insoluble Fe-P and Al-P complexes, or is organically bound in humus. It is particularly interesting that the amount of extractable P in the mineral soil of the snow bed at Hardangervidda is the highest observed (Table 2). Irrespective of the slightly different methods of analysis used, the low mineral values of the Finnish sites in relation to Norwegian soils are mainly attributable to strong fixation with sesquioxides in the more podzolized soils. VALMARI (1921) pointed out earlier that P in the humus layer of Finnish forests has no correlation with other soil fertility factors.

The extractable amounts of P, Ca, Mg, and K are calculated down to 30–35 cm in the profiles (Table 3) because this depth was, in IBP tundra studies, found to be the approximate lower level of the biogeochemical cycles comprising the bulk of the rhizosphere. Even if the contents of this zone show the potential nutrient pool for the plants, this does not mean that they are all available for primary production. Apart from capacity (quantity), the intensity (activity, ten-

Table 3. Bioelement pools at IBP sites. Extractable amounts in g m^{-2} 35 cm^{-1} or 30 cm^{-1}

	Tot. C	P	Ca	Mg	K	Tot. N
Mineral soils (0–35 cm)						
KEVO:						
Pine forest	5682	0.86	9.65	3.47	5.17	—
Birch forest	9337	1.30	15.73	8.64	12.93	—
Low alpine heath	7663	0.94	6.24	1.81	1.99	—
HARDANGERVIDDA:						
Birch forest	9764	6.24	13.96	6.18	14.95	1099
Lichen heath	6020	4.54	10.55	2.05	4.03	245
Dry meadow		3.97	420.49	11.14	19.42	1030
Snow bed		12.74	21.80	2.52	11.83	384
Organic soils						
HARDANGERVIDDA:						
Wet meadow (0–35 cm)	32506	0.71	325.32	5.38	15.43	1762
STORDALEN						
Oligotrophic mire (0–30 cm)	12862	n.d.	49.48	12.32	6.31	342

Table 4. Recalculation of values for comparison of soil analysis calculated from Norwegian soil data. "Readily soluble" nutrients analyzed by the ammonium acetate-lactate method (EGNER et al., 1960), and "exchangeable" nutrients by the ammonium acetate method partly after SCHOLLENBERG and SIMON (1945)

1. Bulk density (y kg/l) $= 0.687 - 0.0057 \cdot$ Loss on ignition (x %)
   $n = 102, r = -0.846$.
   This equation accords with the findings of JEFFREY (1970) within the limit of observed results.
2. Total C (y %) $= -0.17 + 0.53 \cdot$ Loss on ignition (x %)
   $n = 235, r = 0.990$.
3. Total N (y %) $= -0.02 + 0.027 \cdot$ Loss on ignition (x %)
   $n = 194, r = 0.91$.
4. $pH_{CaCl_2}(y) = -0.44 + 0.94 \cdot pH_{H_2O}(x)$
   $n = 421, r = 0.904$.
5. "Exchangeable" K (y m.e./100 g) $= 0.06 + 0.03 \cdot$ "Available or readily soluble" K (K–AL) (x mg/100 g)
   $n = 133, r = 0.982$.
6. "Exchangeable" Mg (y m.e./100 g) $= -0.05 + 0.11 \cdot$ "Available or readily soluble" Mg (Mg–AL) (x mg/100 g)
   $n = 133, r = 0.990$.
7. "Exchangeable" Ca (y m.e./100 g) $= 0.75 + 0.09 \cdot$ "Available or readily soluble" Ca (Ca–AL) (x mg/100 g)
   $n = 133, r = 0.950$.

sion) of the chemical factors also affects growth (SJÖRS, 1961). Table 4 shows one way of expressing the relationship between readily soluble or extractable and "exchangeable" mineral amounts for comparison.

The highest concentrations of extractable Ca are observed in the two meadow ecosystems at Hardangervidda (Tables 2 and 3). The content and distribution of

Ca in the podzols at Hardangervidda and Kevo are similar. The Ca content of the oligotrophic mire at Stordalen is about one tenth that of the Norwegian eutrophic meadows, but 5–8 times higher than that of the oligotrophic alpine heaths in Norway and Finland. Phytosociologically, however, these sites are structurally relatively similar. Although some of the differences may be due to the different methods of analysis used, this can only be of minor significance. The high Mg pool in the Stordalen mire can also be noted. The birch forests are rich in K, but the highest content is found in the dry meadow. However, with regard to the major cations, Ca, Mg, and K, the two birch forest sites and the two alpine heath sites, at Hardangervidda and Kevo respectively, are fairly similar.

On the basis of the nutrient pools alone, the fertility of the soils seems generally to be highest for the Norwegian and lowest for the Finnish sites, whereas the Swedish site is in an intermediate position. However, soil fertility is also determined by other interrelated parameters such as moisture, aeration and temperature, and is ultimately shown by the productivity of the ecosystem (JOHANSSON and LINDER, 1975; KJELVIK and KÄRENLAMPI, 1975; WIELGOLASKI, 1975).

References

ASSARSSON, G.: Södra Sveriges torvtillgångar. II Kemiska analyser. (The supply of peat in southern Sweden. II. Chemical analyses). Sveriges Geol. Unders. C **578**, 1–25 (1961).

BONDORFF, K. A.: Om bestemmelse af jordens rumvægt. Planteavl **53**, 449–460 (1950).

BROWN, J.: Soil group. In: HEAL, O. W. (Ed.): Proceedings of the Tundra Biome Working meeting on analysis of ecosystems, pp. 203–215. London: IBP Tundra Biome Steering Committee 1971.

BROWN, J., VEUM, A. K.: Soils properties of the international tundra sites. In: HOLDING, A. J., HEAL, O. W., MACLEAN, S. F., FLANAGAN, P. W. (Eds.): Soils Organisms and Decomposition in Tundra, pp. 27–48. Stockholm: Tundra Biome Steering Committee 1974.

EGNÉR, R., RIEHM, H., DOMINGO, W. R.: Untersuchungen über die chemische Bodenanalyse als Grundlage für Beurteilung des Nährstoffzustandes der Böden. Kungl. Lantbr. högsk. Ann. **26**, 199–215 (1960).

HINNERI, S.: Podzolic processes and bioelement pools in subarctic forest soils at the Kevo Station, Finnish Lapland. Rep. Kevo Subarctic Res. Stat. **11**, 26–34 (1974).

JEFFREY, D. W.: A note on the use of ignition loss as a means for the approximate estimation of soil bulk density. J. Ecol. **58**, 297–313 (1970).

JOHANSSON, L.-G., LINDER, S.: The seasonal pattern of photosynthesis of some vascular plants on a subarctic mire. In: WIELGOLASKI, F. E. (Ed.): Fennoscandian Tundra Ecosystems, Part 1: Plants and microorganisms, pp. 194–200. Berlin-Heidelberg-New York: Springer 1975.

KALLIO, P., SUHONEN, S., KALLIO, H.: The ecology of nitrogen fixation in *Nephroma arcticum* and *Solorina crocea*. Rep. Kevo Subarctic Res. Stat. **9**, 7–14 (1972).

KJELVIK, S., KÄRENLAMPI, L.: Plant biomass and primary production of Fennoscandian subarctic and subalpine forests and of alpine willow and heath ecosystems. In: WIELGOLASKI, F. E.(Ed.): Fennoscandian Tundra Ecosystems, Part 1: Plants and microorganisms, pp. 111–120. Berlin-Heidelberg-New York: Springer 1975.

MATTSON, S., SANDBERG, G., TERNING, P.-E.: Electrochemistry of soil formation, VI. Atmospheric salts in relation to soil and peat formation and plant composition. Kungl. Lantbr. högsk. Ann. **12**, 101–118 (1944).

POLYNTSEVA, O. A.: Soils of the southwestern part of the Kola Peninsula. Jerusalem 1962 (orig. in Russian, Moskva 1958).

REKSTAD, J.: Fra høifjeldsstrøget mellem Haukeli og Hemsedalsfjeldene. N.G.U. Aarbog for 1903 No. 4, (1903).
RODE, A. A.: Podzol forming process. Jerusalem 1970 (orig. in Russian, Moskva 1937).
ROSENDAHL, H.: The geology of the parts of Southern Norway. 4. Finse Proc. Geol. Ass. **45**, (1934).
SCHOLLENBERGER, C. F., SIMON, R. H.: Determination of exchange capacity and exchangeable bases in soil. Ammonium acetate method. Soil. Sci. **59**, 13–24 (1945).
SJÖQVIST, T., ROSSWALL, T.: Protein mineralization in acid peat. Progress Report 1972. Swedish IBP Tundra Project Tech. Rep. **14**, 167–173 (1973).
SJÖRS, H.: Some chemical properties of the humus layer in Swedish natural soils. Bull. Royal School of Forestry, Stockholm, Sweden **37**, 1–51 (1961).
SKARTVEIT, A., RYDÉN, B. E., KÄRENLAMPI, L.: Climate and hydrology of some Fennoscandian tundra ecosystems. In: WIELGOLASKI, F. E. (Ed.): Fennoscandian Tundra Ecosystems Part 1: Plants and microorganisms, pp. 41–53. Berlin-Heidelberg-New York: Springer 1975.
SONESSON, M.: Studies on mire vegetation in the Torneträsk area, northern Sweden. IV. Some habitat conditions of the poor mires. Botan. Notiser **123**, 67–111 (1970).
SONESSON, M.: Some chemical characteristics of the Stordalen site. In: Progress Report 1972. Swedish IBP Tundra Project Tech. Rep. **14**, 31–43 (1973).
SONESSON, M., WIELGOLASKI, F. E., KALLIO, P.: Description of Fennoscandian tundra ecosystems. In: WIELGOLASKI, F. E. (Ed.): Fennoscandian Tundra Ecosystems, Part 1: Plants and microorganisms, pp. 3–28. Berlin-Heidelberg-New York: Springer 1975.
TAMM, O.: Studier över jordmånstyper och deras förhållande till markens hydrologi i Nordsvenska skogsterränger. Medd. Stat. Skogsförsökanst. **26**, 163–408 (1930).
TEDROW, J. C. F.: Soils of the subarctic regions. In: Ecology of the subarctic regions, Proceedings of the Helsinki symposium, pp. 188–295. UNESCO 1970.
VALMARI, J.: Beiträge zur chemischen Bodenanalyse. Acta Forest Fenn. **20**, 1–20 (1971).
VEUM, A. K.: Jordundersøkelser og nedbryting av organisk materiale. (Soil investigations and decomposition of organic matter). In: IBP i Norge Årsrapport 1972 (IBP in Norway Annual Report 1972), pp. 29–74. Oslo: IBP Norwegian National Committee 1973.
WIELGOLASKI, F. E.: Primary productivity of alpine meadow communities. In: WIELGOLASKI, F. E. (Ed.): Fennoscandian Tundra Ecosystems, Part 1: Plants and microorganisms, pp. 121–128. Berlin-Heidelberg-New York: Springer 1975.

Climate and Hydrology of Some Fennoscandian Tundra Ecosystems

A. Skartveit, B. E. Rydén, and L. Kärenlampi*

Climatic Review

By far the most outstanding topographical feature influencing the climate of Scandinavia is the mountain range running close to the west coast of Norway along the entire Scandinavian Peninsula. Hardangervidda (Norway) is a mountain plateau within this mountain range, while Abisko (Sweden) and Kevo (Finland) are located at the eastern side of it (Sonesson et al., 1975). Polar days last for about 7 weeks at the Swedish (Abisko) and Finnish (Kevo) IBP tundra study areas. The Norwegian IBP tundra study area, Hardangervidda, is located about 6° south of the Polar Circle, which results in marked differences in average sunshine duration between Hardangervidda and Abisko (Fig. 1). Kevo is considered to have a sunshine duration fairly close to, or somewhat lower than that of Abisko.

Scandinavia owes its rather variable weather and its relatively mild maritime (oceanic) climate to the prevailing warm, moisture-bearing westerlies and southwesterlies of the North Atlantic (Johannessen, 1970). Locally, this zonal flow of air is considerably influenced by the topography, elevation and the radiation balance of the earth's surface. Southdirected polar or arctic air masses sometimes play an important role as producers of long-lasting cloud situations east of the mountain range. All these factors make for a considerable diversity of climate within Scandinavia. The altitude differences are essential to the temperature distribution. Figure 1 shows a decrease in temperature and an increase in continentality from Hardangervidda to Abisko, and further on to Kevo. Within Hardangervidda the annual mean temperature decreases about 0.6° C per 100 m. The precipitation regime of Scandinavia is to a great extent determined by the amounts of water released within the westerlies and southwesterlies under the influence of the topography. The heaviest precipitation thus occurs on the western slopes of the Norwegian coastal mountains, while a leeward effect of the Scandinavian mountain chain markedly reduces cloud cover and precipitation eastward in Scandinavia.

* The authorships are divided as follows: A. Skartveit and B. E. Rydén are responsible for "The Climatic Review", L. Kärenlampi for the part concerning Kevo, B. E. Rydén concerning Abisko/Stordalen and A. Skartveit for Hardangervidda.

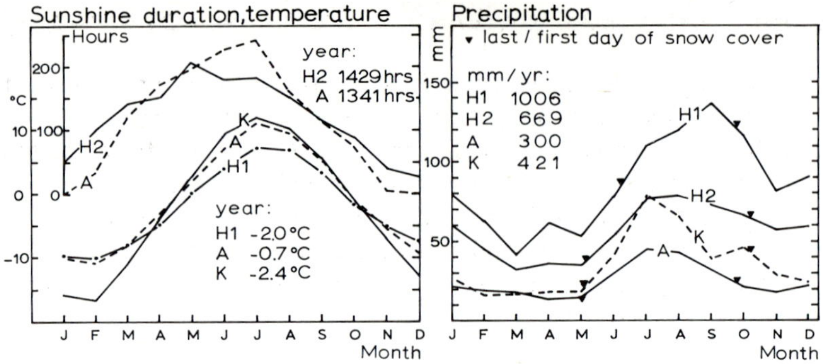

Fig. 1. Monthly and annual values of sunshine duration (Kevo and Abisko are supposed to have the same radiation climate), air temperature (standard level) and precipitation with dates of first and last snow cover at the following stations according to BRUUN, 1967; BRUUN and HÅLAND, 1970; DNMI, in manus; SMHI, 1969; KÄRENLAMPI, 1972:

Abbr.	Site name	Position	Altitude
H1[a]	Hardangervidda W = Slirå	60° 37′ N, 7° 25′ E	1300 m a s l
H2[a]	Hardangervidda E = Haugastøl	60° 31′ N, 7° 52′ E	988 m a s l
A	Abisko	68° 22′ N, 19° 03′ E	345 m a s l
K	Kevo	69° 54′ N, 27° 00′ E	107 m a s l

[a] Representing the more oceanic (H1) and continental (H2) Hardangervidda climate, respectively.

Figure 1 shows an eastward decrease in precipitation amounts across Hardangervidda, and comparatively low precipitation amounts at Abisko and Kevo. The proportion of summer precipitation (June, July, August) is 31%, 39%, and 44% of the yearly total at Hardangervidda, Abisko and Kevo, respectively. The duration of permanent snow cover is of about the same order at Abisko, Kevo and the eastern parts of Hardangervidda, while snow covers the western parts of Hardangervidda about one month longer in the spring (Fig. 1 and for Norwegian sites also Table 2). Furthermore, short periods of snow cover may occur at Hardangervidda throughout the growing season.

Special characteristics of the IBP tundra sites in Fennoscandia are presented below. The energy budget of three Norwegian sites are based on adequate measurements and may illustrate the budget of corresponding surfaces of the sites at Kevo and Stordalen, where the recording instrumentation as to energy budget estimation was not complete. Stordalen is the only site with developed permafrost. These permafrost conditions have been investigated, and show some fairly important hydrological features of the tundra biome. The ecological aspects of a high latitude birch forest seem to be best described by the Kevo studies.

Energy Flow and Microclimate at Hardangervidda, Norway

Most living organisms, especially in high mountain areas, are restricted to the proximity of the meteorologically active earth's surface. Thus, the microclimate at and near this surface, in general being substantially different from the macroclimate, is of great interest both to meteorologists and biologists.

Considering this active surface as an idealized boundary between the atmosphere and the underlying ground, the incoming and outgoing energy fluxes must be in balance. Irrespective of the kind of surface, the main components in this energy balance can be written:

$$Q + B + H + LE = 0 \tag{1}$$

where Q = net radiation balance, B = heatflux out of or into the absorbing subsurface soil, H = transfer of sensible heat from or to the air, and LE = exchange of latent heat by condensation or evaporation at the surface. All these fluxes are considered positive when adding energy to the surface. Terms of energy exchange due to advection, precipitation, freezing, melting, and biotic processes are ignored here.

The net radiation balance at the earth's surface, Q, may be expressed in the form:

$$Q = (R_g - r \cdot R_g) + (R_H - R_F) \tag{2}$$

where the short-wave terms in the first bracket consist of R_g = global radiation, and $r \cdot R_g$ = radiation reflected from the surface, while the long-wave terms in the last bracket consist of R_H = long-wave back radiation from the atmosphere and R_F = long-wave radiation from the surface, see e.g. GEIGER (1966, p. 10 and p. 13).

Half hourly energy balance data from three alpine IBP sites (1225–1320 m.a.s.l.) at Stigstuv, Hardangervidda (60°17′N, 7°40′E) are available for the periods 26.6–8.9 1970, 4.6–14.9 1971, and 8.6–15.9 1972. These periods are hereafter called P707172. The three sites are the lichen heath, the dry meadow and the wet meadow (SONESSON, WIELGOLASKI and KALLIO, 1975), hereafter called *L*, *D*, and *W* respectively. Due to topography and varying permeability of the underground *W* is poorly drained, *D* is well drained, and *L* is excessively drained (HINNERI et al., 1975). At *D* and *W* the transpiring vegetation is relatively deep-rooted grasses, while lichens at *L* are rather disconnected to the underlying soil. In spite of practically the same amount of precipitation the surfaces at *L*, *D*, and *W* are therefore points on a gradient, hereafter called "gradient *LDW*", along which the availability of water increases markedly.

Q, R_g and $r \cdot R_g$ were obtained by direct measurements at *D*. The long-wave term (R_H–R_F) was then calculated from equation (2). At *L* and *W* these measured values of Q were used with appropriate corrections for differences in albedo and surface inclination. No corrections were made for differences in surface temperature. Assuming a solar constant of 1.95 ly min^{-1}, daily totals of solar radiation, Q_e, on a horizontal surface outside the atmosphere have been calculated according to MCCULLOGH (1968). The monthly mean values of Q_e at Stigstuv are 972,

Table 1. Mean values in ly day^{-1} of transfer of sensible heat from or to the air *(H)*, calculated actual evapotranspiration *(LE)*, and potential evapotranspiration *(PLE)* during P707172. The average of three values, namely the extreme diurnal value during each of the three growing seasons, in brackets. Note that a latent heat flux *(LE, PLE)* of −58 ly day^{-1} is equivalent to an evaporation rate of 1 mm day^{-1}. Mean evaporation rates, *WE*, from 0.25 m^2 water pans are also included (mm day^{-1})

Site	*H*	*LE*	*PLE*	*WE*
L	−129 (−332)	− 98 (−191)	−167 (−352)	2.6
D	−120 (−295)	−115 (−243)	−164 (−352)	2.2
W	− 94 (−224)	−127 (−237)	−153 (−315)	—

922, 741, and 501 ly day^{-1} for June, July, August and September, respectively (1 ly = 1 cal cm^{-2}).

The following evaluations were carried out on data from *L*, *D*, and *W*: *B* was evaluated from measurements of temperatures in the top soil layer. *H* and *LE* were evaluated from measurements of temperature and humidity in the near-surface air layer, using the Bowen ratio method. The potential evapotranspiration, *PLE*, was evaluated according to the Penman formula (PENMAN, 1956).

The observed mean value of R_g during P 707172 is 364 ly day^{-1}, which gives a mean value of 0.44 for R_g/Q_e (the transmissivity of the atmosphere to solar radiation). The daily values of R_g/Q_e range between 0.10 and 0.77. The observed mean value of R_g/Q_e is 0.17 on days without bright sunshine, and 0.75 on cloudless days. These mean values are 25–30% and 12–14% higher, respectively, than the corresponding mean values observed at Bergen (60°24′N, 5°19′E, 45 m above m.s.l.) on the western coast of Norway.

At *D* the albedo *r* increases from 0.10–0.15 just after snow melt in early June to 0.20–0.22 about a month later. Then no further increase is observed until 15th September. The albedo at *W* is nearly identical to that at *D*, while the albedo at *L* is some 0.03–0.04 lower than that at *D*.

The observed mean value of the long-wave term $(R_H–R_F)$ during P 707172 is −76 ly day^{-1}. $(R_H–R_F)$ is found to vary mainly with cloudiness. The following regression equation is obtained between daytime mean values of $(R_H–R_F)$ in ly min^{-1} and daily values of R_g/Q_e:

$$(R_H - R_F) = -0.32 \cdot (R_g/Q_e - 0.1)^2 - 0.005\,; \tag{3}$$

correlation coefficient = 0.92 (n = 188).

The lowest observed daytime mean value of $(R_H - R_F)$ is −0.18 ly min^{-1}, while the lowest observed nighttime mean value is −0.14 ly min^{-1}. The lowest observed half hourly mean value of $(R_H–R_F)$ is −0.34 ly min^{-1}. Comparison with data from various other sites confirms the conclusion of GEIGER (1966, p.443) that the long-wave radiation budget is virtually independent of height.

During P707172, the mean precipitation rate (2.0 mm day^{-1}) was less than the mean rates of potential evapotranspiration, PLE, at all sites, while fairly equal to the mean rates of actual evapotranspiration, *LE* (Table 1).

During daytime, the gradient in availability of water induces considerable differences in H and LE along "gradient LDW", while Q and B are quite similar. From observations of the ground water table at W the days with the highest ground water table (23% of total) and the days with the lowest ground water table (17%) are selected, called "wet" and "dry" days, respectively. During daytime on "wet" days the mean values of LE are 86%, 91%, and 99% of the respective mean values of PLE at L, D, and W, respectively. During daytime on "dry" days the mean value of LE is 40%, 57%, and 74% of PLE at L, D, and W, respectively.
During fair weather periods, measurements of surface temperature were carried out by means of an infrared radiation thermometer. Every 15 min readings of surface temperatures were taken at several selected spots at one of the sites, and this measuring program was alternated between L, D, and W. For measurements carried out under cloudless conditions between 11:00 h and 17:00 h the following regression equations between half hourly mean values were obtained:

$$L:(T_0 - T_2) = 18.0 \cdot (1 - r) \cdot R_g + 1.7 \tag{4}$$

$$D:(T_0 - T_2) = 15.7 \cdot (1 - r) \cdot R_g - 1.5 \tag{5}$$

$$W:(T_0 - T_2) = 15.3 \cdot (1 - r) \cdot R_g - 4.6 \tag{6}$$

The correlation coefficient ranged from 0.50 to 0.85 ($n = 50$–60). Here T_0 = mean surface temperature (° C) of several selected spots at the site in question, and T_2 = air temperature (° C) at 2 m level. These regression equations are based on values of $(1-r) \cdot R_g$ ranging between 0.3 ly min^{-1} and 1.2 ly min^{-1}, while mean wind speeds at 2 m level are 3.5 m s^{-1}–4.0 m s^{-1}. The maximum rate of $(1-r) \cdot R_g$ at L, D, and W is a little above 1.1 ly min^{-1}. When $(1-r) \cdot R_g = 1.1$ ly min^{-1}, the differences between surface temperature and the air temperature at 2 m level $(T_0 - T_2)$ are under cloudless conditions as an average 21.5° C, 15.8° C, and 12.2° C at L, D, and W, respectively, according to the Eqs. (4)–(6). The corresponding maximum differences observed under cloudless conditions are 25° C, 18° C, and 14° C. The highest half hourly mean surface temperatures observed under cloudless conditions are 42° C, 35° C, and 27° C at L, D, and W, respectively. These differences in fair weather temperature regimes, at practically identical macroclimatic conditions, exist mainly because of the evapotranspiration, LE, consuming an increasing part of the available energy along "gradient LDW". Furthermore, the wind force being practically the same, the increase along "gradient LDW" of water vapor added to the air from the surface tends to reduce the evaporating power of the air along "gradient LDW". Thus the evaporation from a body of constant degree of wetness should undergo a decrease along "gradient LDW". This conclusion is supported by data from Piche evaporimeters at 2 m level. The mean evaporation rates for July and August 1970, 1971, and 1972 were 4.3 ml day^{-1}, 4.0 ml day^{-1}, and 3.4 ml day^{-1} at L, D and W, respectively. The corresponding rate within the Norwegian IBP birch forest site at Maurset (60°25′N, 7°21′E, 780 m.a.s.l.) was as low as 1.9 ml day^{-1}, probably because of shading and low wind speeds in the forest.

Differences in the amount of radiant energy input to the surface are another fundamental cause of microclimatic variations. Such differences may arise from a

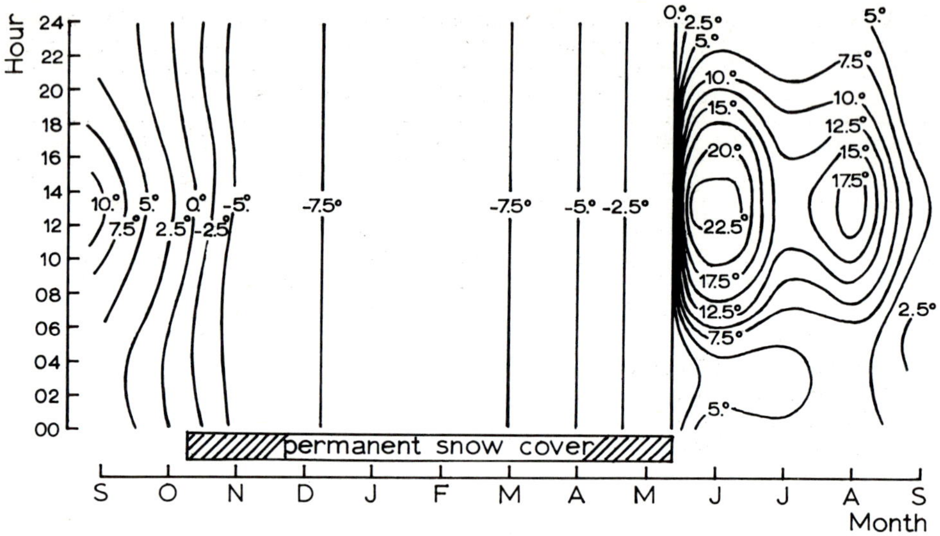

Fig. 2. Isopleth diagram of the monthly means of hourly temperatures within the vegetation at the lichen heath, Hardangervidda, Norway, for the period September 1969 to September 1970

variety of different factors. The effects of a few factors are illustrated in the isopleth diagram of the monthly means of hourly temperatures of a sensor at the soil's surface, completely buried in the vegetation at *L* for the period September 1969–September 1970 (Fig. 2). At *L* snow cover which typically varies in thickness from 30 to 50 cm, usually stays permanent from late October to late May. Under this snow cover a slow day to day variation, but almost no diurnal cycle in ground temperature, is observed. Moreover, the melting of this snow cover brings an almost momentary change from a typical winter to a typical summer temperature regime. This clearly demonstrates the importance of a radiation screening and thermal insulating snow cover to the microclimate within the vegetation.

Furthermore, differences in thickness of snow cover caused by topography, snow-drifting and varying amounts of snowfall, account for considerable differences in ground temperatures during winter. Thus the vegetation on the almost snow-free ridges is subjected to temperatures close to air temperature, including spells of hard frost, while the vegetation in the depressions, where snow may accumulate up to several meters, is rarely subjected to temperatures below $-1°$ C. This is actually the case at *D*, *W* and the birch forest at Maurset, where the snow depth typically reaches about 1–2 m in late winter (Table 2), and even more in a typical *Salix herbacea* snow bed studied at Finse, Hardangervidda.

The influence of varying radiation conditions upon ground temperature during the snow-free period is clearly demonstrated in Fig. 2. For the months June, July, and August 1970, the mean transmissivity of the atmosphere to solar radiation was 0.52, 0.33, and 0.40.

Thus it may be concluded that during the snow-free period, varying amounts of radiant energy input to the surface, together with differing exchanges of this

Table 2. First (D1) and last date (D2) of permanent snow cover together with the observed maximum of snow depth *(H)* in meters during 3 winters. D1 and D2 are given for the spots where litter temperatures were measured. Data of snow depth are obtained from personal communication with T. SOLHØY and S. KJELVIK

Winter:	1969–1970			1970–1971			1971–1972		
Site[a]:	D1	D2		D1	D2	*H*	D1	D2	*H*
W	24.10	10.6	1.9	18.10	5.6	1.6	12.10	8.6	1.5
D	24.10	4.6	1.8	18.10	29.5	1.1	12.10	3.6	1.2
L	24.10	27.5	0.6	18.10	29.5	0.8	12.10	23.5	0.5
B.W.	26.10	26.5	1.3	25.10	25.5	1.4	3.11	12.5	1.0

[a] *W*: Wet meadow at Stigstuv.
D: Dry meadow at Stigstuv.
L: Lichen heath Stigstuv.
B.W. Birch wood at Maurset.

energy, are main factors governing the microclimate and the deviations between microclimate and macroclimate. During winter the microclimate of the alpine plant cover is mainly governed by the thickness of the snow cover.

The three main areas studied at Hardangervidda, Stigstuv, Maurset and Finse have a somewhat different climate. Highest precipitation is generally found at Finse and Maurset, while the values may be estimated to be at least 25% lower at Stigstuv than at Maurset. The lowest summer precipitation (July + August) in the years studied was found in 1971 in all three areas (Table 3), while the annual sum was highest this year. The air temperature (2 m) is generally highest in the lowest elevated Maurset area, and lowest at the wet meadow at Stigstuv, at the highest elevation. In the summers studied, highest mean temperatures of July–August are found in 1969 and 1972, and lowest in 1970 (Table 3). The winter 1969–1970 was also very cold, and the annual mean temperature was therefore lowest in 1970.

Permafrost and Hydrology at Stordalen, Abisko, Sweden

Among the IBP Tundra Biome sites within Fennoscandia, the Stordalen area is of special interest, due to its permafrost situation. The occurence of perennial and seasonal frost is described briefly by SONESSON (1970). The recently obtained data confirm the picture.

The soil situation is studied at Stordalen mire by means of temperature probes recording at 8 levels between surface and 150 cm depth. The boundary between frozen and unfrozen soil is observed weekly with a frost penetration meter of the GANDAHL (1957) type, reaching about 200 cm depth. On several occasions during the growing season, a great number of soundings have been made by a geological stick.

It has been found that the whole mire contains frozen soil throughout the growing season. The seasonal thawing-freezing cycle is shown by the generalized diagram in Fig. 3. The cycle can be regarded as starting during the earliest days of

Table 3. Monthly values of: T_D = daily temperature (°C), T_x = maximum temperature (°C), T_N = minimum temperature (°C), and R = precipitation (mm) from the Norwegian IBP tundra areas Stigstuv (wet meadow), Finse (meteorological station close to the snow bed) and Maurset (birch wood). Station levels in meters are included in brackets. Data from Finse and precipitation data from Maurset (outside the birch wood) are obtained from personal communication with DNMI[c]

	Stigstuv (1320 m)				Finse (1224 m)				Maurset (780 m)			
Period	T_D	T_x	T_N	R	T_D[a]	T_x	T_N[a]	R	T_D	T_x	T_N	R[b]
Jan. 69					-7.2		-10.2					43
Feb. 69					-14.9		-18.8					16
May 69					1.0		-1.8					17
Jun. 69					7.3		3.3					25
Jul. 69	6.9	10.2	3.7	77	7.0		3.9					105
Aug. 69	9.4	13.8	5.3	37	10.0		6.1					13
Sep. 69	3.5	6.5	0.8		4.0		2.0		6.6	10.5	2.8	184
Year 69					-2.0		-5.5					721
Jan. 70	-13.5	-9.1	-17.1		-11.3	-7.5	-15.8	42	-9.8	-6.4	-13.5	30
Feb. 70	-15.3	-8.0	-20.9		-14.7	-10.4	-19.6	18	-13.4	-8.3	-19.5	11
May 70	-0.3	3.1	-3.3		1.1	5.3	-2.1	49	4.2	8.6	-0.2	46
Jun. 70	8.6	13.5	2.6		7.9	13.4	3.3	84	11.8	18.7	3.7	33
Jul. 70	5.1	8.3	1.9	109	5.5	8.8	3.4	180	8.3	12.3	4.1	150
Aug. 70	7.4	11.0	4.2	32	7.8	11.2	4.6	51	10.3	15.4	5.4	35
Sep. 70	2.4	6.2	-0.3		2.9	5.6	0.9	138	5.4	10.2	1.6	91
Year 70	-3.3	1.1	-7.2		-2.6	1.1	-6.2	882	-0.1	4.6	-5.1	826
Jan. 71	-7.3	-4.5	-9.9		-6.8	-3.5	-10.3	104	-4.5	-1.3	-8.3	61
Feb. 71	-5.9	-4.5	-9.9		-6.6	-4.1	-9.2	54	-3.8	-0.7	-7.7	80
May 71	0.4	3.3	-2.5		1.3	3.9	-0.9	45	4.7	8.8	0.6	35
Jun. 71	3.9	7.4	0.8		3.8	7.3	1.2	47	7.3	12.3	2.3	44
Jul. 71	6.8	10.3	3.4	66	6.7	10.2	3.7	80	9.9	14.8	4.6	91
Aug. 71	6.3	10.3	2.7	39	6.8	10.4	4.1	78	9.3	15.0	3.6	44
Sep. 71	3.5	7.4	0.5		3.9	7.3	1.4	121	6.0	10.9	1.5	97
Year 71	-1.5	2.0	-5.0		-1.3	1.9	-4.3	1197	1.5	5.8	-2.8	1200
Jan. 72	-11.7	-9.1	-14.8		-12.1	-8.0	-16.4	32	-8.7	-6.0	-12.0	66
Feb. 72	-9.6	-5.3	-13.0		-10.3	-6.5	-14.2	25	-6.6	-3.2	-10.3	101
May 72	0.9	4.6	-1.8		-1.4	5.2	-1.6	37	5.4	10.0	0.8	13
Jun. 72	4.2	7.5	1.2		4.4	8.2	1.7	117	8.7	13.6	3.8	59
Jul. 72	9.5	13.7	4.9	73	9.3	13.9	5.5	98	12.2	18.4	5.6	47
Aug. 72	6.2	10.1	2.7	79	6.3	10.6	3.3	191	8.6	13.9	3.5	90
Sep. 72					2.0	5.5	-0.6	33				57
Year 72					-1.6	2.0	-4.7	884				902
Jan. 73					-5.2	-2.3	-7.9	47				90
Feb. 73					-8.6	-5.3	-11.7	79				154
May 73					0.0	3.5	-2.6	65				66
Jun. 73					3.5	8.1	0.1	59				49
Jul. 73					8.6	13.1	5.0	85				57
Aug. 73					5.4	9.5	2.6	134				68
Sep. 73					2.1	4.6	0.4	140				85
Year 73					-1.8	1.6	-4.7	989				1125

[a] Temperature data from Finse for 1969 are estimated from Haugastøl data, applying monthly corrections in accordance with the corresponding differences between monthly standard normals.

[b] Up to July 1970 precipitation measurements were carried out at Fet, 735 m a.s.l. and 3 km WSW of Maurset.

[c] Det Norske Meteorologiske Institutt, Oslo.

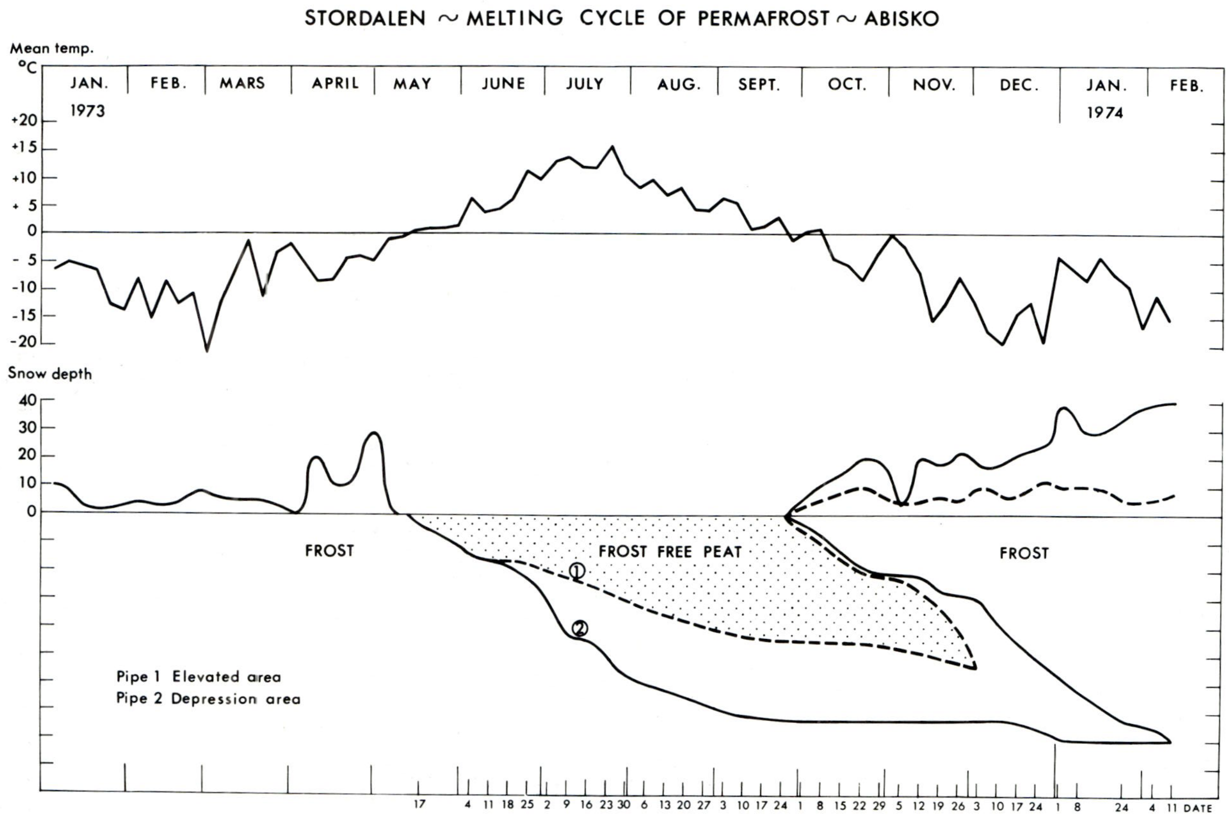

Fig. 3. Seasonal thawing and freezing of peat at Stordalen. Frost penetration meters are installed two meters apart: in an elevated area (1) and in a depression area (2), respectively. In addition, the diagram shows snow cover, snow depth in cm and air temperature as five-day-means. The yearly cycle of thaw ends in February in the depressions, but in December in the elevated area. While peat in the elevated area shows a frost table at 45–54 cm, this is in the depression area found at 73–80 cm depth.

May, when the frost disappears from the surface. The very slow diffusion of heat through the peat layers results in an average thawing depth of 60 cm during the latest part of the growing season (SONESSON and BERGMAN, 1972). The maximum depth of thaw occurs in November, and seems to be not more than some few centimeters greater. But during the middle of October the first freezing starts at the surface, and normally penetrates the whole unfrozen layer of elevated areas within 3–4 weeks. The latest time of temperatures above the freezing point is recorded during early January at 50 cm depth below a depression area (Fig. 3). The general feature of the thawing-freezing cycle thus found coincides well with the mean yearly variation of soil temperature at Abisko Natural Science Station (ÅNGSTRÖM, 1958). The occurrence of permafrost at the Abisko station, however, is found to be much more scattered than the probably continuous permafrost layers at Stordalen mire.

The topography of the frost table is very irregular (RYDEN, 1973; MADSEN and WIDELL, 1974), and reflects i.a. the heterogeneity of the peat. As a mean, the permafrost surface conforms with the surface, but is lower in depressions than in elevated areas of the mire. This is particularly the case in palsas. Thus, the permafrost table shows as a whole greater height differences than the ground surface.

The permafrost plays an important role in the water budget of the mire. In principle, the Stordalen area is a groundwater outflow area fed by an appreciable flow from the mountain rising from the southern border of the mire. However, the permafrost acts as an obstruction layer to part of the groundwater outflow, which is thus restricted to places of unfrozen ground within the area. The small lake east of the mire is a receiver of groundwater, while much of the mire acts as a confining layer above artesian groundwater, which on occasions can be seen welling up through the bottom of one of the pools. This indicates that a permafrost body may be non-continuous.

Surface runoff is not observed within the mire itself at any other time than in connection with snowmelt during May. Vertical exchange of water in the peat is the normal case through the rest of the growing season (RYDÉN op. cit.), i.e. evaporation balances rainfall. Horizontal flow of soil water does not occur until the water has percolated to the permafrost table, upon the irregular surface of which the water finds its way to the depression areas of the permafrost table. Thus, the depressions store much more water and energy than the elevated areas, which should be one of the reasons for the permafrost table being normally lower under depressions.

Some climatic variables from Stordalen are listed in Table 4.

On Climate at Kevo, Finland

On the basis of the average air temperature course, it can be estimated that the temperature is above 0° for approximately 160 days (period from 3rd May through 10th Oct.), and above 5° for about 115 days (from 27th May through 17th Sept.) in the Kevo area. These figures hold for the pine forest and the birch forest in the valley, but both periods are estimated to be 15 days shorter on the low alpine heath (SONESSON et al., 1975; KÄRENLAMPI, 1972).

Table 4. Some climatic variables at Stordalen 1972–1973. Monthly mean temperatures and precipitation

	Jan.	Feb.	Mar.	Apr.	May	Jun.	Jul.	Aug.	Sep.	Oct.	Nov.	Dec.	Year
Air temp. 200 cm °C	−9.3	−12.1	−7.4	−5.8	+1.8	+9.4	+12.7	+8,7	+ 3.3	−4,3	−7.5	−4.7	−1.3° C
Soil temp. −10 cm °C	−3,0	−4,2	−3,7	−1.2	−0.4	+1.9	+ 5.3	+4,2	+ 1.8	−0.4	−2.4	−7.1	−0.8° C
Precipitation mm	[a]	[a]	[a]	[a]	9,5	41.5	68.7	57,4	46.5	[a]	[a]	[a]	[a]

[a] Not available at Stordalen.

A presentation of daily summer data and 5-day-period values throughout the year at Stordalen is found in the Swedish Tundra Biome Project Progress Reports of 1972 and 1973.

The air temperature maxima at standard height on the low alpine heath are 2–3.5° lower in summer than in the valley. Conversely, during a short period in the very middle of the winter (January) they are 1° higher on the mountain summit. This is due to continuous temperature inversion caused by the polar night. The corresponding temperature minima on the low alpine heath are about 1° lower than those in the valley during the summer months, but during the winter months they are, on an average, 2–5° higher. This is mostly due to the drainage of cold air down the slopes. The standard air temperature on the low alpine heath is thus lower during summer, higher during winter and has a slightly higher yearly mean and a range which is several degrees narrower than at the meterological station in the birch forest. The conditions and differences during summer are, naturally, of most interst to ecologists.

The snow cover is thickest in the birch forest. It is also fairly even, having a low standard deviation. The pine forest has a thinner snow cover. Only some spots in openings between trees have snow as thick as at the meterological station. The snow depth is least on the low alpine heath, and shows the greatest variation. This is caused by the wind conditions. During the winter of the observations (1968–69), the snow depth was fairly close to the mean over several years, though slightly less in spring. The thickest snow cover occurs at the end of March, when it is about 70 cm in open places. The mean of the number of days with snow cover is 213 for several years. Thus the snow-free period is about five months.

The development of low forest in the birch forest sample area in the valley is permitted by the often relatively high temperatures. Microtopographically the site is, however, relatively open and windy, so that the forest is sparse and low. The soil is fairly moist, owing to the northern exposure and its fairly good water-holding capacity (small grain size). The moist soil is one of the main factors contributing to the occurence of relatively luxuriant vascular vegetation. The shade from trees and mesic ground vegetation increases the wetness and coldness of the soil, and heat exchange at the soil surface is thus hindered.

Acknowledgement. The Norwegian author, A. SKARTVEIT, wishes to express his thanks to Professor K. UTAAKER for valuable suggestions and criticism.

References

ÅNGSTRÖM, A.: Sveriges klimat. (The Climate of Sweden) 2. ed., Stockholm: 1958.

BRUUN, I.: Standard normals 1931–60 of the air temperature in Norway. Oslo: Det Norske Meteorologiske Institutt 1967.

BRUUN, I., HÅLAND, L.: Standard Normals 1931–60 of number of days with various weather phenomena. Oslo: Det Norske Meteorologiska Institutt 1970.

GANDAHL, R.: Bestämning av tjälgräns i mark med enkel typ av tjälgränsmätare. (Estimation of the frost table by means of a simple frost table meter.) Grundförbättring **10** (1957).

GEIGER, R.: The climate near the ground. Cambridge: Harvard University Press 1966.

JOHANNESSEN, T. W.: The Climate of Scandinavia. In: LANDSBERG, H. E. (Ed.): World Survey of Climatology, Vol. 5, pp. 23–63. Amsterdam-London-New York: Elsevier Publ. Comp. 1970.

KÄRENLAMPI, L.: Comparisons between the microclimates of the Kevo ecosystem study sites and the Kevo Meteorological Station. Rep. Kevo Subarctic Res. Stat. **9**, 50–65 (1972).

MADSEN, I.-L., WIDELL, S.: A vegetation map of the Stordalen site. Progress Report 1973. Swedish IBP Tundra Biome Project Tech. Rep. **16**, 3–15 (1974).

MCCULLOGH, E. C.: Total Daily Radiant Energy Available Extraterrestrically as a Harmonic Series in the Day of the Year. Arch. Met. Geoph. Biokl. B **16**, 129–143 (1968).

Det Norske Meteorologiske Institutt (DNMI): Foreløpige nedbørnormaler 1931–60 (in manus).

PENMAN, H. L.: Estimating Evaporation. Trans. Amer. Geophys. Union **37**, 43–56 (1956).

RYDÉN, B. E.: Markvattenrörelse i torv inom tundramyr—ett experiment. (Soil water movement in peat within a tundra mire—an experiment.) In: Nordiske IHD Feltsymposium. Myrhydrologi. Hydrologiske modeller, pp. 59–64. Trondheim: Den Norske Komité IHD, rap. 4, 1973.

SMHI: Sveriges Meteorologiska och Hydrologiska Institut: Normalnederbörd, månadsvärden för perioden 1931–1960. Stencilerade tabeller från Klimatbyrån. Utgåva 2. Stockholm: SMHI 1969.

SONESSON, M.: Studies on mire vegetation in the Torneträsk area, Northern Sweden. III: Communities of the poor mires. Opera Botanica **26**, 31–35 (1970).

SONESSON, M., BERGMAN, H.: Phytomass changes between two samplings stordalen 1970. Swedish IBP Tundra Biome Project Tech. Rep. **2** (1972).

SONESSON, M., WIELGOLASKI, F. E., KALLIO, P.: Description of Fennoscandian tundra ecosystems. In: WIELGOLASKI, F. E. (Ed.): Fennoscandian tundra ecosystems, Part 1: Plants and microorganisms, pp. 3–28. Berlin-Heidelberg-New York: Springer 1975.

3. Primary Producers

Introduction

F. E. Wielgolaski and P. Kallio

Great variation of plant structure can be seen in different regions and at different elevations of Fennoscandian tundra (Dahl, this section). Plant biomass and primary production are close to zero at the highest peaks, but sometimes also at lower mountain tundra elevations, particularly under nutrient poor conditions, where the ground is covered by large rocks or is free of snow for only a short period during summer, a period generally decreasing northwards towards the real arctic. Yet under the most favorable conditions in Fennoscandian tundra, biomass and production may be even higher than is often found in the lowlands of the region. The structure and mosaic of Fennoscandian tundra means that several plant communities have to be studied to obtain an overall functional view of each area. Typical points on gradients (e.g. from poor to rich communities) must be studied so that intermediate values on the gradient could be estimated later from the observed values.

Before the IBP studies, it had not been decided which methods should be used for estimation of primary producers in the mosaic. The least laborious method is an inventory (semiquantitative) between different communities (Hesjedal, this section), somewhat more detailed within important communities (Lye, this section). Tundra areas are well known to be very fragile (Willard and Marr, 1971; Bliss and Wein, 1972; Rickard and Brown, 1974) having a slow recovery. Diaspore dispersal (reproductive parts) is important for this recovery. The establishment of new vegetation can be studied relatively easily along a gradient from a retreating glacier (Elven and Ryvarden, this section). Also cultural influence e.g. grazing by domestic animals or fertilizing may change the natural plant communities as strongly in Fennoscandian tundra ecosystems as elsewhere and is discussed in this section by Wielgolaski. All these plant successions may be studied by structural technique (phytosociology), but more accurately by harvesting technique.

Primary production is mostly studied by repeated harvestings. However, the variation among samples is great even within relatively homogeneous areas. In the Fennoscandian IBP tundra studies, 10–15 replicates are found to be needed to obtain a production estimate based on destructive sampling with about 10% standard error, when groups of plant species and plant compartments are compared. Frequent harvestings are necessary if relatively realistic estimates of the production are to be obtained from the differences in biomass. The method is therefore very time-consuming and laborious, even if the samples are small (15–

20 full time workers were necessary during summer and fall to sort out the plants from 5–6 intensively studied sites and 3–4 other sites within the Norwegian IBP tundra group studies.)

When primary production is estimated from seasonal changes in biomass, considerable underestimates may arise if consumption by herbivores is not taken into account. It also is very important in the Fennoscandian tundra studies to take into consideration the death of plant parts during the growing season, and decomposition between two harvestings. Material observed as already dead at the first of two measurements, must be allowed to decompose to estimate the correct primary production (WIELGOLASKI, this section). Translocation of material from storage organs in spring (BERG, KJELVIK and WIELGOLASKI, this section) also influences the real primary production inferred from harvestings. (The Norwegian project leader, WIELGOLASKI, who is resposible and author of the last mentioned paper, wishes to thank BERG and KJELVIK for their valuable help to field and laboratory work on the ^{14}C technique.)

In tundra areas, primary production is sometimes studied by the so-called single species approach. Individuals of the main species in the community are separated into age classes (FLOWER-ELLIS, this section) which are measured and weighed. The life cycle of the species is studied and the number of individuals counted on an area basis. Then the production may be calculated both for cryptogams and seed plants (phanerogams) (CALLAGHAN and COLLINS, 1975). This method is less time consuming than the common harvesting method if there are relatively few species at a given site, and if the annual growth is easily observed.

Physiological methods for measuring the primary production, mainly based on CO_2 exchanges per g dry weight of green biomass per ground area, are normally even less time consuming than the single species approach. The conversion of CO_2 measured to dry matter will, however, give only approximate values; a problem is the more or less artificial conditions under which the measurements are carried out. This is of course most pronounced in laboratory studies, but is the case even in physiological field studies. Another practical problem is the transport of sophisticated equipment out to remote tundra areas. Despite the various problems, physiological studies with different CO_2 exchange methods produced some of the most interesting results in Fennoscandian tundra studies.

Utilization of solar energy forms a basis for the dynamic function of the ecosystem, and we ask whether the photosynthetic efficiency in tundra plants differs in a specific way in comparison with other zones. At least there seems to be a positive correlation between the caloric values of the plants and the elevation and/or the increase in latitude (WIELGOLASKI and KJELVIK, this section), and it is possible that the adaptation mechanism takes advantage of the long day light period ("the everlasting" day).

The floristic spectrum typically reflects the influence of a northern situation and its changes in the abiotic factors. Lichens and mosses, very typical components of many tundra ecosystems, seem to be best adapted to low temperatures in the growth period. There are several species in these groups having highest net photosynthesis rates within the temperature range 0° to 5° C, while most vascular plants have highest values at temperatures approximately 10° higher. The minimum temperature with a still positive energy equilibrium is at $-5°$ or $-8°$ C.

Adaptation is partly of an ecotypic origin, but there are good examples of "preadaptation". The acclimatization of photosynthesis to rapid changes of temperature is important in tundra because temperatures may sink below the active range of well adapted plants even during the "growth period." In vascular plants the acclimatiziation takes more time than in lichens and mosses (and this had to be taken into consideration also in the experimental work). In the latter groups, e.g. thawing after an exposure to $-30°$, a positive net assimilation may be seen a few minutes after $0°$ is reached. This type of response is very important from the viewpoint of evolution of life in polar conditions. This has been a special problem in the Finnish IBP tundra work (KALLIO and HEINONEN, this section).

The processes of photosynthesis and respiration determine the energy budget. Photosynthesis, dark respiration and photorespiration are components of the system. These problems are of central concern in the Norwegian IBP work (NYGAARD, this section; SKRE, this section). Respiration during the long winter period, when photosynthesis has no possibilities of functioning, is of particular importance in the north.

The most characteristic feature of the Arctic's abiotic environment is the everlasting day. There may be some non-adaptability involved in this. In higher plants the translocation of the photosynthates may demand a rhythm, and the lack of rhythm may cause disturbances in the photosynthetic apparatus of the cryptogams. Examples are given in this section by KALLIO and VALANNE.

The chlorophyll content is apparently correlated to the photosynthetic capacity. Other factors affecting net production include the leaf-area angle towards the sun. These characteristics may also have some effect in the alpine and subarctic ecology in Fennoscandia (BERG, this section; BERG et al., this section).

Similarly, the primary production of nitrogen indicates some zonal characteristics. It is, however, notable that nitrogenase activity is not so well adapted to low temperature as photosynthesis. It seems that this factor, peculiar to subarctic and arctic zones, coupled with the different features in the adaptation of energy yield, gives the polar regions an ecological image of their own.

The chapters in the present section do not claim to give a full picture of the various methods used in primary production studies in Fennoscandian tundra ecosystems, or to cover all problems of primary production. The principal aim is to show that different useful methods have been evolved and that relatively simple methods and equipment often can and have to be used in remote rural areas. More sophisticated methods may be preferable and possible in solving similar problems in other cases.

The total live below and above ground plant biomass of vascular plants and cryptogams is generally of the order 400–800 g dry-weight per m^2 in the Fennoscandian tundra ecosystems. Usually the highest proportion of vascular plants (particularly monocotyledons) is below ground (KJELVIK and KÄRENLAMPI, this section; WIELGOLASKI, this section). The total annual primary production of the dominating communities in the ecosystems, is on a average about 200–400 g dry weight per m^2 per year (average growing season approximately three months) when death and decomposition are compensated for between harvestings. The highest total live plant biomass is found to be even above 2000 g dry weight per m^2 in wet, relatively, eutrophic meadows. In rich and moist willow thickets approximately 200 m above the tree line in southern Norway (1250 m a.s.l. at 60° N) annual production was above 800 g m^{-2}. Similar biomass values were found in subarctic Finnish oligotrophic birch forests (130 m a.s.l. 69.5° N) and a biomass twice as high was found in subalpine Norwegian oligotrophic birch forests (800 m a.s.l. 60° N) when the tree trunks are included (KJELVIK and KÄRENLAMPI, this section). The lowest ratio between total primary production and total live biomass is, as expected, found in communities with a high proportion of woody plants and/or cryptogams, less than 0.2; the highest ratio, approx. 0.75, is found in a Norwegian relatively eutrophic alpine dry meadow with few shrubs and cryptogams.

In considerable areas of the Fennoscandian tundra region, the biomass of lichens is very high, higher than the above- and belowground biomass of vascular plants. Wet communities, mires, are well developed in some parts of the region with high bryophyte (mostly moss) biomass (e.g. the Swedish IBP Tundra Biome site). Surprisingly high production values are often found for cryptogams when compensated for death and decomposition, in spite of generally lower assimilation rates than for herbaceous vascular plants (SKRE, this section). The main reason for this is probably the good adaptation of cryptogams to tundra conditions mentioned earlier. At the IBP lichen heath site in relatively oceanic alpine southern Norway, an annual production of about $^1/_4$ of the total average live lichen biomass (determined by radioisotopes) is found during the growing season (KJELVIK and KÄRENLAMPI, this section). The production of mosses is found to be about $^3/_4$ of the average live biomass (observed by ^{14}C) in wet Norwegian alpine tundra communities when death and decomposition are compensated for. This is also a high production-biomass ratio, but does not seem to be unrealistic when compared to Finnish laboratory experiments (KALLIO and HEINONEN, this section).

The estimated primary production by harvesting and physiological methods in the Fennoscandian tundra studies is of the same order as found in other tundra areas of the world (review by WIELGOLASKI, 1975). The structural variation is closely related to the variation found elsewhere with latitude, elevation and particularly soil moisture conditions (WIELGOLASKI and WEBBER, in preparation). A high structural similarity for instance, is found between Norwegian alpine wet meadows and high arctic wet meadows at Devon Island, Canada, and also between alpine lichen heaths in southern Fennoscandia (60° N) and beach ridges at Devon Island at 75.5° N. It is easily recognized that even in international comparisons the local nutrient and soil moisture status are often more important for structural plant types developed (based on life forms and not directly on species) than latitude, elevation and precipitation. Internationally, however, there seem to be more heath communities with relatively high amounts of lichens, cushion and rosette forbs, low shrubs and possibly caespitose monocotyledons and bryophytes in the real arctic areas and the highest alpine regions. The tundra areas of the world thus offer an opportunity to test the hypothesis that similarities in environment produce similarities in the plant structure and function just as in the Mediterranean climatic areas, as stressed by MOONEY in Volume 7 of the Ecological Studies series.

The present Fennoscandian and recent international comparisons of primary producers in tundra regions have been made possible through world wide cooperation of the IBP Tundra Biome. Coordinated studies have been carried out in various regions and countries approximately throughout the same time period.

References

BLISS, L. C., WEIN, R. S.: Plant community responses to disturbances in the western Canadian Arctic. Canad. J. Botan. **50**, 1097–1109 (1972).

CALLAGHAN, T. V., COLLINS, N. J.: Life cycles, population dynamics and the growth of tundra plants. In: MOORE, J. J. (Ed.): Tundra and related habitats, in prep. Cambridge: University Press 1975.

RICKARD, W. E., BROWN, J.: Effects of vehicles on Arctic tundra. Environm. Cons. **1**, 55–62 (1974).

WIELGOLASKI, F. E.: Primary production of tundra. In: COOPER, J. P. (Ed.): Photosynthesis and productivity in different environments, pp. 75–106. Cambridge: University Press 1975.

WIELGOLASKI, F. E., WEBBER, J.: A comparison of circumpolar arctic and alpine tundra vegetation. I. Classification, II Ordination. In manuscript, about 30 pp.

WILLARD, B. E., MARR, J. W.: Recovery of alpine tundra under protection after damage by human activities in the Rocky Mountains of Colorado. Biol. Cons. **3**, 181–190 (1971).

3.1. Plant Structure

Flora and Plant Sociology in Fennoscandian Tundra Areas

E. Dahl

Altitudinal Belts

In the alpine tundra region of Fennoscandia four altitudinal belts are recognized (Sjørs, 1967).

1. *A subalpine belt* limited downwards by the boreal coniferous forests, and upwards by the climatic timber line. The climatic timber line is the highest level where forests can grow under optimal, local conditions. The highest level of forests within a somewhat larger area is taken as an estimate of the climatic timber line. The altitude is slightly above 1200 m in the most continental parts of the Jotunheimen mountains in southern Norway; it is at about 1050 m at Hardangervidda in southwestern Norway, at about 700 m at Abisko in northern Sweden, and slightly less than 350 m at Kevo in northern Finland. The limit drops northwards and westwards; the northernmost fringe of Norway lies north of the polar timber line.

The only forest-forming tree is birch (*Betula pubescens* ssp. *tortuosa*). Besides birch forests there may be extensive treeless areas, either heath plant communities or mire communities.

2. *A low alpine belt* dominated by shrub communities with *Betula nana*, *Juniperus communis* or various *Salix* species, or ericaceous heath communities dominated by *Empetrum hermaphroditum*, *Vaccinium myrtillus* and *V. uliginosum* or *Phyllodoce coerulea*. The upper limit of the low alpine belt is placed where *Vaccinium myrtillus* disappears. *Vaccinium myrtillus* requires a snow cover for survival, but in a cold climate a too heavy snow cover might mean a vegetation too short for the species. Ascending the mountains one sees how the width of the *Vaccinium* zone in the snow beds gradually diminishes, until at some level *Vaccinium* is "squeezed" out of existence. This level is taken as the upper limit of the low alpine belt.

Besides shrubs and ericaceous heath communities also grass heaths, i.e. vegetation dominated by xerophilous graminoids, are common. Mires are also common. Snow bed communities on solifluction soils cover small areas at lower levels, but become more extensive upwards.

The low alpine belt is about 250–300 m high, reaching 1500 m in the Jotunheimen mountains.

3. *A mid alpine belt* which may be 400 m high in Jotunheimen (Nordhagen, 1943), and about 200 m in northern Sweden becomes narrower towards the coast

and northwards. No mid alpine belt is present in northern Finland (KALLIOLA, 1939). The vegetation on stable soils consists of grass heaths dominated by *Juncus trifidus*, *Carex bigelowii* or *Festuca ovina;* the only ericaceous species commonly occurring is *Vaccinium vitis-idaea.* Plant communities typical of solifluction soils cover extensive areas, and the percentage increases upwards until the last communities on stable soils disappear. This is taken as the upper limit of the mid alpine belt. The climate is too cool for peat accumulation and consequently mire formation.

4. *A high alpine belt* extends up to the summits of the highest mountains. All soils are unstable, and the vegetation is scattered and patchy. Vascular plants have been found up to about 100 m below the summit of Galdhøpiggen, Norway, 2469 m.

Altitudinal Limits of Species; Life Forms and Phytogeographic Elements in the Vascular Plant Flora

Considerable information is available on altitudinal limits of vascular plant species in southern Scandinavia. The highest level of occurrence of a plant species may be taken to indicate the species' ability to survive in a cold climate, since temperatures decrease with altitude.

In Table 1 the species found in southern Scandinavia occurring above 1000 m have been stratified into groups with altitudinal limits within 200 m interval. The species are also stratified according to life forms in the Raunkiaer system, as understood e.g. by CLAPHAM et al. (1962). Above 2000 m only chamaephytes and hemicryptophytes occur. The percentage of chamaephytes drops downwards, confirming the old observation by RAUNKIAER (1911) that the colder the climate, the higher the percentage of chamaephytes. Apparently it is of positive survival value to a species that its overwintering buds are placed close to, but not in direct contact with the soil surface. The percentage hemicryptophytes is fairly constant with altitude, while the percentage of therophytes, geophytes, and helophytes increases downwards. Annual species may have difficulties in carrying out an entire life cycle within one season in a cold climate. Most of the perennial alpine

Table 1. Species of vascular plants in southern Norway stratified according to altitudinal limit and life form according to RAUNKIAER (1911)

Altitudinal limit	Number of species	Life form						
		Phanerophyte	Chamaephyte	Hemicryptophyte	Geophyte	Helophyte	Hydrophyte	Therophyte
>2000 m	29		44.8%	55.2%				
1800–1999 m	39		25.6%	66.7%	2.6%	2.6%		2.6%
1600–1799 m	75	8.0%	32.0%	48.0%	4.0%	5.3%		2.7%
1400–1599 m	63	4.8%	9.5%	65.1%	11.1%	1.6%	3.2%	4.8%
1200–1399 m	138	6.5%	10.1%	51.4%	10.1%	10.1%	2.9%	8.7%
1000–1199 m	109	2.8%	5.5%	55.0%	10.1%	11.9%	5.5%	9.2%

Table 2. Species of vascular plants in southern Norway stratified according to altitudinal limit and phytogeographic element

Altitudinal limit	Number of species	Phytogeographic element			
		Amphiatlantic	Circumpolar	Eurasiatic	Endemic
> 2000 m	29	51.7%	37.9%	10.3%	0
1800–1999 m	39	35.9%	53.8%	10.3%	0
1600–1799 m	75	32.0%	58.7%	8.0%	1.3%
1400–1599 m	63	23.8%	47.6%	27.0%	1.6%
1200–1399 m	138	14.5%	37.0%	48.6%	0.7%
1000–1199 m	109	11.0%	36.7%	52.3%	0

species have a period of several seasons of bud differentiation, as shown by RESVOLL (1917). Geophytes cannot easily survive on soils subject to strong frost action. The helophyte flora grows mainly in mires which hardly occur above the low alpine belt.

In Table 2 the species have been stratified according to altitudinal limits and phytogeographic element. The amphiatlantic element comprises species occurring along both shores of the North Atlantic Ocean, but with large distribution gaps either east of the area in Europe or west of their area in North America. The circumpolar element comprises species having a fairly continuous distribution around the North Pole. The eurasiatic element comprises species occurring more or less continuously eastwards from Scandinavia, but absent from the eastern parts of North America. A very small endemic element restricted to Fennoscandia is also present. Information about species distribution in Table 2 is taken from HULTÉN (1958; 1962; 1971), supplemented by data from various floras.

It appears from the table that the percentage amphiatlantic species drops from about 50% in the element growing above 2000 m to about 10% in the element with altitudinal limits between 1000 and 1199 m. The amphiatlantic species are believed to have survived one or several glacial ages in unglaciated refuges along the Norwegian coast. Such species must be able to endure severe conditions, and therefore are also capable of growing at high altitudes today. The percentage eurasiatic species is highest among species with low altitudinal limits, while the circumpolar species have a maximum percentage among species with altitudinal limits between 1400 and 2000 m. The percentage of species with wide distribution areas in the circumpolar tundra areas is high. Hence, the possibilities of direct comparison of Fennoscandian vegetation types with corresponding vegetation types in other tundra areas around the North Pole are good (WIELGOLASKI and WEBBER, in prep.).

Vegetation and Environment

Probably nowhere in the world do environmental conditions for plant life change more abruptly within short distances than in alpine tundra areas. In the mountainous regions there is naturally a large-scale relief pattern. In the glaciated regions, a more small-scale relief pattern created by glacier erosion or developed in unconsolidated sediments left by the ice is superimposed on the large-scale relief pattern. In the absence of forest, the solar radiation is absorbed by a low vegetation or the soil, creating very different microclimatic conditions depending upon slope, aspect and exposure to winds (SKARTVEIT et al., 1975). Also, soil parent materials vary as to texture and mineral content (HINNERI et al., 1975). Drainage conditions vary from water-logging in the depressions to very dry conditions on gravelly hillsides. Important, too, is the small-scale relief pattern which creates very different conditions as to snow accumulation in winter. The thickness of the snow cover may vary from zero on exposed ridges to several metres of snow in the depressions. The unequal snow cover in turn results in very different temperature conditions experienced by the low plants in winter, in different times of snowmelt in spring and in a different duration of the vegetative season in summer (SONESSON et al., 1975). Within a horizontal distance of 20 m, mean annual temperatures in the soil may vary as much as 4° C, or about as much as the mean annual difference between Oslo and Paris (DAHL, 1956). Associated with differences in snow conditions are also differences in soil stability.

The large variation in environmental conditions over short distances results in correspondingly large differences in vegetational composition. The vegetation becomes a mosaic of very different vegetation types, each with a floristic composition and ecological niche of its own, but each patch of a small size. In vegetation mapping, the units mapped in alpine tundra areas in the scale of 1:10000 correspond to alliances in the phytosociological hierarchy (HESJEDAL, 1975), while in forested areas the mapping units on the same scale are associations and subassociations (WIELGOLASKI, 1971; MARKER, 1973).

In ecological studies, vegetational heterogeneity may be an advantage as well as a disadvantage. If the object of study is a comparison of plant life under very different conditions within a limited area, it is an advantage. Classical studies of this kind include the influence of snow cover in winter on tundra vegetation (VESTERGREN, 1902; NORDHAGEN, 1943; GJÆREVOLL, 1956; DAHL, 1956). Much phytosociologic research has been done on Fennoscandian tundra areas, and in general there is agreement on how to classify tundra vegetation (MARKER, 1973), although there are still differences in detail, as can be seen by a comparison of the papers by LYE (1975) and HESJEDAL (1975). But in other respects the vegetational and environmental heterogeneity may be a disadvantage. In ecological studies which analyze the functioning of an entire ecosystem including animals with considerable mobility, the heterogeneity is a complicating factor. To obtain a reliable average productivity figure, more observations are required in heterogeneous vegetation than in areas with more limited variation. For more precise studies it is necessary to concentrate on areas with relatively uniform conditions and vegetation, but sites for such studies must be carefully selected.

The vegetation types of the Fennoscandian tundra areas can be arranged along a set of environmental gradients. Besides an overall climatic gradient expressed in the altitudinal belts, the following gradients are recognized as important:

1. *A soil fertility gradient* related to the amount of cations, especially calcium ions, released by weathering from the soil parent minerals. Species or plant com-

munities capable of growing on the poorest and most leached soils are termed oligotrophic, those requiring somewhat more fertile soils are termed eutrophic, while calcicolous species or communities are restricted to soils high in calcium. On stable soils, profiles in oligotrophic communities tend to be podzols, while soils in calcicolous communities are often brown earths. Associated with high base saturation is usually low loss on ignition in the A_0 or A_1 layer, and high nitrogen content of the humus. Oligotrophic species are commonly xeromorphic, while calcicolous or eutrophic species are more commonly mesomorphic (DAHL, 1956).

The nutrient-rich soils are usually derived from more or less metamorphosed cambro-silurian limestones and schists (often phyllitic), while sandstones (sparagmites) and highly metamorphosed precambrian rocks usually produce oligotrophic soils.

2. *A snow cover gradient* related to the amount of snow accumulating in winter and time of snow-melt in spring. Species or communities restricted to areas with little snow in winter are termed chionophobous, those to areas with a good snow cover are termed chionophilous. An area where snow accumulates in winter is termed a snow bed.

Chionophobous communities are often dominated by lichens, especially in continental areas where reindeer grazing has not been too destructive. In snow beds with a limited amount of snow accumulation, many lowland species ascend high into the mountains e.g. *Vaccinium myrtillus*, *Solidago virgaurea* and *Trientalis europaea*. In areas with late snow-melt in summer, soils tend to become unstable solifluction soils, and the vegetation is dominated by mosses with few vascular plants e.g. *Salix herbacea*.

3. *A drainage gradient*. On well-drained and leached soils the communities are usually ericaceous or grass heaths, and the soil profile an iron podzol. In soils with somewhat impeded drainage and humus podzols, *Nardus* heaths commonly occur in the subalpine and low alpine belts. In areas with a permanently high water table, mires are formed in the subalpine and low alpine belts. In addition there is a regional gradient from oceanic to continental areas. In continental areas chionophobous lichen heaths cover large areas, while the chionophobous communities in oceanic areas cover less area, and are also richer in mosses.

References

CLAPHAM, A.R., TUTIN, T.G., WARBURG, E.F.: Flora of the British Isles. Second edition. Cambridge: University Press 1962.

DAHL, E.: Rondane. Mountain Vegetation and Its Relation to the Environment. Oslo. Skr. Norske Vid. Akad. Oslo I. Mat.-Nat. Kl. **3** 1956.

GJAEREVOLL, O.: The Plant Communities of the Scandinavian Alpine Snowbeds. Skr. Kgl. Norske Vid. Selsk. Trondheim **1**, 1956.

HESJEDAL, O.: Vegetation mapping at Hardangervidda. In: WIELGOLASKI, F.E. (Ed.): Fennoscandian tundra ecosystems, Part 1: Plants and microorganisms, pp. 74–81. Berlin-Heidelberg-New York: Springer 1975.

HINNERI, S., SONESSON, M., VEUM, A.K.: Soils of Fennoscandian IBP tundra ecosystems. In: WIELGOLASKI, F.E. (Ed.): Fennoscandian tundra ecosystems, Part 1: Plants and microorganisms, pp. 31–40. Berlin-Heidelberg-New York: Springer 1975.

HULTÉN, E.: The Amphi-Atlantic Plants. Stockholm. K. Sv. Vet. Akad. Handl. 4 Ser. **7**, No. 1, 1958.

HULTÉN, E.: The Circumpolar Plants. I. Stockholm. K. Sv. Vet. Akad. Handl. 4 Ser. **8**, No. 5, 1962.

HULTÉN, E.: The Circumpolar Plants II. Stockholm. K. Sv. Vet. Akad. Handl. 4 Ser. **13**, No. 1, 1971.

KALLIOLA, R.: Pflanzensoziologische Untersuchungen in der alpinen Stufe Finnisch-Lapplands. Helsinki. Ann. Botan. Soc. Zool.-Bot. Fenn. Vanamo **13** (1939).

LYE, K. A.: Survey of the main plant communities on Hardangervidda. In: WIELGOLASKI, F. E. (Ed.): Fennoscandian tundra ecosystems, Part 1: Plants and microorganisms, pp. 68–73. Berlin-Heidelberg-New York: Springer 1975.

MARKER, E. (Ed.): IBP-CT symposium om vegetasjonskartlegging. IBP i Norden **11**. Oslo: Nordic IBP committees 1973.

NORDHAGEN, R.: Sikilsdalen og Norges fjellbeiter. Skr. Bergens Museum **22**, 1943.

RAUNKIAER, C.: Det arktiske og det antarktiske Chamaefytklima. Biol. Arbeider tilegnede Eug. Warming. pp. 7–27. København 1911.

RESVOLL, T. R.: Om planter som passer til kort og kold sommer. Oslo. Arch. Mat. og Nat. vid. **35**, No. 6 (1917).

SJØRS, H.: Nordisk växtgeografi. 2. ed. Stockholm: Scandinavian University Books 1967.

SKARTVEIT, A., RYDÉN, E. B., KÄRENLAMPI, L.: Climate and hydrology of some Fennoscandian tundra ecosystems. In: WIELGOLASKI, F. E. (Ed.): Fennoscandian tundra ecosystems, Part 1: Plants and microorganisms, pp. 41–53. Berlin-Heidelberg-New York: Springer 1975.

SONESSON, M., WIELGOLASKI, F. E., KALLIO, P.: Description of Fennoscandian tundra ecosystems. In: WIELGOLASKI, F. E. (Ed.): Fennoscandian tundra ecosystems, Part 1: Plants and microorganisms, pp. 3–28. Berlin-Heidelberg-New York: Springer 1975.

VESTERGREN, T.: Om den olikaformiga snöbetäckningens inflytande på vegetationen i Sarekfjällen. Lund. Botan. Not. 241–268 (1902).

WIELGOLASKI, F. E. (Ed.): Forslag til vegetasjonsklassifisering (Preliminary classifications of Scandinavian mountain, forest, wetland and seashore vegetation). IBP i Norden **7**. Oslo: Nordic IBP committes 1971.

WIELGOLASKI, F. E., WEBBER, P. J.: A comparison of circumpolar arctic and alpine tundra vegetation. I. Classification, II. Ordination. In prep.

Survey of the Main Plant Communities on Hardangervidda

K. A. Lye

Introduction and Methods

During the years 1967–1968 and 1972, about 400 vegetation samples were investigated by qualitative plant-sociological methods at Hardangervidda. Homogenous samples were chosen subjectively and usually two samples, each 10 m^2, were investigated in each locality. Cover in percent and frequency were noted for all species. Soil samples were collected and analyzed chemically.

Localitites were investigated from the subalpine birch forest to the high alpine zone with only sparse vegetation from about 1600 m. In the intensive Norwegian IBP tundra localitites (Sonesson et al., 1975), a larger number of samples was analyzed than in the other areas. Permanent plots were fenced in a subalpine birch forest, and three alpine vegetation types (oligotrophic lichen heath, eutrophic dry meadow and eutrophic wet meadow) and changes in the vegetation noted at the meadows after three years without grazing. Taxonomic mushroom studies were carried out by Gulden (1975) at these plots in early August 1971. She found more species in the subalpine area than at higher elevation, and only very few of these were found in the alpine region. The fruit-bodies were considerably larger in the subalpine forest (often 10–15 cm diameter) than in alpine plots (mostly 0.5–2.5 cm diameter).

Vegetation Types

In the following condensed survey of the plant communities the classification system of Dahl et al. (1971) is accepted; this is largely based on Dahl (1956).

A. Subalpine Birch Forest. On both its eastern and western side, Hardangervidda is fringed by rather extensive birch forests dominated by *Betula pubescens* ssp. *tortuosa*, *Juniperus communis*, *Vaccinium myrtillus*, *Deschampsia flexuosa*, *Maianthemum bifolia*, *Melampyrum pratense*, *Solidago virgaurea*, *Trientalis europaea* and the bryophytes *Hylocomium splendens* and *Pleurozium schreberi*. Several species of the fungi *Cortinarius*, *Russula* and *Lactarius* and e.g. *Leccinum versipellis* and *L. scabrum* were particularly abundant in the forests.

This forest occurs even on soils very poor in plant nutrients. On better soils, however, a richer facies develops; it has the following additional species constantly present: *Anthoxanthum odoratum*, *Melampyrum silvaticum* and *Dryopteris linnaeana*. On the richest soils, the composition of the forestfloor changes remarkably and *Aconitum septentrionale*, *Geranium silvaticum* and *Potentilla erecta* often dominates.

The poorest types belong to the *Vaccinio-Piceion* alliance (KIELLAND-LUND, 1971), while the richest types belong to the *Lactucion alpinae* alliance (DAHL et al., 1971).

B. Lichen-Heaths. Most lichen heaths on Hardangervidda are oligotrophic, and occur in localities with very sparse snow cover (maximum snow cover during winter usually less than 50–60 cm). This community is dominated by *Empetrum hermaphroditum*, *Vaccinium vitis-idaea* and the lichens *Alectoria ochroleuca*, *Cetraria nivalis*, *Cladonia mitis*, *C. rangiferina* and *Ochroleucia frigida*. The last-mentioned species tends to be more common when other lichens have been severely damaged by overgrazing and throdding (LYE and LAURITZEN, 1975). Only few fungi, e.g. *Inocybe lacera* and *Laccaria laccata var. montana*, were found on lichen heaths. These lichen heaths are best developed in the eastern parts of Hardangervidda, and are very important as wintergrazing habitats for reindeer. In more oceanic facies of this community, the moss *Rhacomitrium lanuginosum* occasionally dominates.

In habitats with more accumulated snow during winter, *Alectoria ochroleuca* and *Cetraria nivalis* disappear and *Cladonia alpestris*, together with *C. rangiferina* and *C. mitis*, tend to dominate. This lichen heath is, however, more closely related to oligotrophic heaths and grasslands of intermediate snowcover.

C. Oligotrophic Heaths and Grasslands of Intermediate Snowcover. In habitats with a snow cover of 1–3 m during winter, a number of characteristic species appear, e.g. *Vaccinium myrtillus*, *Deschampsia flexuosa*, *Festuca ovina*, *Carex bigelowii*, *Juncus trifidus*, *Hieracium alpinum*, and on slightly better soils also *Anthoxanthum odoratum* and *Alchemilla alpina*.

This is a rather heterogeneous group of communities, which in some localities forms a continuous variable series, while in other places distinct discontinuities are found. The communities dominated by *Festuca ovina* and *Juncus trifidus* are found especially on soils very poor in organic matter and plant nutrients, while the *Vaccinium myrtillus* heath is sometimes found on better soils. These communities are frequently placed in different alliances, e.g. *Phyllodoco-Vaccinion myrtilli*, *Nardo Caricion bigelowii* and *Deschampsia-Anthoxanthion*.

D. Eutrophic Heaths and Meadows. In areas with cambro-silurian phyllitic rocks (rarely also limestones), different plant communities which are much richer in species develop. In eutrophic habitats where water is not scarce, e.g. near streams and on edges of mires, a meadow dominated by *Geranium silvaticum*, *Alchemilla glomerulans*, *Deschampsia caespitosa*, *Salix glauca*, *S. lapponum* and occasionally also *Aconitum septentrionale* and *Ranunculus platanifolius* is found. This community is found scattered throughout the area, but is more common in the subalpine birch forest zone than in the alpine zone. In this community lichens are very rare or absent, while many demanding bryophytes are found. Generally, the willow thickets contain a variety of different mycorrhizal fungi, particularly species of *Inocybe*, *Cortinarius* and *Hebeloma*. In taxonomic plant sociology, this community is usually placed in the *Lactucion alpinae* alliance (DAHL et al., 1971).

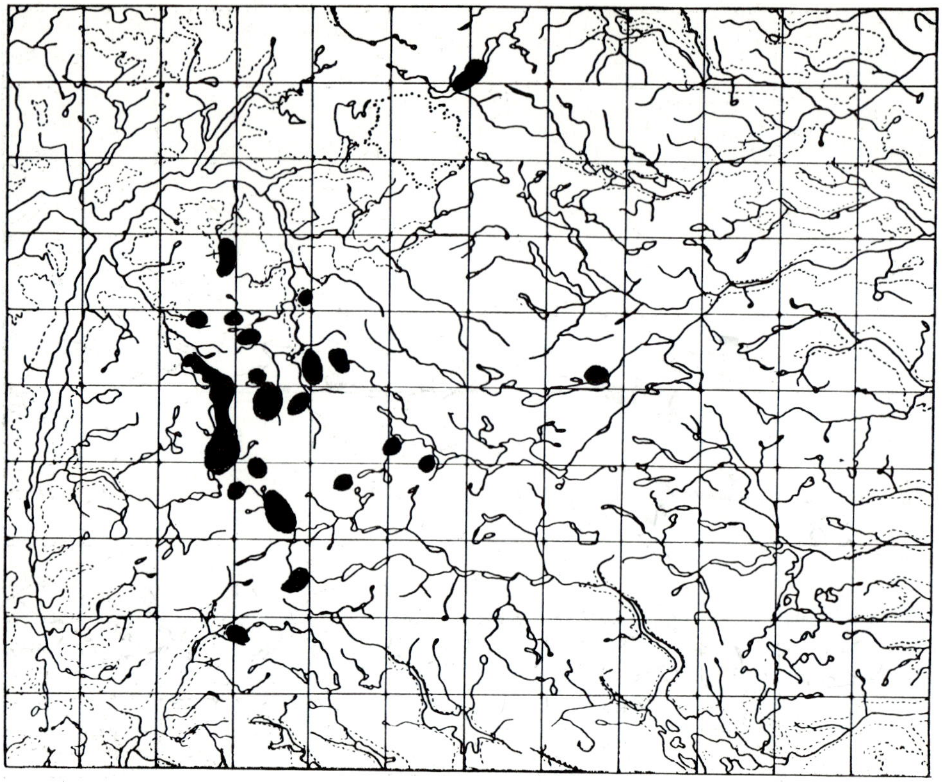

Fig. 1. The distribution of the most demanding society of the *Kobresio-Dryadion* alliance (i.e. eutrophic dry meadows) on Hardangervidda. This society is characterized by the occurrence of one of the following three species, viz. *Carex glacialis*, *Kobresia myosuroides*, and *Potentilla nivea*. The society is only recorded as smaller patches within the marked area

In eutrophic dry habitats, a very different low-growing community rich in cushion plants, herbs and grasses develops. This community, popularly known as the *Dryas* heath and in plant sociology as the *Kobresio-Dryadion* alliance, is dominated by *Dryas octopetala*, *Silene acaulis*, *Salix reticulata*, *S. herbacea*, *Carex bigelowii*, *Alchemilla vulgaris* (coll.), *Polygonum viviparum*, and many others (Lye, 1972), and is also relatively rich in mushrooms, e.g. *Inocybe* and *Cortinarius* species. In certain localitites, especially on the westernmost parts of Hardangervidda, a richer facies occurs where such rare plants as *Carex glacialis*, *Kobresia myosuroides* and *Potentilla nivea* are found. The distribution of this rare society on Hardangervidda is shown in Fig. 1.

E. Eutrophic Mires. The distribution of eutrophic mires on Hardangervidda corresponds well to that of eutrophic dry heaths or meadows. Such mires are frequently dominated by *Carex nigra*, *Comarum palustre*, *Leontodon autumnalis*, *Polygonum viviparum*, *Ranunculus acris*, and the bryophytes *Bryum pseudotriquetrum*, *Campylium stellatum*, *Drepanocladus exannulatus*, *D. revolvens*, *Mnium pseudopunctatum*, *Paludella squarrosa* and *Philonotis fontana*. The soils of such mires

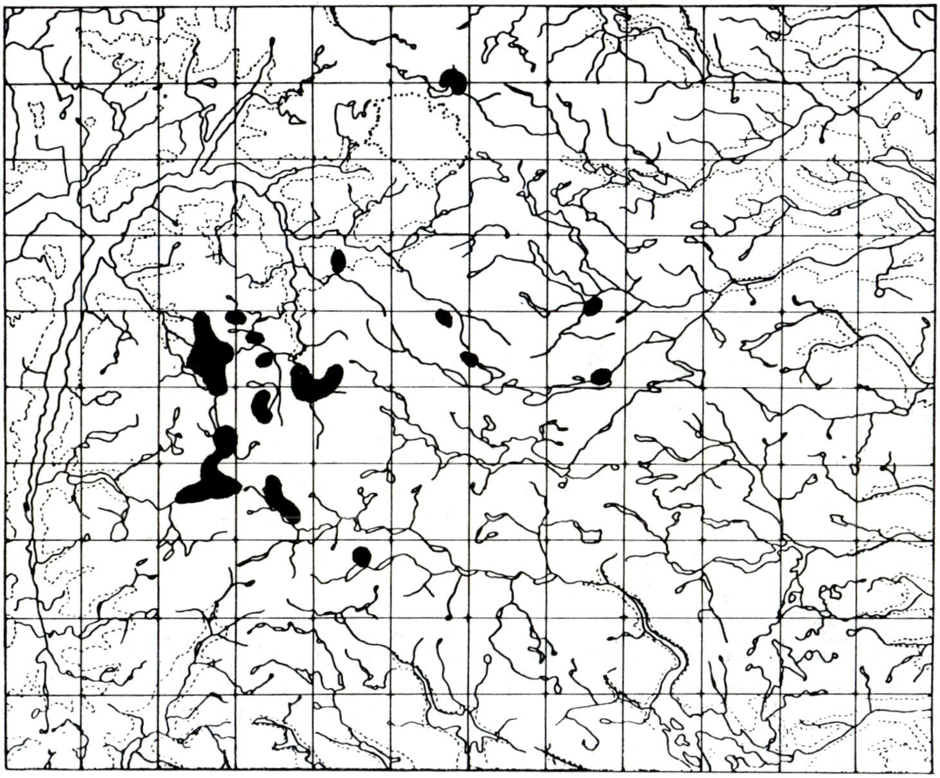

Fig. 2. The distribution of the most demanding society of the *Salicion myrsinitae* alliance (i.e. eutrophic mires) on Hardangervidda. This society is characterized by the occurrence of one of the following three species, viz. *Carex microglochin*, *C. capitata* and *Kobresia simpliciuscula*. The society is only recorded as smaller patches within the marked area

frequently have a nutrient-content as high as that of the eutrophic dry meadow. These societies belong to the *Salicion myrsinitae* or the *Caricion canescentis-nigrae* alliances (DAHL et al., 1971). Only relatively few mushroom species are common in this community. At the IBP wet meadow, *Galerina heterocystis* seems to dominate among the mushrooms. In scattered localitities, a slightly different society with even more demanding species, such as *Carex microglochin*, *C. capitata* and *Kobresia simpliciuscula* (LID, 1959) is found. The distribution of this society is shown in Fig. 2.

F. Oligotrophic and Mesotrophic Mires. The majority of mires on Hardangervidda are mesotrophic, and are usually dominated by *Eriophorum angustifolium*, more rarely *Carex rostrata* or *Eriophorum scheuchzeri*. There is, however, a continuous variation from the most eutrophic to the oligotrophic plant communities, and a phytosociological splitting of this group of societies is highly arbitrary. These societies belong to the *Leuco-Scheuzerion* (mainly below 900 m) or *Stygio-Caricion limosae* alliances (DAHL et al., 1971).

The most oligotrophic societies are dominated by *Rubus chamaemorus*, *Eriophorum vaginatum*, *Empetrum hermaphroditum*, *Sphagnum fuscum*, *S. robustum*, and at lower altitudes also *Betula nana*. The nutrient content of the soil in such communities is very low, and most of the minerals are brought down by rain and snow. In alpine regions, however, there is always some wind-blown sand, soil or organic matter from the environment, so fully ombrotrophic bogs do not exist here. This society belongs to the *Oxycocco-Empetrion hermaphroditi* alliance.

G. Water Communities. In the alpine region, macrophytic water communities are rare, although *Carex rostrata* and *Equisetum fluviatile* sometimes form pure stands in open water. Other plants occasionally forming water communities are *Alopecurus aequalis*, *Potamogeton alpinus*, *P. gramineus*, *P. pusillus*, *P. filiformis* and *Sparganium affine*. Such societies can be classified within the *Potamogeton eurosibiricum* or *Magnocaricion* alliances, but are very poorly developed.

H. Snowbed Communities. The greater majority of snowbeds in the area are of the oligotrophic type. In habitats with 4–10 m deep snow cover, a community dominated by *Salix herbacea*, *Dicranum starkei*, *Gymnomitrium concinnatum*, *Marsupella varians* and *Solorina crocea*, or more rarely *Polytrichum norvegicum* and *Conostomum tetragonum* develops. This society is much grazed by reindeer at midday, especially during hot periods in summer. Among plant sociologists, this society is known as the *Cassiopo-Salicion herbacea* alliance (DAHL et al., 1971). In this society mushrooms, especially of the genera *Cortinarius*, *Inocybe* and *Hebeloma*, are common.

In similar localities on substratum rich in phyllites a slightly different society, often known as the *Polarion*-alliance, develops. This community differs in having *Salix polaris* as a dominating species, instead of *S. herbacea*. Also, such demanding plants as *Saxifraga oppositifolia*, *S. rivularis*, *Taraxacum croceum* and *Leontodon autumnalis* occasionally occur, but when many herbs are encountered, the society has a thinner snow cover and is probably better classified among the eutrophic meadows.

I. The Vegetation of Springs. The vegetation in and near springs and streams from springs and melting ice is dominated by the bryophytes *Philonotis fontana* and *Scapania uliginosa*, with a more scattered occurrence of *Saxifraga stellaris*, *Cerastium cerastoides* and *Epilobium* spp. Among plant sociologists, this is known as the *Montio-Epilobion hornemanni* alliance.

Near water with a higher nutrient content, other vascular plants such as *Saxifraga aizoides* and *Juncus biglumis* tend to take over, and more rarely a demanding moss such as *Cratoneuron* sp. dominates the spring vegetation. In this case, the society is known as the *Cratoneuro-Saxifragion aizoidis* alliance (DAHL et al., 1971).

J. Micro-Communities. In addition to the plant communities mentioned above, a number of other characteristic communities are developed. These communities are, however, generally found in minute areas only, e.g. lichen and moss communities on exposed rocks and boulders, fungus and moss communities on droppings of reindeers and other mammals (especially *Splachnum ovatum* and *S. vasculosum* societies). Also, the communities on dead carcases (especially *Tetraplodon mnioides* societies) are interesting. On Hardangervidda, the ornithocoprofilous communities on rocks and boulders where birds of prey have their resting places are of considerable interest, since they are unusually common and well-developed in this area. In the early facies, such communities are dominated by crustaceous lichens, *Ramalina* spp. and *Parmelia* spp., but when soil has started to accumulate, vascular plants such as *Carex bigelowii*, *Eriophorum vaginatum* and *Rubus chamaemorus* dominate.

References

DAHL, E.: Rondane. Mountain vegetation in South Norway and its relation to the environment. Skr. Norske Vid.-Akak. Oslo. I. Mat.-Nat. Kl. **3**, 1956.

DAHL, E., KALLIOLA, R., MARKER, E., PERSSON, Å.: Nordisk vegetationsklassifisering för kartläggning. IBP Norden **7**, 3–12 (1971).

GULDEN, G.: Mushroom inventory at Hardangervidda, Autumn 1971. In: VIK, R. (Ed.): IBP in Norway. Methods and results. Sections PT-UM Grazing project, Hardangervidda. Botanical investigations, pp. 373–380. Oslo: Norwegian National IBP Committee 1975.

KIELLAND-LUND, J.: A classification of Scandinavian forest vegetation for mapping purposes (draft). IBP Norden **7**, 13–43 (1971).

LID, J.: The vascular plants of Hardangervidda, a mountain plateau of southern Norway. Nytt Mag. Botan. **7**, 61–128 (1959).

LYE, K. A.: Vegetation of selected localitites for IBP vegetation in Hardangervidda, Southern Norway. IBP Norden **8**, 101–111 (1972).

LYE, K. A., LAURITZEN, E. M.: Effect of grazing in alpine vegetation on Hardangervidda, South Norway. Norw. J. Botan. **22** (1975) (in press).

SONESSON, M., WIELGOLASKI, F. E., KALLIO, P.: Description of Fennoscandian tundra ecosystems. In: WIELGOLASKI, F. E. (Ed.): Fennoscandian tundra ecosystems, Part 1: Plants and microorganisms, pp. 3–28. Berlin-Heidelberg-New York: Springer 1975.

Vegetation Mapping at Hardangervidda

O. HESJEDAL

The aim of the project was to organize the plant communitites at Hardangervidda into a "vegetation type" classification for mapping at a scale of 1:10000, and to survey the distribution of these vegetation types, by estimating the areal coverage of the mosaic, in order to get an idea of the total coverage of different types at Hardangervidda. The project served as a methodological basis for studies of the use of air photos for vegetation mapping at different scales (HESJEDAL and LARSSON, 1975).

Vegetation Classification and Mapping

The plant sociological classification of Norway's mountain vegetation was carried out primarily by NORDHAGEN (1943), GJAEREVOLL (1956), and DAHL (1956). Common guidelines for Fennoscandian vegetation classification, e.g. of mountain vegetation, were recently given in WIELGOLASKI (1971) and MARKER (1973). Dominating and characteristic plant species of the main plant communities at Hardangervidda are shown in LYE (1975). The classification system presented in Table 1 is a combination of sociologically defined communitites, combined into "vegetation types" which can be used as mapping units (HESJEDAL, 1973). This system is a result of several research projects, conducted by IBP and DLR. These projects have been mutually beneficial, and have successively modified the mapping procedure. It is obvious that many of the vegetation types are only found in the subalpine zone, while others are found in the altitudinal belts all the way from subalpine to mid-alpine zones. The high alpine zone covers a relatively small area at Hardangervidda (Table 3), and various snow bed communities are the only vegetation types of extension to be classified in the present system, as seen in Table 1. Other areas in the high alpine zone consist of rock and boulders without vegetation cover. Snow cover and distinction in eutrophic and oligotrophic areas are characteristics of the various vegetation types.

The areas mapped at Hardangervidda are shown in Fig. 1. The intensive alpine IBP sites are within plot P on the map (willow thicket site within plot B), while the subalpine birch forest IBP site is within plot H.

Paper prints of air photos are used as base maps for all mapping. Most of the photos used on this project have been taken on panchromatic film (approximately) at 1:10000 scale.

Table 1. "Vegetation types" at Hardangervidda

Type code	Type name	Communities included [all formal names marked N are sensu NORDHAGEN (1943); marked G are sensu GJAEREVOLL (1956)]	Trophy. Oligo-trophic: O Eu-trophic: E	Snowcover Sparse: 1 Interm.: 2 Medium longlasting: 3 Longlasting: 4 Extremely longlasting: 5	Zone Subalpine: I Low alp: II Mid alp: III High alp: III
1a	*Loiseleuria* heath	Loiseleurio-Arctostaphylion N, except Betuletum-empetro-cladinosum	O	1	(I)–II
1b	*Cetraria nivalis-Juncus trifidus* heath	Chionophobous forms of Juncion trifidi scandinavicum N.	O	1	III
2a	*Vaccinium myrtillus* heath	Phyllodoco-Vaccinion myrtilli N, except Betuletum empetro-hylocomiosum and Betuletum myrtillo-hylocomiosum	O	2	(I)–II
2b	*Cetraria islandica-Juncus trifidus* heath	Chionophilous forms of Juncion trifidi scandinavicum N.	O	2	III
3	*Dryas* heath	Kobresieto-Dryadion N	E	1	(I)–II–III
4a	*Nardus-Carex bigelowii* heath	Deschampsio-Anthoxanthion G and Ranunculo-Anthoxanthion G	O	3	(I)–II–III
4b	*Salix herbacea* snowbed	Herbaceon G and Stellaro-Oxyrion G	O	4	II–III–(IV)
4c	*Polytrichum norvegicum* snowbed	Polytrichion norvegici G	O	5	II–III–(IV)
5a	*Salix reticulata* snowbed and *Ranunculus acris* snowbed	Reticulato-Poion alpinae G, Ranunculo-Poion alpinae G and Potentillo-Polygonion vivipari N	E	3	(I)–II–III
5b	*Salix polaris* snowbed	Polarion G and fragments of Oppositifolio-Oxyrion G	E	4	II–III–(IV)
6	Vegetated rock and scree	Veronico-Poion glaucae N, Saxifragion cotyledonis N and related communities	O–E	1–2	I–II
7a	*Lactuca alpina-Geranium silvaticum* meadow, typical	Typical Lactucion alpinae N, unwooded	E	2–3	I–II–III
7b	*Lactuca alpina-Geranium silvaticum* meadow, along flooding rivers and brooks	Salicetum ulmariosum alpicolum N and Salicetum deschampsiosum alpicolum N	E	2	I–II–(III)
8	*Empetrum-Cladonia-Betula tortuosa forest*	Betuletum empetro-cladinosum N and Betuletum empetro-hylocomiosum N	O	1–2	I
9	*Vaccinium-Dryopteris-Betula tortuosa* forest	Betuletum myrtillo-hylocomiosum N	O	2	I
10a	*Aconitum-Betula tortuosa* forest, typical form	Typical Betuletum geraniosum subalpinum N	E	2–3	I

Table 1 (continued)

Type code	Type name	Communities included [all formal names marked N are sensu NORDHAGEN (1954); marked G are sensu GJAEREVOLL (1956)]	Trophy Oligo-trophic: O Eu-trophic: E	Snowcover Sparse: 1 Interm.: 2 Medium longlasting: 3 Longlasting: 4 Extremely longlasting: 5	Zone Subalpine: I Low alp: II Mid alp: III High alp: IV
10b	*Aconitum-Betula tortuosa* forest, along flooding rivers and brooks	Forest communities close to Salicatum ulmariosum alpicolum N and Salicetum deschampsiosum alpicolum N	E	2	I
11	"Pastures", conditioned by heavy grazing	Heavily grazed, strongly modified, meadow-like areas. Potential vegetation information is given on the maps	O–E	2–3–4	I–II–III
12	Spring communities	Mniobryo-Epilobion hornemanni N, and fragments of Cratoneuro-Saxifragion aizoides N	O–(E)	2–3–4–5	I–II–III
13	*Oxycoccus-Empetrum*	Oxycocco-Empetrion hermaphroditi N	O	(1)–2–3	I–II–(III)
14	*Carex limosa-Sphagnum lindbergi* fen	Leuco-Scheuchzerion N	O	(1)–2–3	I–II–(III)
15a	*Carex chordorrhiza* fen	Stygio-Caricion limosae N	O–E	2–3	I–II–III
15b	*Eriophorum angustifolium* fen	Exceptionally long snowcovered forms of Caricion canescentis-fuscae N and Caricion atrofuscae-saxatilis N	O–E	4–5	II–III
16a	*Carex canescens* fen	Oligotrophous forms of Caricion cannescentis-fuscae N	O	2–3	I–II–III
16b	*Carex fusca* fen	Eutrophous forms of Caricion canescentis-fuscae N	E	2–3	I–II–III
17	*Salix myrsinites* fen	Caricion atrofuscae-saxatilis N	E	2–3–(4)	I–II–III
18	Oligotrophous *Betula-Salix* forest	Wooded forms of oligotrophous Caricion canescentis-fuscae N and closely related communities	O	2	I
19	Eutrophous *Salix-Betula* forest	Wooded forms of eutrophous Caricion canescentis-fuscae N and Caricion atro-fuscae-saxatilis N, and closely related communities	E	2	I
20	*Carex rostrata* community	Oligotrophous forms of Magno-caricion	O	2–3	I–II–III

Other mapping units are:
Open water: Minor water bodies within the mapped plots.
Bare rocks: Areas without soil or higher vegetation of any significance.
Cultivated areas: Fully or semi-cultivated areas.
Others: Unvegetated rockfalls and boulders, roads etc.

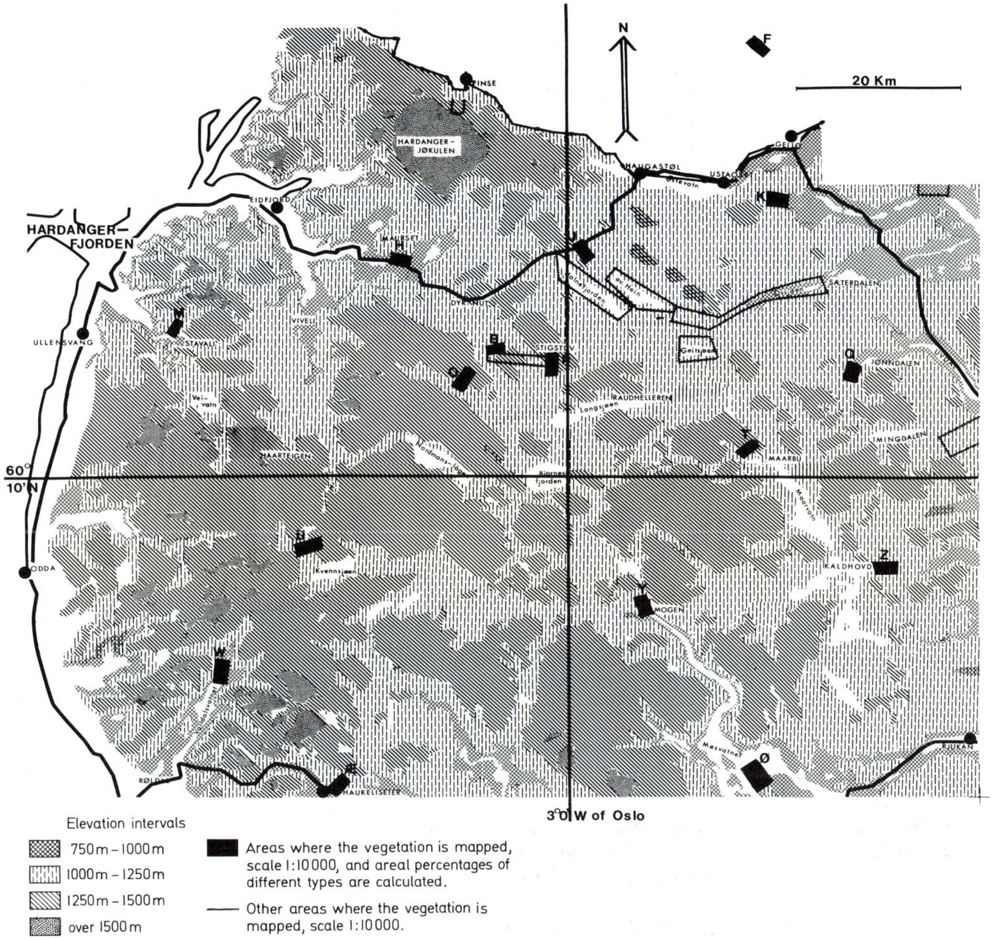

Fig. 1. Elevation zones and distribution of mapped areas at Hardangervidda

Using a 2X field stereoscope, mappers viewed the terrain stereoscopically at c. 1:5000. The mappers ink-typed boundaries directly on the photos. Many vegetation types have unique tone/texture signatures on the photos, and these can be outlined after a brief field inspection. Other types have vague or non-photoidentifyable boundaries which require more field work to delineate.

Area Measurements

Sixteen plots were selected for areal measurements of various vegetation, each plot to include as great an internal range of factors as possible in an area of 3–6 km^2. Location of the plots are shown in Fig. 1. The plot areas cover about 0.85 percent of the total area of Hardangervidda (about 8300 km^2). Delineation

Table 2. Elevation and area of 16 sample plots at Hardangervidda

Sample plot (Fig. 1)	Maximum elevation (m)	Minimum elevation (m)	Average elevation (m)	Calculated area (1000 m²)
J	1260	1180	1225	3290
E	1190	1020	1110	3345
F	1115	1020	1070	4490
Q	1140	995	1110	3275
Z	1210	1060	1120	5755
O	1350	1230	1280	3255
U	1380	1175	1250	5355
H	1000	775	820	3935
Y	1225	965	1070	4905
Ø	1020	920	950	6360
T	1385	1140	1270	3915
K	1150	900	1060	3035
M	1120	900	1000	3655
P	1390	1215	1300	5325
W	1085	745	920	3615
B10	1275	1155	1210	3305
All plots:	1390	745	1110	66815

Table 4. Area percentages of different vegetation types within plots, and average percentages for combined plots

Plot code \ Type code	1a	1b	2a	2b	3	4a	4b	4c	5a	5b	6	7a	7b	8	9
J	6.0	2.2	29.6	12.3	—	23.2	16.1	2.7	1.0	—	—	—	—	—	—
Æ	9.9	—	35.6	—	—	23.0	15.8	0.8	—	—	—	1.2	—	—	—
F	10.6	—	34.7	—	—	0.1	—	—	—	—	—	3.2	—	1.2	28.4
Q	17.3	—	40.1	—	—	4.7	—	—	—	—	—	7.8	—	1.9	10.0
Z	5.9	—	63.3	0.5	—	0.8	—	—	0.1	—	—	9.5	0.6	—	0.3
O	—	2.7	2.1	7.1	—	39.0	5.8	0.9	0.3	—	—	0.1	0.3	—	—
U	—	29.3	—	0.2	0.7	32.6	16.7	0.8	2.6	0.1	—	—	0.4	—	—
H	7.4	—	19.7	—	—	0.1	0.3	—	—	—	—	0.4	—	15.0	10.8
Y	5.5	—	38.5	—	—	0.2	—	—	—	—	0.8	4.6	0.9	9.5	10.3
Ø	0.3	—	28.4	—	—	0.3	—	—	—	—	—	0.8	—	15.0	2.5
T	16.3	1.6	44.7	6.3	—	6.3	1.5	—	—	—	—	8.4	0.1	—	—
K	9.2	—	29.7	—	—	3.4	2.2	0.2	—	—	—	9.5	0.1	1.8	14.5
M	15.9	—	55.3	—	—	9.3	0.8	—	—	—	—	0.4	1.2	0.7	—
P	0.5	8.6	4.6	4.4	—	45.8	9.5	0.7	3.1	—	—	0.5	—	—	—
W	1.4	—	36.0	—	—	22.1	2.5	—	—	—	—	0.2	—	—	20.5
B10	8.0	4.4	33.8	0.8	—	13.7	1.5	—	4.2	—	—	3.1	—	—	—
Average of 16 plots	7.1	3.0	31.0	2.0	<0.1	14.0	4.5	0.4	0.7	<0.1	<0.1	3.1	0.2	2.8	6.1

Table 3. Calculated area of Hardangervidda and of sample plots in 4 elevation zones

Elevation in m	750–1000	1000–1250	1250–1500	1500–(1875)	Sum
Area of Hardanger-vidda in km^2:	825	4075	3195	220	8310
Percentage of H.'s total area:	10	49	38.5	2.5	100
Area of sample plots im km^2:	13.9	35.5	17.4	0	66.8
Percentage of total area of sample plots:	21	53	26	0	100

and elevation intervals of Hardangervidda are also shown in Fig. 1. Hardangervidda's average elevation is estimated to be slightly less than 1200 m, while the average elevation of the 16 plots is 1110 m (Table 2). From Table 3 it is clear that the two lowest elevation zones (generally corresponding to subalpine and low alpine zones) are over-represented in the sample plots, the zone 1250–1500 m (approximately mid alpine zone) is underrepresented, and the highest (high alpine) zone is not represented at all.

10a	10b	11	13	14	15a	15b	16a	16b	17	18	19	20	Water	Bare rock	Cul-tiv-ated	Oth-ers
—	—	—	—	—	—	0.9	1.2	0.7	0.8	—	—	—	3.2	—	—	—
—	—	0.5	—	—	—	0.2	1.0	—	—	—		—	11.1	—	—	1.1
5.2	—	1.4	0.3	—	—	—	5.4	0.9	5.5	—	—	0.3	2.2	0.4	0.2	—
1.8	—	0.2	0.1	—	0.1	—	11.4	1.6	2.6	—	0.2	—	0.2	—	—	—
0.4	—	0.4	0.1	—	1.0	—	8.0	1.6	3.8	—	—	—	2.8	0.1	—	0.6
—	—	—		2.0	—	—	37.4	2.3	—	—	—	0.1	0.1	—	—	—
—	—	—	—	—	—	0.6	4.4	0.9	0.1	—	—	—	2.4	2.2	—	6.0
5.5	—	—	14.6	1.5	0.4	—	11.4	—	—	5.0	0.7	—	3.2	0.5	1.6	1.7
0.5	0.2	—	0.4	—	0.6	0.1	7.9	5.0	0.6	0.1	2.2	—	—	1.3	—	4.9
2.2	—	—	3.2	0.4	1.1	—	28.7	10.7	2.3	0.2	3.0	0.1	0.4	—	0.4	—
—	—	1.4	—	1.0	1.2	—	9.5	—	—	—	—	0.3	1.1	0.1		0.1
5.0	—	0.8	—	0.1	—	—	5.9	2.2	8.7	—	2.9	0.1	1.9	—	1.6	—
0.5	—	—	0.7	—	0.3	0.8	9.0	—	—	—	—	—	2.8	0.7	—	1.3
—	—	—	—	—	1.5	0.7	0.9	14.3	2.3	—	—	0.1	2.1	—	—	
0.2	0.3	0.5	—	—	—	0.2	5.0	1.8	0.3	—	0.4	—	3.0	0.5	0.3	4.2
—	—	—	0.1	—	0.1	0.1	8.3	11.1	11.7	—	—	—	—	—	—	—
1.7	<0.1	0.3	1.2	0.3	0.4	0.2	9.7	3.3	2.4	0.3	0.6	<0.1	2.3	0.4	0.3	1.2

All areas are measured by electronic planimeter. Measurements of sample plot areas are carried out directly on transparencies, on to which original photo drawings have been transferred. An air photo is a central projection of the terrain, and consequently it does not show the true area at the actual scale. Distortion is most serious towards the photo edges, but can be reduced by using only the central part of each photo. Other sources of error are: Average scale of the photos, listed by the photo company, is incorrect; camera axis moves off the vertical; relief displacement within the photo area, unstable photo-base and inaccurate planimeter operation. Of these sources of error, scale incorrectly listed is the only systematic one. Most other errors tend to compensate. Table data are therefore only corrected for average scale incorrectly listed. The procedure for this correction is omitted in this short paper.

In spite of these possible errors and the bias of the plots selected, the percentages of the vegetation types given in Table 4 are the best estimates at present accessible for the average at Hardangervidda. From field experience, however, it seems that the average percentages (Table 4) of the Types 1a, 2a, 8–10 and 13–17 may be somewhat overestimated, whereas 1b, 2b and 4a–c may be somewhat underestimated in relation to their total occurrence at Hardangervidda.

The most important vegetation type at Hardangervidda is obviously the oligotrophic *Vaccinium myrtillus* heath (Type 2a). Within plot P (with the intensive alpine IBP sites) this vegetation type covers a smaller area than the average, and is partly replaced by the more chionophilous oligotrophic *Nardus-Carex bigelowii* heath (Type 4a). Both in plot P and plot B 10 (including the IBP willow thicket), there are overrepresentations of the eutrophic snowbed-meadow vegetation Type 5a, in plot B 10 also of the eutrophic *Lactuca alpina*—*Geranium silvaticum* meadow (Type 7a). The eutrophic *Carex fusca (C. nigra)* fen is also represented in both these plots. Generally, the plots including the alpine IBP sites have more eutrophic vegetation than the average at the Hardangervidda plateau. As could be expected plot H, including the IBP birch forest site, shows higher percentages than average for subalpine vegetation types, especially the oligotrophic-mesotrophic Types 8–9, but also the eutrophic forest Type 10a.

The vegetation types and their coverages given in Tables 1 and 4 may be grouped more roughly according to nutrient status and snow cover. This is done in Table 5, showing that about 83% are oligotrophic and 13% eutrophic

Table 5. Vegetation types mapped at Hardangervidda grouped according to nutrient and snow cover

	Oligotrophic	Eutrophic
Chionophobous heaths vegetated rocks etc.	10.1	<0.1
Chionophilous heaths	33.0	<0.1
Moderate snowbeds	14.1	4.2
Birch woods	9.0	2.3
Mires (bogs and fens)	11.8	5.8
Extreme snowbeds	4.9	<0.1

vegetation types; the rest being mostly bare rocks and water. Most types are found under a moderately lasting snow cover (above 80%). Based on the area percentages in Table 5 and the primary production values estimated by WIELGOLASKI (1975) and KJELVIK and KÄRENLAMPI (1975) for the vegetation types studied at Hardangervidda by IBP, there might be a possibility to calculate roughly the total annual primary production of the plateau.

Acknowledgement. The project was made possibly by the cooperation between the Department of Agriculture, Division of Land Registration (DLR), and IBP, PT/UM Norway. P. K. BJØRKLUND, J. HOFSTEN, J. LARSSON and the author took part on behalf of DLR, E. MARKER and R. ELVEN on behalf of IBP.

References

DAHL, E.: Rondane mountain vegetation and its relation to the environment. Skr. Vid. Akad. Oslo, I. Mat.-Nat. Kl. **3**, 1956.

GJAEREVOLL, O.: The plant communities of the Scandinavian alpine snowbeds. Skr. Kgl. Norske Vid. Selsk. Trondheim **1**, 1956.

HESJEDAL, O.: Vegetasjonskartlegging. ISBN 82-557-0010-2. Ås-NLH: Landbruksbokhandelen 1973.

HESJEDAL, O., LARSSON, J.: Remote sensing of vegetation in conservation of tundra landscapes. In: WIELGOLASKI, F. E. (Ed.): Fennoscandian tundra ecosystems, Part 2: Animals and systems analysis. Berlin-Heidelberg-New York: Springer 1975 (in press).

KJELVIK, S., KÄRENLAMPI, L.: Plant biomass and primary production of Fennoscandian subarctic and subalpine forests and of alpine willow and heath ecosystems. In: WIELGOLASKI, F. E. (Ed.): Fennoscandian tundra ecosystems, Part 1: Plants and microorganisms, pp. 111–120. Berlin-Heidelberg-New York: Springer 1975.

LYE, K. A.: Survey of the main plant communities on Hardangervidda. In: WIELGOLASKI, F. E. (Ed.): Fennoscandian tundra ecosystems, Part 1: Plants and microorganisms, pp. 68–73. Berlin-Heidelberg-New York: Springer 1975.

MARKER, E. (Ed.): IBP-CT symposium om vegetasjonskartlegging. Rapport IBP i Norden **11**. Oslo: Nordic IBP committees 1973.

NORDHAGEN, R.: Sikilsdalen og Norges fjellbeiter. Skr. Bergens Museum **22**, 1943.

WIELGOLASKI, F. E. (Ed.): Forslag til vegetasjonsklassifisering (Preliminary classifications of Scandinavian mountain, forest, wetland and sea shore vegetation). IBP i Norden **7**, Oslo: Nordic IBP committees 1971.

WIELGOLASKI, F. E.: Primary productivity of alpine meadow communities. In: WIELGOLASKI, F. E. (Ed.): Fennoscandian tundra ecosystems, Part 1: Plants and microorganisms, pp. 121–128. Berlin-Heidelberg-New York: Springer 1975.

Dispersal and Primary Establishment of Vegetation

R. Elven and L. Ryvarden

A retreating glacier leaves large areas of open and virgin soil. Very soon the first plants appear, but there are indications that it takes hundreds of years to establish a mature vegetation. It is of interest to know which are the decisive factors in the development of the vegetation in the first phase, and also what happens when new plant communities replace the preceeding ones in a succession. With the purpose of gaining some basic insight into these problems, seed dispersal and vegetation were studied in front of the glacier Hardangerjøkulen at Finse, Central Norway. R. Elven was responsible for studies on establishment of vegetation while L. Ryvarden was in charge of the dispersal experiments.

It is important to know what sorts of diaspores and seeds are transported into the area. It can be observed in front of the glacier that some species are far more abundant than others. The reasons may be manifold. These species may have been more successful in dispersing their diaspores into the new areas than others. However, it is possible that they are only more successful in coping with the harsh climatic conditions and the disturbed soil prevailing in the area near the glacier, and that an equal number of other species are also dispersed into the area without any possibilities of growth there.

Methods

No experiments seem to have been performed with trapping of diaspores in the alpine zone, and a new sort of trap had to be developed (Ryvarden, 1971). The traps were filled with water, and all diaspores coming to rest on the water were trapped. Later they were collected, counted, and identified. The traps were plastic vessels 60 × 40 × 10 cm with a hole of about 3 cm in a corner, below which was a smaller plastic vessel with a perforated bottom, acting as a filter. The whole trap was dug into the ground so that the upper edge of the larger vessel was level with the soil. The trap was slightly tilted, causing the water to cover the bottom of the larger vessel with the exception of a small area around the hole. The water was drained twice a week, and trapped diaspores were collected on the filter paper in the lower vessel. They were in operation from the end of July to mid-September during three seasons, and were placed along gradients perpendicular to the glacier front. The lines started some meters from the ice and ended a kilometer away in mature vegetation. The reader is referred to Ryvarden (1971) for the exact position of the traps. To obtain an estimate of the winter-dispersal, snowbeds and snow on the glacier were scraped down to 5 cm in plots of 2 and 4 m^2. The snow was melted, drained, and the diaspores collected and identified. This was done several times during two summers (for details, see RYVARDEN, 1975).

In front of Hardangerjøkulen, there are several moraine ridges reflecting different stages of retreat. The age of these ridges was partly determined by lichenometry (ANDERSEN and SOLLID, 1971, our own datings 1972–1973), and partly by local Finse inhabitants, who had visited the glacier regularly since 1934. The investigations of establishment of vegetation were performed from the eldest terminal moraine, dated to A.D. 1750, to the glacier front. Three lines were drawn perpendicularly on the glacier, and along these lines plots of 3 m^2 were investigated on each moraine ridge. The plots were placed on inner and outer slopes, on the top, and in the depression between the ridges. In each plot all species and their coverage were noted (except fungi and algae), and the number of individuals counted. The depth of snow was measured for each plot during two seasons (for details, see ELVEN, 1975).

Results

Trapping of diaspores with the water-filled traps showed that each square meter in front of the glacier for a distance of about 1 km in average received approx. 650 diaspores in the summer season. The number of diaspores trapped was lowest close to the glacier, increasing steadily with an increasing cover of vegetation. All the 57 species recorded from the traps are known from the Finse area, and no long-distance dispersed diaspores were found. To obtain an estimate of the actual distances of dispersal, all species around the traps were noted. Assuming that diaspores of a species had come from the nearest representatives, all diaspores dispersed less than 5 m were excluded. This left only 13% of the diaspores, and indicates the importance of the first individuals which can act as centra for new establishment.

When the diaspores were classified according to morphology (DAHL, 1958), there was no qualitative overrepresentation in the samples of anemochorous species, compared with the indigenous flora of Finse. Nothing is known of the production of diaspores by the different species, and a quantitative comparison cannot be made. However, even smooth, heavy diaspores without any apparent morphological adaptation to any agency of dispersal, were found to have dispersed several hundreds of meters, clearly indicating that the morphological classification to assumed type of dispersal may be quite misleading.

The scraping of snowbeds and glacier snow showed that very few diaspores were dispersed into the area by action of wind in winter. On the glacier on average one diaspore was deposited per m^2, while the snowbeds in front of the glacier received 5–7 diaspores per m^2. The results seem to indicate that most diaspores have already been dispersed when the first snow arrives, and only a very few are released later for chionochor dispersal. All the 34 species represented in the snowbed samples were known from the Finse area. The maximum distance of dispersal (under the same assumptions as mentioned above) is about 4 km for diaspores of *Betula tortuosa*. If the diaspores are classified on morphology, there is

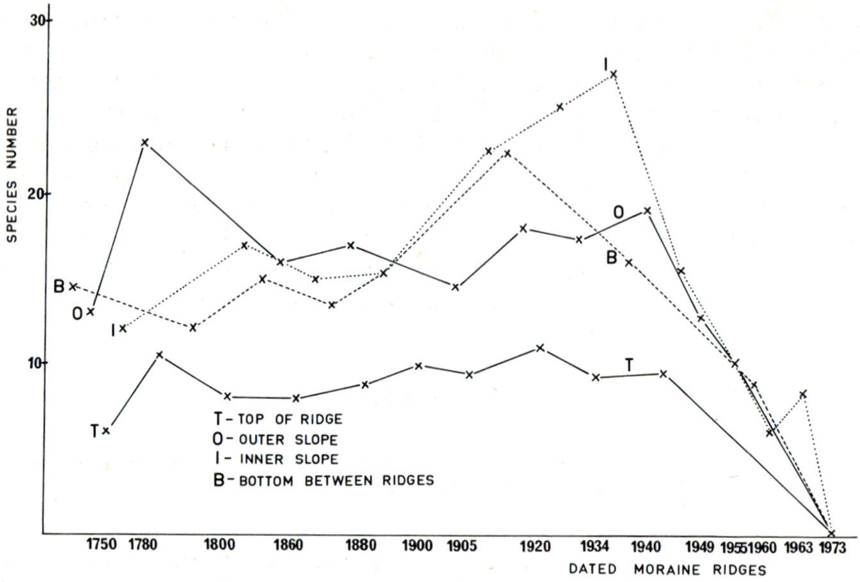

Fig. 1. Variation in species number (phanerogams), with time and position on moraine ridges

again a very close qualitative correspondence with the indigenous flora. There is no overrepresentation of anemochorous species, and there are several representatives of species with heavy and smooth diaspores. This again makes precaution advisable in assuming a certain way of dispersal on the basis of diaspore morphology. Many species without any apparent morphological adaptations seem to be easily dispersed by wind, at least in the alpine zone.

In the first 30 years after deglaciation of soil, there is a steady increase in the number of species established. In the pioneer group, the following phanerogams were the most common ones (percentage frequency given in parenthesis): *Deschampsia alpina* (52), *Arabis alpina* (34), *Saxifraga groenlandica* (34) and *Poa alpina* f. *vivipara* (25). The mean distance from the glacier in this group are from 180 to 280 m. From this stage the snow cover, and to a lesser degree, the age, are responsible for the development, and two main vegetation types can be observed. On the snow-free areas on top of the dry moraine ridges, *Empetrum hermaphroditum* and chionofobous lichens dominate (LYE, 1975), while *Carex lachenalii* and *Deschampsia alpina* dominate in wet snowbeds. *Salix herbacea* is an important species, but shows no special preference at this stage. Later, when the competition becomes more severe, e.g. after some hundred years and in mature vegetation, this species becomes a typical snowbed species.

The area between the glacier and the "1750" moraine ridge is about 1 km wide, and a total of 274 species have been noted here (fungi and algae omitted). Of the 78 dicotyledons *Salix herbacea* is the most frequent of the woody species and *Gnaphalium supinum* among the perennials, while *Euphrasia frigida* is the only annual abserved; 32 monocotyledons have been observed, *Trisetum spicatum* being the most frequent. Of the 164 cryptogams recorded, *Lycopodium selago*,

Gymnomitrion concinnatum, Pohlia drummondii and *Stereocaulon alpinum* are the most frequent species in the different systematic groups.

Besides the marked qualitative difference in species composition between the top, the slopes and the bottom of the moraine ridges, there was also a certain quantitative difference. This is shown in Fig. 1. The figure demonstrates that there is a constant increase in species number per plot until a peak of diversity is reached 30 to 50 years after the first invasion. Later, the more competitive species slowly "squeeze" the less competitive ones out, leaving the first invaders to restricted areas like riverbanks, solifluction slopes, small screes etc. The result is a decreasing number of species and an approach towards a mature vegetation.

References

ANDERSEN, J. L., SOLLID, J. L.: Glacial chronology and glacial geomorphology in the marginal zones of the glaciers, Midtdalsbreen and Nigardsbreen, South Norway. Norsk Geogr. Tidsskr. **25**, 1–38 (1971).

DAHL, E.: Amfiatlantiske planter. Blyttia **16**, 93–121 (1958).

ELVEN, R.: Plant communities on recently deglaciated moraines at Finse, southern Norway. In: VIK, R. (Ed.): IBP in Norway. Methods and results. Sections PT-UM Grazing project, Hardangervidda, Botanical investigations, pp. 381–467. Oslo: Norwegian National IBP Committee 1975.

LYE, K. A.: Survey of the main plant communities on Hardangervidda. In: WIELGOLASKI, F. E. (Ed.): Fennoscandian tundra ecosystems, Part 1: Plants and microorganisms, pp. 68–73. Berlin-Heidelberg-New York: Springer 1975.

RYVARDEN, L.: Studies in seed dispersal I. Trapping of diaspores in the alpine zone at Finse, Norway. Norw. J. Botan. **18**, 215–226 (1971).

RYVARDEN, L.: Studies in seed dispersal II. Trapping of winter-dispersed diaspores in the alpine zone at Finse, Norway. Norw. J. Botan. **22** (1975) In press.

Comparison of Plant Structure on Grazed and Ungrazed Tundra Meadows

F. E. WIELGOLASKI

Introduction and Methods

A review of literature of grazing influence on plant succession of rangelands, from the prairie to the subalpine zones in USA, was given by ELLISON (1960). Grazing in mountain areas has been studied by BRANSON and PAYNE (1958) in the mountains of Montana, by ELLISON (1954) in Central Utah, and by PAULSEN (1960) in the central Rocky Mountains. In Great Britain, studies of grazing influence on plant succession have been carried out in the mountains of Wales (HUGHES et al., 1964), in the highlands of Scotland (KING and NICHOLSON, 1964) and to above 800 m in Northern England (WELCH and RAWES, 1965; RAWES and WELCH, 1966). The importance of various plant communities for grazing in Norway was shown by NORDHAGEN (1943).

The present sheep grazing study in the Norwegian IBP tundra project at Hardangervidda, southern Norway (SONESSON et al., 1975) lasted four years in the dry meadow and three years in the wet meadow. The abiotic factors, both soil and climate, were studied in detail at both grazed and ungrazed plots. Influence of grazing was studied on both plants and invertebrates. Inventory studies on vegetation were carried out before the research started (LYE, 1972).

The driest of the two meadows may be said to be a mixture of the relatively eutrophic phytosociological alliances Ranunculeto-Oxyrion digynae, important for summer grazing by both reindeer and sheep (NORDHAGEN, 1943 p. 287), including Reticulata-Poion alpinae (DAHL et al., 1971), Potentilleto-Polygonion vivipari (WIELGOLASKI, 1972) and Kobresio-Dryadion, but mostly within the suballiance Alchemillion glabrae chionophilum sensu NORDHAGEN (1943) of the first alliance. The wet meadow is phytosociologically a mixture dominated by the alliance Caricion canescentis-nigrae. In wet depressions, however, *Eriophorum angustifolium* dominates (LYE, 1972) while, as expected, *Ranunculus acris*, *Taraxacum croceum* and *Thalictrum alpinum* are more in evidence in the drier parts of this site. The grazing pressure by large mammals had been minimal in the areas for several years before the research started. In the dry meadow, sheep and cattle had grazed occasionally. In the wet meadow, reindeer may occasionally have had some grazing and trampling effect before the present investigations.

One sheep (ewe) and two lambs (variety "Dalasau") were allowed to graze within enclosures of about one hectar during each of the research years, both in the dry and wet meadows. Snow usually melts in early-mid June at the sites, a few days earlier in the dry meadow than in the wet meadow. This means that the new vegetation was relatively well developed before grazing started in the first part of July. Grazing ended each year in mid-September, just before new snow could be expected.

Plant cover was estimated after three years of research in early July 1972, both in the enclosures grazed by sheep and in enclosures where all large mammal grazing was excluded. The values were compared with the vegetation analyses before the research started. Plant harvestings within the enclosures grazed by sheep followed the wire cage method (10 parallels, each 0.5 m^2) outlined in the IBP Grassland Handbook (MILNER and HUGHES, 1968). Plant dry weight is expected to be the most objective measurement for changes in vegetation structure (ELLISON, 1960) and therefore, only changes in weight are discussed in the detailed herbaceous plant studies.

Results and Discussion

Changes in plant cover were found to have taken place by comparing the cover before and after the studies of enclosures covering three years, even in areas ungrazed by sheep (Table 1). The relative shrub cover in the dry meadow was reduced by about 50%, mostly caused by a reduction in the cover of *Salix herbacea* from about 9% of the total amount of vascular plants to about 2%. A possible explanation may be an increase of small mammals in the fall of 1970. In spring 1971, it could be seen that especially stems of *Salix herbacea* had been grazed and cut off during the winter by these animals. The monocotyledons increased during the three years, even at plots ungrazed by sheep. This may be caused by less competition, e.g. from the shrubby vegetation after destruction by small mammals, but also by more available nutrients released from the dead *Salix herbacea* stems. The better nutrition conditions, and thus more extensive herbaceous vascular plant growth, may also be the reason for the somewhat lower lichen cover at the end of the study.

The dry meadow was a more favourable habitat for the small mammals than the wet meadow. Therefore, no reduction in *Salix herbacea* was found in the wet meadow, in the study areas where there had been no sheep grazing. On the contrary, because grazing and trampling by reindeer was excluded during the three year period, a somewhat higher shrub cover was found than before the studies started. This resulted in a tendency (not significant) to reduced monocotyledonous cover at the ungrazed plots during the same period at this site (Table 1). A higher moss cover at the same plots in 1972 than in 1969, and a tendency to an increase in dicotyledonous herbs, may have been caused mostly by the exclusion of the trampling effect by reindeer for three years.

The shrub cover was strongly reduced by sheep grazing during the three years of study, both in the dry meadow and the wet meadow (Table 1), which is in accordance with the results, e.g. from Great Britain (TANSLEY, 1953). This reduction occurred although shrubs were generally found to account for less than 50% of the herbivory by sheep, including woody parts (WIELGOLASKI, 1976). One rea-

Table 1. Percentage total cover of vascular plants and cryptogams before grazing, at ungrazed fenced plots and at plots grazed by sheep for three years (estimated before grazing 1972)

	Dry meadow						Wet meadow					
	Before grazing		Ungrazed		Grazed 1969–71		Before grazing		Ungrazed		Grazed 1969–71	
	1969		1972		1972		1969		1972		1972	
	% Cover	Rel. % Cover	% Cover	Rel. % Cover	% Cover	Rel. % Cover	% Cover	Rel. % Cover	% Cover	Rel. % Cover	% Cover	Rel. % Cover
Shrubs (including *Dryas*)	25.1	18.1	15.1	9.4	3.8	2.6	21.7	13.2	33.1	15.0	4.7	4.2
Monocot. herbs	33.7	24.3	63.0	39.1	99.2	66.6	34.5	21.1	31.7	14.3	36.0	32.0
Dicot. herbs	24.5	17.7	29.0	18.0	18.6	12.5	14.4	8.8	19.2	8.7	4.4	3.9
Sum vascular	83.3	60.1	107.1	66.5	121.6	81.7	70.6	43.1	84.0	38.0	45.1	40.1
Mosses	44.5	32.2	46.8	29.0	23.6	15.8	92.6	56.5	136.4	61.7	66.9	59.6
Lichens	10.6	7.7	7.3	4.5	3.7	2.5	0.6	0.4	0.7	0.3	0.4	0.4
Total	138.4	100.0	161.2	100.0	148.9	100.0	163.8	100.0	221.1	100.0	112.4	100.0

son is that nanophanerophytes (RAUNKIAER, 1934) and chamaephytes (such as *Salix herbacea*, *Salix reticulata* and *Vaccinium vitis-idaea*) with aboveground perennating buds are highly sensitive to grazing. Even a relatively low grazing intensity may cause many plants to die during the following winter. Similar effects could be expected from trampling by the sheep. The reduction in cover of mosses and lichens in the sheep-grazed areas reflects mostly a trampling effect by the sheep at both sites, but more detailed studies were not carried out for the cryptogams in the grazed enclosures.

There was also a reduction in the dicotyledonous herb cover (including *Pteridophyta*) in the enclosures, after three years of sheep grazing at both sites (Table 1). The monocotyledonous cover was, however, found to increase after sheep grazing, especially in the dry meadow and on a relative basis. This is in accordance with results found by FENTON (1949) and GIMINGHAM (1964), and is also often found in the U.S. IBP Grassland studies (LEWIS, 1971). ELLISON (1954) says that a subalpine area dominated by grasses reflects overgrazing by sheep. This is also seen from practical experience in Norway.

The strong increase in monocotyledons in the dry meadow resulted in no reduction of the total vascular plant cover at this site by grazing, as is often found in grazing studies (PEARSON, 1965). In the wet meadow, however, cover was strongly reduced by three years of sheep grazing (Table 1).

Table 2 shows (in the first four columns) that all herbaceous vascular species in the ungrazed enclosures at the dry meadow had the same or higher weight in 1972 than in 1969, possibly to some degree caused by favorable growth conditions at the site in 1972. The dry weight sum of forbs increased more in the ungrazed enclosure than the cover of this plant group at the site (Table 1). On a relative basis, however, the weight of all dicotyledonous herbs decreased in this area without large herbivore grazing, while both the absolute and relative weight of the monocotyledons (as for the cover) increased in the ungrazed area, except for the species *Carex bigelowii* and *Deschampsia flexuosa*.

The increase in total plant dry weight of monocotyledons in the dry meadow after three years of heavy sheep grazing is caused by a weight increase in just these two species, and also in *Poa alpina*, while the other monocotyledons showed lower dry weight in grazed than in plots ungrazed by large herbivores in the same period (Table 2). *Carex bigelowii* and *Deschampsia flexuosa* are normally less favored by sheep (on an average about 45% grazed) than most of the other monocotyledons in the dry meadow. *Poa alpina*, however, is highly palatable for sheep (60–70% of available biomass grazed). A reason for the higher biomass of the heavily grazed *Poa alpina* may be high regeneration ability after grazing, and invasion of the bare soil found after sheep trampling. Seeds of the species may also have good germination ability after passing through the animals, and the general fertilization effect of the sheep may also be of importance.

The dicotyledonous herbs usually had, as expected from the cover values, lowest aboveground biomass at the sheep-grazed plots, except *Ranunculus acris* and *Thalictrum alpinum* at both sites. Normally *Ranunculus acris* is a bad forage plant, but the species was grazed at a medium rate (50%) in the dry meadow, and at an even higher rate in the wet meadow. The regeneration of the two species must be good, either from old roots or from seed germination by the reduced

Table 2. Dry meadow. Dry weight in gm^{-2} of aboveground plant material and relative dry weight in relation to total herbaceous vascular aboveground biomass before grazing, at ungrazed fenced plots and at plots grazed by sheep for three years

Spezies	August 1969 Before grazing		August 1972 Ungrazed exclosure 1969–1972		August 1972 Grazed 1969–1971, Ungrazed 1972	
	Total above-ground biomass	Relative weight	Total above-ground biomass	Relative weight	Total above-ground biomass	Relative weight
Anthoxanthum alpinum	3.5	6.6	7.6	9.2	6.0	7.8
Carex bigelowii	4.7	9.0	7.2	8.7	12.9	16.7
Deschampsia flexuosa	6.5	12.5	9.3	11.3	14.4	18.6
Festuca ovina	0.8	1.6	3.0	3.7	1.4	1.8
Luzula frigida	0.2	0.4	1.9	2.3	0.4	0.5
Phleum commutatum	0.3	0.5	2.1	2.5	0.4	0.6
Poa alpina	0.9	1.7	4.0	4.8	14.6	18.8
Other monocotyledons	0.2	0.4	1.7	2.0	0.3	0.4
Sum monocotyledons	17.1	32.7	36.8	44.5	50.5	65.2
Selaginella selaginoides	0.4	0.7	0.6	0.7	0.2	0.2
Alchemilla vulgaris coll.	3.0	5.8	4.4	5.3	2.3	2.9
Antennaria dioica	3.5	6.7	5.0	6.1	2.6	3.4
Astragalus alpinus	2.2	4.3	2.3	2.7	1.1	1.4
Bartsia alpina	0.8	1.5	1.1	1.3	0.4	0.5
Campanula rotundifolia	0.8	1.5	1.2	1.4	0.6	0.8
Euphrasia frigida	0.3	0.5	0.3	0.4	0.2	0.2
Hieracium alpinum	0.2	0.3	0.2	0.3	0.1	0.2
Leontodon autumnalis	2.1	4.0	3.1	3.7	1.6	2.1
Polygonum viviparum	9.4	18.2	10.9	13.2	6.2	8.0
Potentilla crantzii	0.4	0.8	0.5	0.7	0.2	0.3
Ranunculus acris	1.9	3.6	2.7	3.3	4.1	5.3
Saussurea alpina	1.6	3.0	2.3	2.8	0.8	1.0
Sibbaldia procumbens	0.3	0.5	0.4	0.5	0.2	0.3
Silene acaulis	2.7	5.2	4.2	5.0	1.3	1.7
Thalictrum alpinum	3.6	6.8	3.8	4.6	3.8	5.0
Veronica alpina	0.4	0.7	0.5	0.6	0.3	0.4
Other dicotyledons	1.7	3.2	2.3	2.9	1.1	1.4
Sum herbaceous dicotyledons and Pteridophyta	35.0	67.3	46.0	55.5	26.9	36.8
Sum herbaceous vascular plants and Pteridophyta	52.1	100.0	82.6	100.0	77.4	100.0

competition and more open soil after heavy sheep grazing. The dicotyledonous herbaceous species most strongly reduced in aboveground biomass by sheep grazing in the dry meadow were the highly palatable species *Astragalus alpinus*, *Saussurea alpina*, and *Silene acaulis*, all to less than half the biomass of the same species at the plots ungrazed by sheep, and before the research started.

The biomass of *Polygonum viviparum* was not decreased at the sheep-grazed area at the end of the study in the relatively densely vegetated wet meadow

Table 3. Wet meadow. Dry weight in gm^{-2} of aboveground plant material and relative dry weight in relation to total herbaceous vascular aboveground biomass before grazing, at ungrazed fenced plots and at plots grazed by sheep for three years

Species	August 1969 Before grazing		August 1972 Ungrazed exclosure 1969–1972		August 1972 Grazed 1969–1971 Ungrazed 1972	
	Total aboveground biomass	Relative weight	Total aboveground biomass	Relative weight	Total aboveground biomass	Relative weight
Anthoxanthum alpinum	0.2	0.3	0.1	0.1	0.4	0.3
Carex canescens	0.5	0.6	0.2	0.2	0.8	0.6
Carex nigra	51.7	61.4	51.2	64.4	99.1	76.1
Carex rariflora	0.6	0.7	0.2	0.3	1.0	0.8
Eriophorum angustifolium	1.5	1.8	0.6	0.7	2.6	2.0
Festuca rubra	16.2	19.2	13.2	16.6	9.3	7.1
Luzula frigida	0.2	0.3	0.1	0.1	0.4	0.3
Phleum commutatum	0.3	0.3	0.3	0.4	0.1	0.1
Poa alpina	1.4	2.1	0.6	0.8	3.3	2.3
Other monocotyledons	2.4	2.8	0.8	1.0	3.9	3.0
Sum monocotyledons	75.4	89.5	67.3	84.6	120.9	92.9
Equisetum palustre	1.2	1.4	2.2	2.8	0.5	0.4
Selaginella selaginoides	0.1	0.1	0.1	0.1	0.1	0.1
Comarum palustre	0.9	1.1	1.8	2.3	1.0	0.8
Epilobium alpinum coll.	0.2	0.3	0.4	0.5	0.3	0.2
Leontodon autumnalis	2.1	2.6	2.8	3.6	1.0	0.8
Polygonum viviparum	2.8	3.4	2.8	3.3	3.6	2.8
Ranunculus acris	0.5	0.6	0.6	0.7	1.4	1.1
Rubus chamaemorus	0.2	0.2	0.4	0.5	0.3	0.2
Taraxacum croceum	0.1	0.4	0.1	0.2	0.2	0.1
Thalictrum alpinum	0.1	0.1	0.1	0.1	0.2	0.1
Veronica alpina	0.4	0.1	0.7	0.9	0.4	0.3
Other dicotyledons	0.2	0.2	0.3	0.4	0.3	0.2
Sum herbaceous dicotyledons and Pteridophyta	8.8	10.5	12.3	15.4	9.3	7.1
Sum herbaceous vascular plants and Pteridophyta	84.2	100.0	79.6	100.0	130.2	100.0

(Table 3), in contrast to this species in the somewhat more open dry meadow. Species of the two grasses *Festuca* and *Phleum* decreased in dry weight as a result of sheep grazing at both sites. The dominating monocotyledon in the wet meadow, *Carex nigra* (about half the aboveground biomass), however, increased considerably in weight from ungrazed to grazed plots. This species was grazed only lightly, about 15% of the aboveground biomass consumed per year (WIELGOLASKI, 1975). The weight of the other sedges in the wet meadow was also found to increase after sheep grazing, possibly because of low grazing intensity and reduced competition with other species, and also fertilization by the sheep. The total daily herbivory was found to be somewhat lower in the wet meadow than in the dry meadow (on an average about 0.7 gm^{-2} and 1.0 gm^{-2}, respectively, for the populations studied), which is possibly caused by the generally less favorable species for sheep grazing in the wet meadow.

One reason why most of the monocotyledons at the Norwegian sites are not destroyed to a great extent by grazing is that they are more or less strongly tufted (e.g. *Poa alpina* and *Deschampsia flexuosa*, and partly *Carex nigra*), or have stoloniferous shoots and rhizomes with several buds (e.g. *Carex bigelowii*). ELLISON (1960) says that the resistance of grasses to cropping is often extolled, usually with reference to the basal meristem as an adaptation to grazing.

During the first summer of grazing, *Astragalus alpinus* was already strongly reduced in aboveground biomass. This mineral-rich species (see N-fixing forbs in Table 1, WIELGOLASKI et al., 1975) was one of the preferable species in the sheep diet, and more than 90% of the available material was grazed in the first summer. The regeneration was not rapid enough to build up new plant material before the next year's grazing. It seems to be common that legumes are decreased by grazing. ELLISON (1960) reports that nearly one-third of the dicotyledonous species found to decrease by grazing were legumes.

Life forms generally seem to be very important for dicotyledonous herbs to withstand grazing (in addition to consumption preference). It is important to note that regeneration can easily take place from adventitious buds on the remaining organs after grazing. Appressed leaf rosettes also save some dicotyledonous herbs from overgrazing. The rosettes, in addition to more or less rhizomatous growth, may e.g. be the reason for the small changes in weight of *Thalictrum alpinum* after sheep grazing.

On the whole, considerable quantitative variation is found in plant structure after three years of heavy sheep grazing; but none of the species present before the studies began were absent, nor were any new species found. This means that if invasion of other species has occurred, they are possibly only very sparsely present. It is obvious that the relative weight increase in grazed areas for many of the monocotyledons is not only caused by low grazing pressure (as for the *Carex* species), but is also a result of the growth form allowing these low grasses to with stand heavy grazing at least for a few years. The root biomass, however, was strongly reduced at the grazed areas compared to the enclosed ungrazed areas (WIELGOLASKI, 1976). This may also cause a reduction in the amount of monocotyledons, if heavy grazing is permitted for many consecutive years.

References

BRANSON, F. A., PAYNE, G. F.: Effects of sheep and gophers on meadows of the Bridger Mountains of Montana. J. Range Management **11**, 165–169 (1958).

DAHL, E., KALLIOLA, R., MARKER, E., PERSSON, Å.: Nordisk vegetations-klassificering för kartläggning. IBP i Norden **7**, 3–12 (1971).

ELLISON, L.: Subalpine vegetation of the Wasatch Plateau, Utah. Ecol. Monog. **27**, 89–184 (1954).

ELLISON, L.: Influence of grazing on plant succession of rangelands. Botan. Rev. **26**, 1–78 (1960).

FENTON, E. W.: Vegetation changes in hill grazings with particular reference to heather (*Calluna vulgaris*). J. Brit. Grassland Soc. **4**, 95–103 (1949).

GIMINGHAM, G. H.: Dwarf shrub heaths. In: BURNETT, J. H. (Ed.): The vegetation of Scotland, pp. 232–287. Edinburgh-London: Oliver and Boyd 1964.

HUGHES, R. E., MILNER, C., DALE, J.: Selectivity in grazing. In: CRISP, D. J. (Ed.): Grazing in marine and terrestrial environments, pp. 189–202. Symp. British Ecological Soc. 1962. Oxford: Blackwell 1964.

KING, J., NICHOLSON, J. A.: Grasslands of the forest and subalpine zones. In: BURNETT, J. H. (Ed.): The vegetation of Scotland, pp. 168–215. Edinburgh-London: Oliver and Boyd 1964.

LEWIS, J. K.: A synthesis of structure and function. In: FRENCH, N. R. (Ed.): Preliminary analysis of structure and function in grasslands, pp. 317–387. Proc. Symp. U.S. Grassland Biome 1971.

LYE, K. A.: Vegetation of selected localities for IBP on investigation in Hardangervidda, Southern Norway. IBP i Norden **8**, 101–111 (1972).

MILNER, C., HUGHES, R. E.: Methods for the measurement of the primary production of grassland. IBP Handbook No. 6. Oxford: Blackwell 1968.

NORDHAGEN, R.: Sikilsdalen og Norges fjellbeiter. Berg. Mus. Skr. **22**, 1–607 (1943).

PAULSEN, H. A., JR.: Plant cover and forage use of alpine sheep ranges in the central Rocky Mountains. Iowa State J. Sci. **34**, 731–748 (1960).

PEARSON, L. C.: Primary production in grazed and ungrazed desert communities of eastern Idaho. Ecology **46**, 278–285 (1965).

RAUNKIAER, C.: The life forms of plants and statistical plant geography. Oxford: Univ. Press 1934.

RAWES, M., WELCH, D.: Further studies on sheep grazing in the Northern Pennines. J. Brit. Grassland Soc. **21**, 56–61 (1966).

SONESSON, M., WIELGOLASKI, F. E., KALLIO, P.: Description of Fennoscandian tundra ecosystems. In: WIELGOLASKI, F. E. (Ed.): Fennoscandian tundra ecosystems. Part 1: Plants and microorganisms, pp. 3–28. Berlin-Heidelberg-New York: Springer 1975.

TANSLEY, A. G.: The British Islands and their vegetation. Vol. 1. Cambridge: Univ. Press 1953.

WELCH, D., RAWES, M.: The herbage production of some Pennine grasslands. Oikos **16**, 39–57 (1965).

WIELGOLASKI, F. E.: Vegetation types and plant biomass in tundra. Arctic Alpine Res. **4**, 291–305 (1972).

WIELGOLASKI, F. E.: Grazing by sheep. In: WIELGOLASKI, F. E. (Ed.): Fennoscandian tundra ecosystems. Part 2: Animals and systems analysis, in press. Berlin-Heidelberg-New York: Springer 1975.

WIELGOLASKI, F. E.: The effects of herbage intake by sheep on primary production, ratios top-root and dead-live aboveground parts. Polish Ecol. Studies **1** (accepted) (1976).

WIELGOLASKI, F. E., KJELVIK, S., KALLIO, P.: Mineral content of tundra and forest tundra plants in Fennoscandia. In: WIELGOLASKI, F. E. (Ed.): Fennoscandian tundra ecosystems. Part 1: Plants and microorganisms, pp. 316–332. Berlin-Heidelberg-New York: Springer 1975.

Plant Phenology of Fennoscandian Tundra Areas

F. E. WIELGOLASKI and L. KÄRENLAMPI

Introduction and Methods

A short review of literature on phenological studies in tundra is given by WIELGOLASKI (1974). Generally, snow cover is found to be the primary factor influencing tundra phenology (HOLWAY and WARD, 1965), with soil temperature also of importance.

Dates of the phenophases budbreak, visible flower stem, visible flower bud, flowering, flower fading, fruit ripening, seed dispersal and autumn coloring were determined during 1970–1973 at the Norwegian sites studied on the mountain plateau area Hardangervidda in southern Norway (SONESSON et al., 1975). Budbreak was defined as 0.5 cm new growth in perennial herbaceous plants and the first visible leaf stalk in woody plants. Flowering was defined (when possible) as the first day the stamens and pistil could be seen in the first flower. The beginning of fruit ripening and autumn color were, however, difficult, to define exactly. The fractions of plants in the various phenophases were found each week. At the same time, the heights of the plants (length of new shoots or leaves of grasses) were measured and the number of new leaves counted, always at the same marked 15 plant specimens and thickness-growth measured on large shrubs and trees.

The Finnish phenology studies (1969–1972) at sites in northern Finland (SONESSON et al., 1975) were concentrated on growth in length on some vascular plants and cryptogams, and growth in thickness on trees. The length-growth of plants at the Finnish sites was measured by small plastic scales attached to the plants. The readings on the millimeter scale were taken at intervals of a few days during the period of apparent growth. For some species, e.g. *Pleurozium schreberi*, the intervals were considerably longer and irregular. Stem thickness-growth measurements were carried out by growth bands on Norwegian birch trees and other woody plants and on Finnish pine trees, and by two cross diameter measurements on Finnish birch trees.

Results and Discussion

The study has shown that there are considerable variations between the species in the date of budbreak in the spring, and also between the sites. In the Norwegian studies, plants at the subalpine birch site started growth first, as expected. Green buds of *Betula tortuosa* and seedlings of *Melampyrum* and breaking buds of *Maianthemum bifolium*, *Trientalis europaea* and grasses were found on May 28th in the earliest year studied (1971) and on June 7th in the latest year (1973), although about 15% snow cover was still present. Generally, grasses and sedges are early in the spring. At all the Norwegian alpine sites except the wet meadow most of the monocotyledons had started new green growth before June 8th in the earliest year, and before June 21st in the latest year. The wet meadow was about one week later. Most specimens of *Dryas octopetala*, *Polygonum viviparum*, *Ranunculus acris* and other forbs also started growth at these dates. The *Salix* species were later. Even at the early Norwegian lichen heath site (about one week earlier than the dry meadow), budbreak was seen in only about 50% of the *Salix herbacea* plants in the two years on June 8th and 22nd, respectively, and in about 5% of the specimens in the dry meadow. None were visible in the wet meadow until mid and end of June, respectively, in the two extreme years.

The latest start of new shoot growth was found for the evergreen shrubs in the Norwegian alpine sites, while these plants were found to be as early as other species (June 5th–June 20th in different years) in the Finnish subarctic studies. *Vaccinium vitis-idaea* began shoot growth on about June 23rd at the early Norwegian alpine lichen heath situated, 200–300 m above the tree line in the earliest year, and about one week later in the latest year, while *Empetrum hermaphroditum* started budbreak on about July 5th in both years at this site. However, flowering of this species started at about the same time as budbreak of the earliest species. The delay in budbreak of the evergreen species may be one form of protection for those species growing on exposed sites with a thin, early melting snow cover.

Based on the dates of the first budbreak, the start of the growing season at the Norwegian sites may on an average be about May 30th in the subalpine birch forest, June 3rd at the lichen heath, June 8th at the dry meadow and willow thicket sites and June 12th in the wet meadow. Similarly, the start of the growing season at the Finnish sites is estimated to be on an average June 7–10th when based on the first budbreak, maybe two to three days delayed at the mountain heath site compared to the forest sites. The annual variation in climate (SKARTVEIT et al., 1975), however, strongly influences the phenophases. The spring phases at the Norwegian sites were generally earlier in 1970, with high June temperatures, than in the other years. More species were, e.g., flowering before about July 1st in 1970 than in the other years studied (*Poa alpina*, *Dryas octopetala* and *Thalictrum alpinum* in the dry meadow, and *Ranunculus acris* in the wet meadow). Especially species preferring relatively eutrophic conditions in the Norwegian alpine areas seemed to be influenced strongly by the warm early spring (June) in 1970. At the Finnish sites, variations in budbreak from June 1st to June 20th are found throughout the years because of the climate.

KÄRENLAMPI (1972) estimated that the snow cover at the sites in northern Finland generally disappeared about May 20th, and the stable snow cover forms on about October 20th. Thus, the snow-free period is 150 days. Apparent growth of the dominant vascular plants at the Finnish sites (KJELVIK and KÄRENLAMPI, 1975) begins roughly 20 days after snow melt and lasts about 40 days (i.e. ends about July 20th), thus ceasing 90 days before the forming of snow cover. The snow cover at the Norwegian sites lasts longer in spring, to about May 28th, at the lichen heath with a thin snow cover, and a couple of days earlier at the lower elevated subalpine site. In the wet meadow snow melts on an average about June 10th, and about one week earlier in the dry meadow and at the willow thicket sites. New stable snow cover forms about October 20th at the Norwegian alpine sites, and about November 1st in the birch forest, 10 days later than in northern Finland (SKARTVEIT et al., 1975). There is a shorter period from snow-melt to budbreak at the Norwegian than at the Finnish sites, causing an earlier growth start at most of the Norwegian sites in spite of later snow-melt. The length-growth ends on about July 25th in the Norwegian birch forest and lichen heath, and about a week later at the eutrophic sites. This longer period of length-growth of the Norwegian compared to the Finnish plants may be caused by lower summer temperatures (KÄRENLAMPI, 1972; SKARTVEIT et al., 1975) at Hardangervidda. The thickness-growth of the trees in both the Norwegian and Finnish birch forests started some days after the length-growth, and was most active in the period June 20th–July 20th.

Figure 1 suggests that the late snow-melting in 1972 generally caused later spring phenophases at the Norwegian sites than in years with earlier snow-melt (e.g. 1971, as shown by WIELGOLASKI, 1974). This is in agreement with the statement by e.g. HOLWAY and WARD (1965) that snow cover is a very important factor in alpine phenology. The warm July of 1972 resulted in mostly shorter periods for each of the phenophases studied in Norway than found in the other years studied. The flowering period of *Astragalus alpinus* is, for instance, strongly influenced; it lasted about 50 days in 1971, but only about 23 days in 1972.

Low temperatures at the Norwegian sites in the important growth-month July in 1970 usually delayed development of the later phenophases, particularly in the meadows that year, compared to the other study years. This delay in 1970 was to some degree also found for phenophases in September, in spite of relatively high August temperatures in 1970. Later ripening and diaspore dispersal in August–September was e.g. found in the meadows in 1970 compared to 1971, 1973, and particularly 1972 (up to three weeks earlier than 1970) for the species *Anthoxanthum alpinum*, *Thalictrum alpinum* and *Polygonum viviparum* (bulbils). The difference between 1972, which was warm, and the other years was more pronounced in the cold, wet Norwegian meadow than in the warmer, dry meadow. At the extremely dry and warm Norwegian lichen heath, however, the phenophases occurring in August were earlier after the cold, wet July of 1970 than in the other three years, and latest in 1972 and 1973 as found, e.g., for the first ripe seeds of *Salix herbacea*. This was possibly caused by the better moisture conditions in 1970 (particularly in July) than in the other years studied (SKARTVEIT et al., 1975).

The autumn colors at the Norwegian sites started later in the dry summer of 1971 than in the other years. The same species started to show color about one week later in the dry than in the wet meadow, and a few days later at the protected willow thicket site (e.g. *Polygonum viviparum* and *Anthoxanthum alpinum*). *Arctostaphylos alpina* was the first species really showing beautiful autumn colors (at the Norwegian lichen heath), starting on about August 10th in 1972, one week later in 1970, and still some days later in 1971 and 1973. By the end of August in 1972 and one week later in 1970, 1971 and 1973, most species were showing

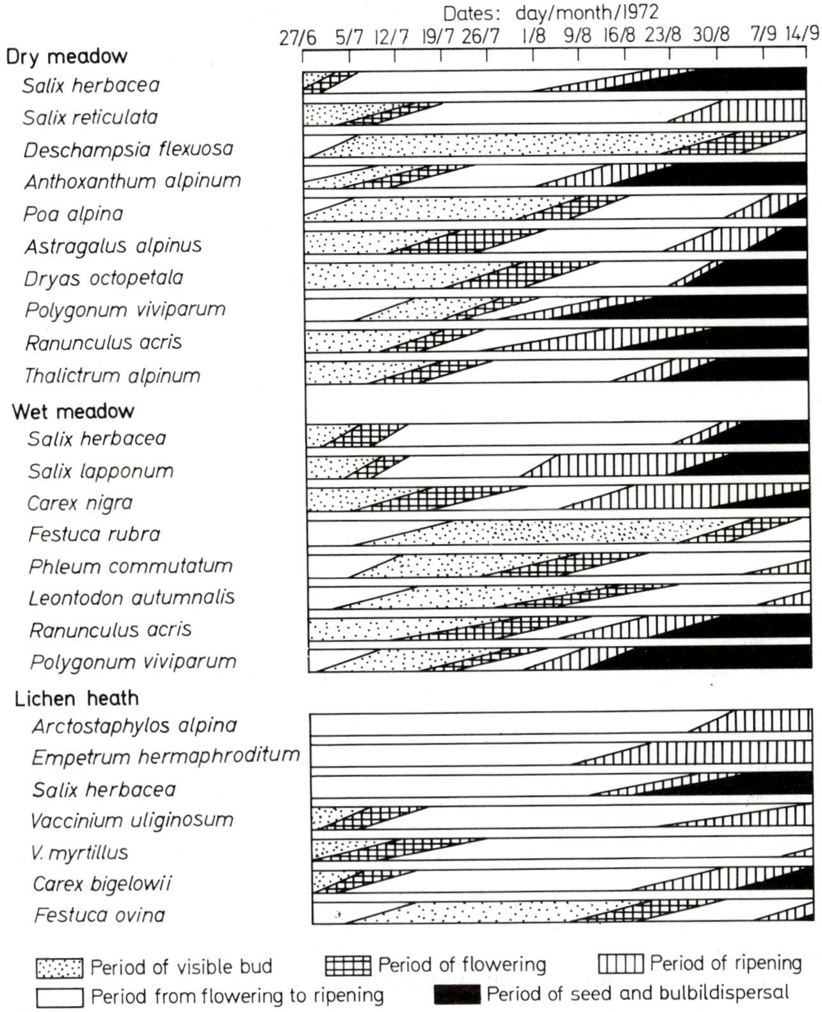

Fig. 1. Reproductive development in alpine tundra sites at Hardangervidda, southern Norway, 1972

autumn colors on the leaves at all sites, even in the birch forest (except evergreen species and *Astragalus alpinus* and most specimens of *Salix lapponum*, *Equisetum palustre* and *Deschampsia caespitosa*). The time of the autumn colors clearly shows when decreasing photosynthesis could be expected to take place in various species. The vegetation period at the Norwegian alpine sites ended about September 20th and about two weeks later in the birch forest.

A detailed comparison between the Norwegian sites for the phenology of all main species has been given elsewhere by WIELGOLASKI (1975). The various early summer phenophases for a given species were, as expected, earlier at the subalpine forest site than at the sites above the tree-line, e.g., about 10 days for the flowering

Table 1. Fertility in percentage of total number of shoots at the Norwegian IBP Tundra sites 1973

Species	Birch forest	Lichen heath	Dry meadow	Willow thicket	Wet meadow
Empetrum hermaphroditum	12.2	6.3	—	—	—
Vaccinium myrtillus	18.3	11.7	2.3	—	—
Vaccinium uliginosum	11.6	8.2	—	—	—
Vaccinium vitis-idaea	3.2	0.6	0	—	—
Salix herbacea	—	18.1	11.2	10.3	8.7
Vaccinium lapponum	—	—	—	9.1	1.7
Carex bigelowii (C. vaginata)	—	12.1	8.7	8.5	—
Anthoxanthum alpinum	—	—	9.3	6.6	—
Deschampsia flexuosa	3.7	—	5.4	—	—
Festuca rubra	—	—	—	4.3	1.7
Phleum commutatum	—	—	7.2	—	6.4
Polygonum viviparum	—	—	18.3	16.8	12.7
Ranunculus acris	—	—	15.7	8.4	13.1
Thalictrum alpinum	—	—	6.6	5.3	3.0

of *Vaccinium uliginosum*, but also for the early-flowering species *Empetrum hermaphroditum*. Flowering was five days to one week earlier for *Anthoxanthum alpinum* and *Ranunculus acris* at the willow thicket site than in the dry meadow. The difference was a few more days for *Ranunculus* in the wet meadow, while there was from 6 to 12 days difference between the willow thicket site and the wet meadow for flowering of *Salix lapponum*. Generally, the phenophases in the wet meadow were somewhat later than in the dry meadow, caused by later snow-melting and low temperature (SØRENSEN, 1941). However, the first phenophases studied (e.g. visible flower bud) on *Polygonum viviparum* and partly *Ranunculus acris* were, in both 1971 and 1972, earliest in the wet meadow. This may have been caused by the moisture conditions of the same or the previous year. Lack of water delayed the ripening of the berries of *Vaccinium uliginosum* at the Norwegian lichen heath in 1972 and in part of 1973, compared to the birch forest. Drought is possibly also the reason why seed-ripening (as also mentioned in the comparison between years), as well as seed dispersal, were later for *Salix herbacea* at the lichen heath than in the dry meadow, particularly in 1972, while the low temperature in the wet meadow caused all phenophases of the species to be latest at that site in 1972 (Fig. 1), as well as in other years.

The fertility varied considerably from site to site in the Norwegian studies. Highest fertility was found in the sheltered birch forest. In 1973 about 12% of the shoots of *Empetrum* were fertile at this subalpine site, but only 6% at the alpine lichen heath (Table 1). Similarly, the *Vaccinium* species all showed a higher percentage fertility in the birch forest than at the lichen heath. The fertility also seemed to be lower in the dry meadow and the willow thicket than in the birch forest, for woody plants as well as monocotyledons. The lowest fertility of both these plant groups and of forbs was mostly found in the wet meadow (Table 1). The fertility was generally higher in 1973 than in any of the other years. *Astragalus alpinus* showed more than 70% of fertile shoots in 1973, while only about 10% were

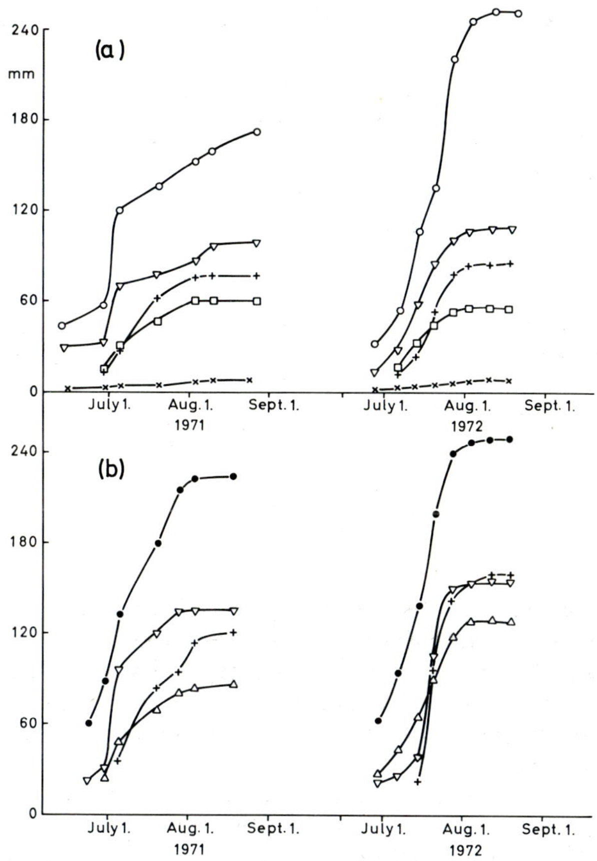

Fig. 2a and b. Shoot growth at Norwegian alpine sites in mm 1971 and 1972. (a) Dry meadow; (b) wet meadow. ○: *Anthoxanthum alpinum.* ▽: *Ranunculus acris.* +: *Polygonum viviparum.* □: *Astragalus alpinus.* × : *Vaccinium vitis-idaea.* △: *Equisetum palustre.* ●: *Carex nigra*

found in 1971. This may be explained as an effect of the July temperature the year before (low in 1970 and high in 1972). The same effect was found e.g. on the species *Dryas octopetala*, *Viola bicolor*, *Arctostaphylos alpina* and *Carex bigelowii.*

The shoots of *Salix lapponum* at the Norwegian wet meadow site were three times larger in early July 1970 (about 25 mm) than on the same date in 1971 and six times larger than in 1972, while the difference in length of the leaves of *Phleum commutatum* in early July was smaller between the years ($1^1/_2$ times longer in 1970 (75 mm) than in 1971 and $2^1/_2$ times longer than in 1972). *Poa alpina* leaves at the dry meadow site were, in early July, about twice as long in 1970 (about 130 mm) as in 1971, and three times longer than in 1972. The shoots of *Thalictrum alpinum* at the same site were also longest in 1970 (about 80 mm, which is five times the length in the other years for early July). At the early snow-melting lichen heath site, the differences between the years were smaller at the beginning of July.

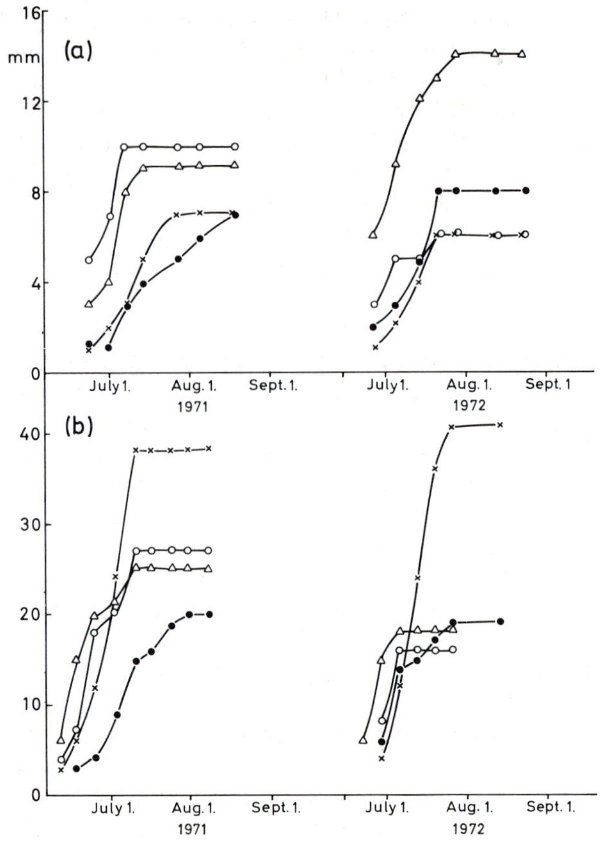

Fig. 3a and b. Shoot growth in mm 1971 and 1972: (a) Norwegian alpine lichen heath; (b) Norwegian subalpine birch forest. ×: *Vaccinium vitis-idaea.* △: *Vaccinium myrtillus.* ○: *Vaccinium uliginosum.* ●: *Empetrum hermaphroditum*

The variation in shoot lengths for some species at four of the Norwegian sites studied in 1971 and in the somewhat warmer summer 1972 are shown in Figs. 2 and 3. For some species, the differences in maximum lengths between the two years are small (e.g. *Astragalus alpinus*), while others have longer shoots and/or leaves in 1972 than in 1971. This was found e.g. for leaves of *Anthoxanthum alpinum* in the dry meadow, and for shoots of *Ranunculus acris* at both meadow sites (Fig. 2). The effect seemed to be most extensive in the wet meadow, again showing the stronger positive influence of high summer temperatures at this cold site than in the warmer dry meadow.

Generally, the plants were higher (maximum length) at the Norwegian sites in 1970 than in 1971, caused by the rapid growth in the warm spring of 1970. They were, however, usually somewhat lower than in 1972 at all sites except the lichen heath, where the wet summer of 1970 caused longer shoots to develop than in the drier years. The maximum shoot length of *Vaccinium uliginosum* was e.g. above 20 mm at the lichen heath in 1970, while only 10 mm in 1971 and 6 mm in 1972.

The high July temperature and moderate precipitation in 1972 also resulted in early stop of growth and shorter shoots than in 1971 for both this species and *Vaccinium myrtillus* in the subalpine birch forest at Maurset (Fig. 3). On the other hand, the drought-resistant species *Empetrum hermaphroditum* showed no difference in maximum shoot length in these two years. In northern Finland, however, it was found that the new shoot growth of *Empetrum* was very small in years with low temperatures (e.g. 1969). The birch *(Betula tortuosa)* had a much stronger shoot growth in 1972 than in 1971 at the Norwegian site (maximum 110 and 60 mm respectively), but the length-growth ended about on July 20th–25th in both years.

The shoot-growth of vascular plant species which were found at more than one of the sites, was generally stronger in the subalpine birch forest in both NORWAY and Finland than at the sites above the tree line, as seen for *Empetrum hermaphroditum* and all the *Vaccinium* species at the Norwegian sites in Fig. 2 and 3. The maximum shoot-length of *Empetrum* seemed to be some millimeters less in the Finnish than in the Norwegian birch forest. At the Finnish mountain site, growth of *Empetrum* shoots was 1–2 mm more than at the Norwegian lichen heath. Longer shoots were found for *Salix lapponum* and *Ranunculus acris* at the willow thicket site in Norway than in the wet meadow, and the leaves of *Anthoxanthum alpinum* were longer at the willow thicket site than in the dry meadow. *Ranunculus acris* shoots were longer in the wet meadow than in the dry meadow, as well as the shoots of *Polygonum viviparum.* The shoots of *Vaccinium vitis-idaea* were somewhat longer in the dry meadow than in the lichen heath. The general variation in plant heights between the Norwegian sites is shown in BERG et al. (1975).

Observations on the moss *Pleurozium schreberi* show that the length-growth continues during the whole snow-free period. There was no certain growth season like many other mosses, and for the vascular plants. The length-growth of this species was, in the Finnish birch forest, about 6–10 mm per year and in the pine forest about 3–4 mm, which was the maximum length-growth found in *Dicranum elongatum* (SONESSON, 1973; KALLIO and HEINONEN, 1975). Dry periods obviously reduce growth of bryophytes. The same was found for reindeer lichens studied. *Cladonia alpestris* increases 1–2.5 mm in length per year dependent on age, with the highest growth rate of the thallus in young specimens (KÄRENLAMPI, 1970). The dominating cryptogams at the Finnish and Norwegian sites grow when temperature, light and moisture conditions are all favorable. Moisture is the most limiting factor in the middle of summer, and light in autumn.

The differences between the species in shoot-length at the various sites, and between the years, reflect the most important minimum factors for growth. It is obvious that temperatures strongly limit the growth of birch as well as *Ranunculus acris* (and also *Phleum commutatum*) in the Norwegian wet meadow, and for *Anthoxanthum alpinum* (and also *Poa alpina*) in the dry meadow, as well as generally for *Empetrum hermaphroditum.* The growth of the species of *Vaccinium*, and especially *V. uliginosum*, possibly also of *Dryas octopetala* and to some degree *Astragalus alpinus* seemed, however, to be more strongly dependent on moisture. The influence of moisture is especially reflected in vascular plants as in cryptogams by the variation in the shoot growth between sites within one year.

Acknowledgement. Much of the phenological field work and many calculations in the Norwegian studies on which this paper is based have been carried out by the technical assistant S. KRIGSVOLD. The senior author wishes to express thanks to KRIGSVOLD for his valuable help.

References

BERG, A., KJELVIK, S., WIELGOLASKI, F. E.: Measurement of leaf areas and leaf angles of plants at Hardangervidda, Norway. In: WIELGOLASKI, F. E. (Ed.): Fennoscandian tundra ecosystems. Part 1: Plants and microorganisms, pp. 103–110. Berlin-Heidelberg-New York: Springer 1974.

HOLWAY, J. G., WARD, R. T.: Phenology of alpine plants in an area of Colorado alpine. Ecology **46**, 73–83 (1965).

KALLIO, P., HEINONEN, S.: CO_2 exchange and growth of *Rhacomitrium lanuginosum* and *Dicranum elongatum*. In: WIELGOLASKI, F. E. (Ed.): Fennoscandian tundra ecosystems. Part 1: Plants and microorganisms, pp. 138–148. Berlin-Heidelberg-New York: Springer 1974.

KÄRENLAMPI, L.: Morphological analysis of the growth and productivity of the lichen *Cladonia alpestris*. Rep. Kevo Subarctic Res. Stat. **7**, 9–15 (1970).

KÄRENLAMPI, L.: Comparisons between the microclimates of the Kevo ecosystem study sites and the Kevo Meteorological Station. Rep. Kevo Subarctic Res. Stat. **9**, 50–65 (1972).

KJELVIK, S., KÄRENLAMPI, L.: Plant biomass and primary production of Fennoscandian subarctic and subalpine forests and of alpine willow and heath ecosystems. In: WIELGOLASKI, F. E. (Ed.): Fennoscandian tundra ecosystems. Part 1: Plants and microorganisms, pp. 111–120. Berlin-Heidelberg-New York: Springer 1974.

SKARTVEIT, A., RYDÉN, B. E., KÄRENLAMPI, L.: Climate and hydrology of some Fennoscandian tundra ecosystems. In: WIELGOLASKI, F. E. (Ed.): Fennoscandian tundra ecosystems. Part 1: Plants and microorganisms, pp. 41–53. Berlin-Heidelberg-New York: Springer 1974.

SONESSON, M.: Studies in production and turnover of bryophytes at Stordalen 1972. Progress report 1972. Swedish IBP tundra project Tech. Rep. **14**, 66–75 (1973).

SONESSON, M., WIELGOLASKI, F. E., KALLIO, P.: Description of Fennoscandian Tundra ecosystems. In: WIELGOLASKI, F. E. (Ed.): Fennoscandian tundra ecosystems. Part 1: Plants and microorganisms, pp. 3–28. Berlin-Heidelberg-New York: Springer 1974.

SØRENSEN, T.: Temperature relations and phenology of the northeast Greenland flowering plants. Medd. Grønland **125**, 1–305 (1941).

WIELGOLASKI, F. E.: Phenology studies in tundra. In: LIETH, H. (Ed.): Phenology and seasonality modelling, pp. 209–214. Berlin-Heidelberg-New York: Springer 1974.

WIELGOLASKI, F. E.: Phenology and population density of Norwegian tundra plants. In: VIK, R. (Ed.): IBP in Norway. Methods and results. Sections PT-UM Grazing project, Hardangervidda, Botanical investigations, pp. 323–367. Oslo: Norwegian National IBP Committee 1975.

Measurement of Leaf Areas and Leaf Angles of Plants at Hardangervidda, Norway

A. BERG, S. KJELVIK, and F. E. WIELGOLASKI

Introduction and Methods

The leaf areas and leaf angles of vascular plants were measured in 1972 and partly in 1971 at Norwegian IBP Tundra sites, lichen heath, dry meadow, wet meadow and willow thicket in the low-mid alpine zones and at the subalpine birch forest site described by SONESSON et al. (1975).

Two methods were compared for the leaf area measurements in the studies. The leaf area by the inclined point quadrate method is estimated from the number of contacts observed with leaves of different plant species when the point quadrate is inserted into the vegetation at specific angles. The method is described in detail by LEVY and MADDEN (1933) and WARREN WILSON (1959, 1960, 1963) and they also discuss the usefulness in practice of the method. In the present studies single quadrates with a 32.5° angle were used for determination of the leaf area index ($m^2 m^{-2}$ ground area) of living and dead attached leaves (on a single leaf surface basis) and the stems (half the appr. circular stem area). The leaf angles were determined by the same method, but with the point quadrate inserted into the vegetation at 0° and 90°. At each measurement about 100 quadrates were analyzed.

The other method is based on optical measurements of leaf areas per unit dry weight (KJELVIK, 1972), and the leaf area index is determined by multiplication with the dry weight of each species per unit ground area. This method was not useful for cylindrical leaves.

Results and Discussion

At all sites studied the maximum LAI of green material is generally found in mid August (Fig. 1), earlier at the exposed dry lichen heath than in the sub-alpine birch forest (later senescence). This tendency to earlier maximum LAI of green material under more extreme conditions is confirmed by the two to three weeks earlier maximum found in Canadian high arctic studies, at Devon Island, especially at dry beach ridges, but also at sedge meadows (SVOBODA, 1972; MUC, 1973), and at U.S. tundra sedge meadows at Point Barrow, Alaska (DENNIS and TIESZEN, 1972) than in the Norwegian studies.

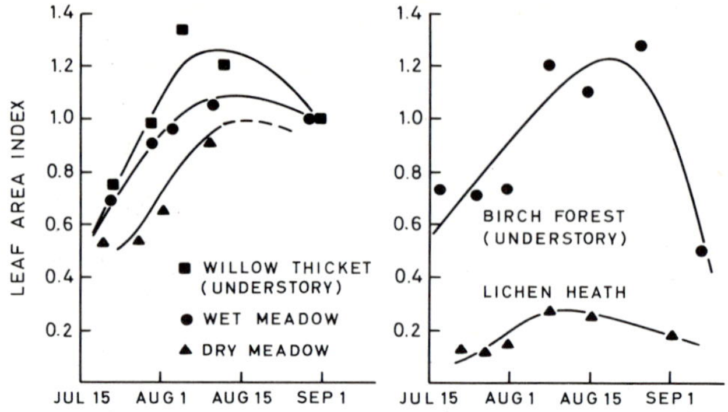

Fig. 1. Leaf area indices ($m^2\ m^{-2}$) of green leaves at different sites at Hardangervidda (single surface) measured by inclined point quadrates, 32.5°

The lowest maximum LAI value is found at the lichen heath, about 0.3 (Fig. 1). This seems to be higher than at the dry Devon Island beach ridge also with sparse vascular plant vegetation and relatively rich in lichens (SVOBODA, 1972). The maximum LAI at the Norwegian wet meadow is slightly above the values at the sedge meadows at Point Barrow (DENNIS and TIESZEN, 1972), and much higher than at the Devon Island sedge meadows (maximum LAI 0.3–0.6, MUC, 1973). The LAI values of the Norwegian alpine meadows are, however, lower than at alpine meadows in Colorado, where the maximum values are found to be of the order of two (dry *Kobresia* site) to three (moist meadows) according to CALDWELL et al. (1972). Highest maximum LAI values are, in Norway, found in the field- and dwarf-shrub-layers of the birch forest and willow thicket sites, about 1.2–1.3.

The relationships between the leaf area indices at the different alpine sites compare well with the production values estimated for the same sites (WIELGOLASKI and KJELVIK, 1973). The production found for the understory in the subalpine birch forest is, however, lower than for the alpine meadows contrary to the results for LAI. When this is especially high for the plants below trees and high shrubs in the birch forest and willow thicket it may be seen as a result of the sheltering effect and shading which causes generally larger leaves. A high correlation is found between the LAI and the chlorophyll amount per ground area at the sites.

The field- and dwarf-shrub-vegetation is also higher in the birch forest and willow thicket than at the other alpine sites studied where there is no protection from overstory plants (Fig. 3). The highest LAI is found between 2 cm and 8 cm above ground in the willow thicket and between 6 cm and 14 cm height in the birch forest. At the last mentioned site relatively few leaves are developed below 4 cm in contrast to the willow thicket site and the other alpine sites. The reason is the low light intensity available to the lowest parts of plants in the birch forest.

The maximum leaf areas in the shrub- and tree-layers at this site are higher than in the field- and dwarf-shrub-layers, LAI about 2.1. The total LAI of green

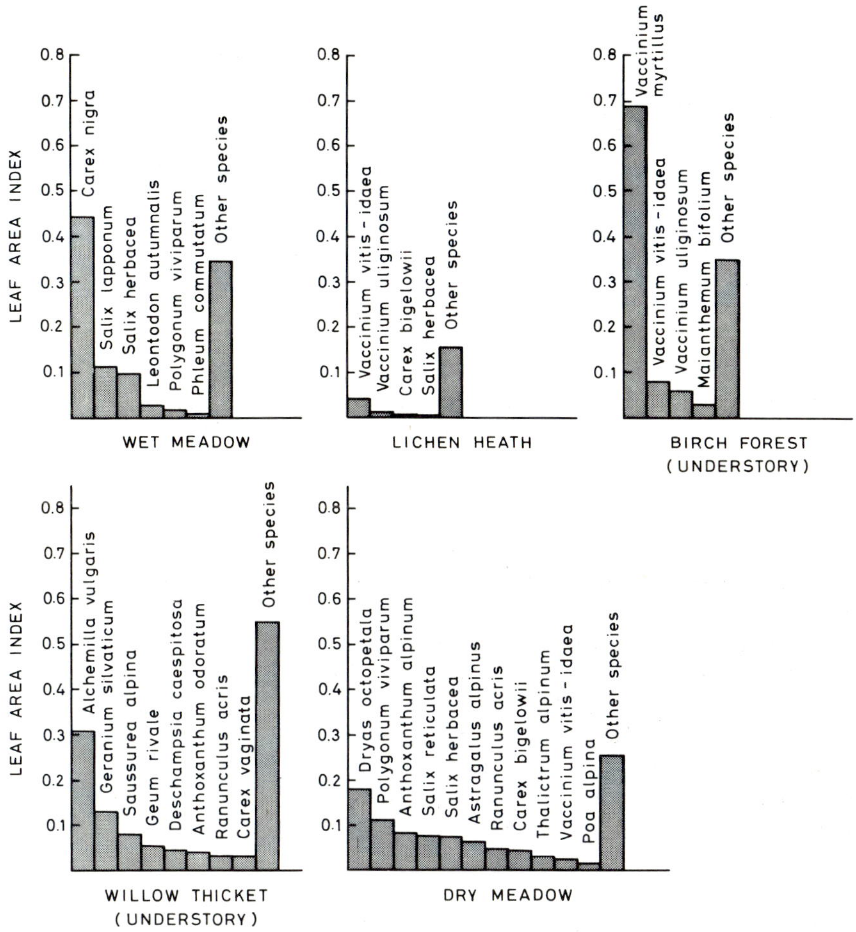

Fig. 2. Leaf area indices ($m^2\ m^{-2}$) of green leaves for some species at different sites in the middle of August (single surface) measured by optical method

vascular plant material in the birch forest is then as high as about 3.2 in mid August. The height of the willows (mostly *Salix lapponum*, 0.5–1.0 m high) in the willow thicket is less than for the birch trees (*Betula tortuosa*, 4–6 m high). The number of willow stems per unit area is higher than the number of birch trees (KJELVIK and KÄRENLAMPI, 1975), but the total maximum LAI in the willow thicket (including the shrubs) is only about 2.0.

At the willow thicket with high species diversity the highest LAI is found in broadleaved, eutrophic forb species such as *Alchemilla vulgaris*, *Geranium silvaticum* (sometimes above 30 cm high), *Saussurea alpina* and *Geum rivale*, and also in some monocotyledons (Figs. 2 and 3). Oligotrophic and drought resisting species are kept out of the willow thicket with oxygen rich water and favorable nutrient conditions in competition with several eutrophic species.

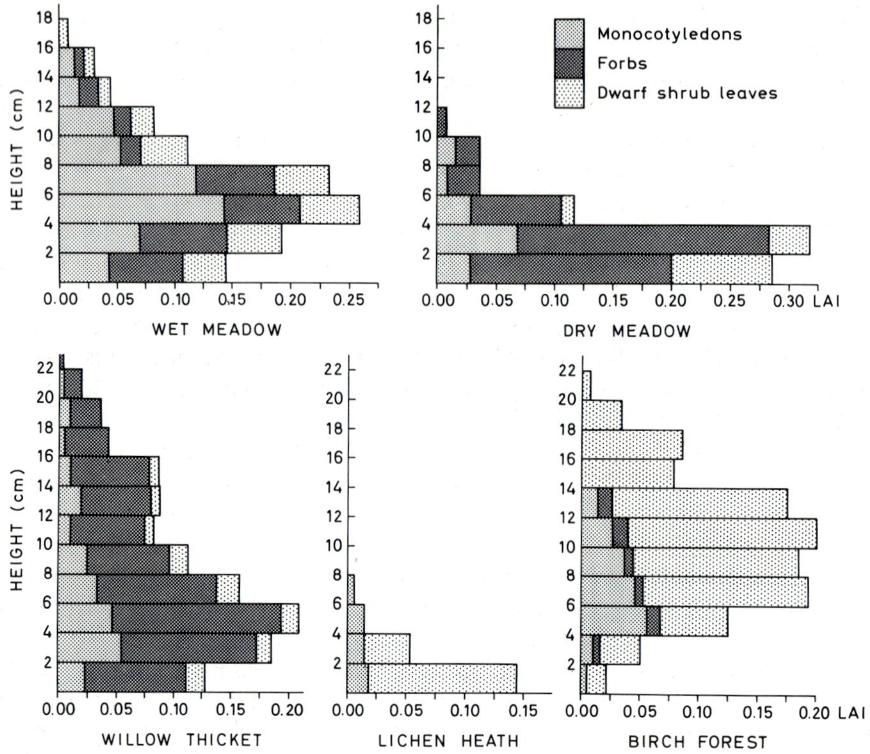

Fig. 3. Maximum leaf area indices ($m^2 m^{-2}$) at different plant heights (single surface) measured by inclined point quadrates, 32.5°

The oligotrophic ericaceous dwarf shrubs dominate in the birch forest (Fig. 3). About $^2/_3$ of the leaf area in the understory is found in the *Vaccinium* species at the site, especially *V. myrtillus* (Fig. 2), and also *Empetrum hermaphroditum* is important. The plants are considerably higher than the same species at the lichen heath.

The average leaf angle with the horizontal is about 55° for both monocotyledons and forbs in the willow thicket at 6, 12, and 17 cm height. The variation in the leaf angle is higher in the birch forest (BERG et al., 1973). The forbs show the same angle as in the willow thicket (55°) both at 6 and 9 cm height, while the angle in the monocotyledons is about 75° 3 cm above the ground, 45° at 6 cm height and 70° at 9 cm. The relatively low leaf angle at 6 cm is caused by *Deschampsia flexuosa* which has nearly horizontal leaves at this height. Only few leaves of this species reach 9 cm, and the dominating grass at this height is the more erect *Nardus stricta*.

The more exposed alpine sites, wet meadow, dry meadow and especially lichen heath have lower vegetation. At the wet meadow the plants are mostly lower than 8 cm, at the dry meadow generally below 4 cm and at the highly exposed lichen heath only very few leaves are found above 2 cm (Fig. 3).

The water deficit and poor nutrient conditions in the glacifluvial sandy soil (HINNERI et al., 1975) at the lichen heath means that only drought resisting species will survive. The few plants which are higher than 2 cm are mostly the monocotyledons *Festuca ovina* and *Carex bigelowii* with stiff leaves. The very low mat-forming dwarf shrub *Empetrum hermaphroditum* (less than 2 cm) dominates together with lichens. Forbs are nearly totally absent at this site (Fig. 3).

At the eutrophic dry meadow, however, forbs dominate completely among the more than 50 species of vascular plants (LYE, 1972). Most important at this site is the calciphilous *Dryas octopetala* (classified as a forb although functionally reacting more like a dwarf shrub) with nearly $^1/_5$ of the total LAI (Fig. 2). It is found that the mean leaf area per live leaf of *Dryas* increases steadily from 0.21 cm^2 in early June just after budbreak (1971) to 0.56 cm^2 in mid August. This indicates that the leaves of *Dryas* are usually not overwintering at Hardangervidda alive as found in e.g. high arctic Canada (SVOBODA, 1972).

The wet meadow is somewhat more sheltered than the dry meadow or the lichen heath, and the ground water table is often close to the surface of the peaty soil. This condition is favorable for *Carex nigra* which makes up more than 40% of the total LAI at the site (Fig. 2).

The variation in leaf angles is high at the three low-mid alpine exposed sites. At the lichen heath the leaf angles of the monocotyledons are nearly 90° at all heights measured, while the dwarf shrub leaves are more horizontal with angles less than 30° (BERG et al., 1973). The leaves of monocotyledons at the dry meadow are softer than at the lichen heath and the wet meadow and therefore more horizontal at higher levels above the ground; 75°, 60°, and 25° at the heights 2, 3, and 4 cm, respectively, at the dry meadow, and 80°, 70°, and 50° at 1, 4, and 7 cm, respectively, at the wet meadow. These monocotyledons seem to be somewhat less erect than those found at dry and moister meadows in alpine Colorado (CALDWELL et al., 1972). The forbs and dwarf shrubs have also somewhat more erect leaves at the Norwegian wet than at the dry meadow. The erect leaf growth will cause increased light intensity to reach the bottom layer and influences the rich bryophyte cover at the wet meadow (LYE, 1972).

The leaf area per unit dry weight of leaves is found to vary strongly during the season. Most of the species show low values in the spring and in the fall, with a maximum value (thinnest leaves) sometimes during the growing season (Table 1). Some species have the maximum value early in the season e.g. *Vaccinium uliginosum* at the alpine lichen heath (Table 1) as well as in the sub-alpine forest. *Salix herbacea* shows a decreasing trend in the leaf area per unit dry weight during the whole growing season studied at both the dry and wet meadow and at the lichen heath. This may mean that an increasing leaf area per unit dry weight for this early species could possibly have been found for a short period in the early spring before the observations began, (maximum value in early June). Other species e.g. *Vaccinium vitis-idaea* (new leaves) at all sites studied, at the birch forest (Table 1), at the lichen heath and at the dry meadow, show a maximum leaf area per unit dry weight at about July 20. Many of the monocotyledonous species also have maximum leaf area values per g (thinnest leaves) in the middle of the growing season, e.g. *Anthoxanthum alpinum* at the dry meadow (Table 1), as well as e.g. *Carex bigelowii* and *Poa alpina*. Some species did not reach the maximum leaf

Table 1. Leaf areas in cm^2 per g dry weight of some plant species measured by optical method

Species	Comp.	Site	June					July					August				
			6	12	18	24	30	6	12	18	24	30	6	12	18	24	30
Vaccinium myrtillus	leaves	lichen heath	— —	374	373	348	—	323	233[a]	—	231	202	—	224[a]	—	280	—
Vaccinium uliginosum	leaves	lichen heath	—	110	121	146	151[a]	119	131[a]	—	107	111	—	102[a]	—	102	—
Vaccinium vitis-idaea	current yr. leaves	birch forest	—	—	96	124[a]	167	—	178	183	211	178	156	160	121[a]	121	—
Salix herbacea	leaves	wet meadow	—	—	—	—	227	224	—	209[a]	220	185	159[a]	151	144[a]	—	128
Anthoxanthum alpinum	leaves	dry meadow	85	—	—	129	139	185	198	235	272	—	284	253	—	215	—

[a] Observations 1972, the others 1971.

area per unit dry weight before the last measurement in late August, but increased throughout the season, e.g. many forbs at the dry meadow (*Astragalus alpinus*, *Polygonum viviparum*, *Ranunculus acris* and *Thalictrum alpinum*). Other species show a minimum leaf area per unit dry weight in the middle of the growing season, e.g. *Vaccinium myrtillus* both at the lichen heath (Table 1) and at the birch forest (Berg et al., 1973) and *Betula tortuosa* at the sub-alpine site.

The leaf area value per unit dry weight for e.g. *Salix herbacea* is higher at the wet meadow than at the lichen heath, for *Carex bigelowii* higher at the dry meadow than at the lichen heath, and for *Vaccinium myrtillus* lower at the lichen heath than within the shaded birch forest. For the forbs *Polygonum viviparum* and *Ranunculus acris* there seems, however, to be a tendency to lowest leaf area per unit dry weight at the wettest sites, in the latter part of the season.

Acknowledgements. The analyses in field and laboratory were carried out mainly by the student K. Hofstad and the technical assistants S. Krigsvold, A. Polder and R. Surdal. We wish to thank them for all their help.

References

Berg, A., Skre, O., Wielgolaski, F. E., Kjelvik, S.: Leaf areas and angles, chlorophyll and reserve carbon in alpine and sub-alpine plant communities, Hardangervidda, Norway. In: Bliss, L. C., Wielgolaski, F. E. (Eds.): Primary production and production processes, Tundra Biome, pp. 239–254. Edmonton-Oslo: IBP Tundra Biome Steering Committee 1973.

Caldwell, M. M., Tieszen, L. L., Fareed, M.: Comparative Barrow and Niwot Ridge canopy structure. In: Bowen, S. (Ed.): Proceedings U.S. 1972 Tundra Biome Symposium, pp. 22–28. Hannover-New Hampshire: USA CRREL 1972.

Dennis, J. G., Tieszen, L. L.: Seasonal course of dry matter and chlorophyll by species at Barrow, Alaska. In: Bowen, S. (Ed.): Proceedings 1972 U.S. Tundra Biome Symposium, pp. 16–21. Hannover-New Hampshire: USA CRREL 1972.

Hinneri, S., Sonesson, M., Veum, A. K.: Soils of Fennoscandian IBP tundra ecosystems. In: Wielgolaski, F. E. (Ed.): Fennoscandian tundra ecosystems, Part 1: Plants and microorganisms, pp. 31–40. Berlin-Heidelberg-New York: Springer 1975.

Kjelvik, S.: Måling av bladareal ved hjelp av lys ved IBP's primærproduksjonsundersøkelser på Hardangervidda. IBP i Norden **8**, 87–90 (1972).

Kjelvik, S., Kärenlampi, L.: Plant biomass and primary production of Fennoscandian subarctic and subalpine forests and of alpine willow and heath ecosystems. In: Wielgolaski, F. E. (Ed.): Fennoscandian tundra ecosystems, Part 1: Plants and microorganisms, pp. 111–120. Berlin-Heidelberg-New York: Springer 1975.

Levy, E. B., Madden, E. A.: The point methods of pasture analysis. N.Z.Y. Agric. **46**, 267–279 (1933).

Lye, K. A.: Vegetation of selected localities for IBP investigation in Hardangervidda, southern Norway. IBP i Norden **8**, 101–111 (1972).

Muc, M.: Primary production of plant communities of the Truelove Lowland, Devon Island, Canada. Sedge Meadows. In: Bliss, L. C., Wielgolaski, F. E. (Eds.): Primary production and production processes, Tundra Biome, pp. 3–14. Edmonton-Oslo: IBP Tundra Biome Steering Committee 1973.

Sonesson, M., Wielgolaski, F. E., Kallio, P.: Description of Fennoscandian tundra ecosystems. In: Wielgolaski, F. E. (Ed.): Fennoscandian tundra ecosystems, Part 1: Plants and microorganisms, pp. 3–28. Berlin-Heidelberg-New York: Springer 1975.

SVOBODA, J.: Vascular plant productivity studies of raised beach ridges (semi-polar desert) in the Truelove Lowland. In: BLISS, L. C. (Ed.): Devon Island IBP Project. High arctic ecosystem. Project report 1970 and 1971, pp. 146–184. Edmonton: University of Alberta, Dept. of Botany 1972.

WARREN WILSON, J.: Analysis of the spatial distribution by two-dimensional point quadrats. New Phytol. **58**, 92–101 (1959).

WARREN WILSON, J.: Inclined point quadrats. New Phytol. **59**, 1–8 (1960).

WARREN WILSON, J.: Estimation of foliage denseness and foliage angle by inclined point quadrats. Aust. J. Botan. **11**, 95–105 (1963).

WIELGOLASKI, F. E., KJELVIK, S.: Production of plants (vascular and cryptogams) in alpine tundra, Hardangervidda. In: BLISS, L. C., WIELGOLASKI, F. E. (Eds.): Primary production and production processes, Tundra Biome, pp. 75–86. Edmonton-Oslo: IBP Tundra Biome Steering Committee 1973.

3.2. Biomass and Production

Plant Biomass and Primary Production of Fennoscandian Subarctic and Subalpine Forests and of Alpine Willow and Heath Ecosystems

S. KJELVIK and L. KÄRENLAMPI

Introduction and Methods

The areas selected for sampling by Norwegian and Finnish IBP are described elsewhere (SONESSON et al., 1975). The Finnish sites are also discussed in KÄRENLAMPI (1972), with maps of the distribution of trees and bushes at the Kevo subsites. All the sample areas are placed in a pattern analysis diagram of the direct gradient analysis (KÄRENLAMPI and KAUHANEN, 1972). Production estimates are given by KÄRENLAMPI (1973). The production and total biomass of the Norwegian alpine lichen heath and the understory vegetation of the willow thicket and the sub-alpine birch forest are reported by WIELGOLASKI and KJELVIK (1973); values for bushes and trees are given by KJELVIK (1972; 1973). The biomass estimation method is similar in the two countries (WIELGOLASKI and KJELVIK, 1972).

Pine litter estimates were obtained in Kevo with litter funnels, which were 1 m high and had a collecting surface of 0.5 m^2. Ten funnels were placed at random and emptied every two weeks during summer. During winter, wooden trays with nylon net were left under the snow and emptied in spring. Birch forest litter was collected in Kevo in wooden trays, which were left out in autumn and emptied in spring. Thus the values refer to material which was fairly decomposed and includes macrolitter only. Measurement of Norwegian birch leaves shows a weight loss of 20% before leaf fall. Any production, therefore, predicted from leaf litter values will tend to be an underestimate.

Sampling in Northern Finland was carried out in August, 8–16 replicates taken at random were used in sampling of the understory. The Norwegian understory biomass values from Hardangervidda are means of 7–8 harvestings of 10 replicates throughout three seasons at the birch forest and lichen heath, and in two seasons at the willow thicket. Studies by $^{14}CO_2$ were carried out to separate living from dead roots at the willow thicket site, and also to separate living and dead lichens at the lichen heath. The separation of live and dead parts of mosses and lichens was carried out by visual inspection in the Finnish investigation, the live biomass values may, therefore, tend to be underestimates.

The production of understory vegetation in the Norwegian studies was estimated from changes in standing crop throughout the season, when death and decomposition of plant material were taken into consideration. The method is described by WIELGOLASKI (1975). Production of trees and bushes (including production of leaves) were obtained in the Norwegian investigations with the regressions used to calculate biomass, and the yearly increase in height and diameter (KJELVIK, 1972; 1973).

Results and Discussion

Trees. The results concerning the trees of pine and birch forest are given in Table 1. The total biomass of pines in the areas studied at Kevo varies from 40 metric tons ha^{-1} (which is close to the maximum of about 50 tons ha^{-1}

Table 1. Live and dead material of pine and birch in the Finnish and Norwegian forest sites. Dry weight (kg ha^{-1}). Biomass = live material. Standing crop = live + dead material

	Pine Kevo[a]	Birch Kevo	Birch Hardanger-vidda
Stem wood	13897	3530	13940
Stem bark	2010	1587	3390
Stem total	15907	5117	17330
Leaves	2314	595	970
Attached dead leaves	427	—	—
Attached dead twigs	1783	122	300
Living branches	8759	1782	4770
Dead branches	1395	85	131
Cones	743	—	—
Dead basal shoots	—	833	399
Roots	7324	(3541)	11340
Total tree biomass	35047	11035	34310
Total tree standing crop	38650	12075	35140

[a] Values given here are mean of pine central sample area and pine measurement area.

Table 2. Biomass of bushes (kg ha^{-1}) at the Finnish alpine heath site and the Norwegian alpine willow thicket site

	Finland birch bush	Norway willows
Leaves	32	752
Living stems and branches	877	6279
Dead stems and branches	1009	522
Roots	349	4245
Total biomass	1261	11276
Total standing crop	2270	11798

Table 3. Average standing crop in g dry weight per m^2 at three IBP sites at Hardangervidda, Norway, for the important species and groups ± standard error (understory vegetation only)

Sites and Species	Above l. root					Roots		Total alive
	Green							
	Min	Max	Mean	Non-green	Dead	Live	Dead	
Birch forest								
Empetrum hermaphroditum	24.5	44.3	35.1 ± 3.9	36.3 ± 2.8	34.0 ± 4.2	62.4 ± 21.9	2.4 ± 0.3	133.8
Vaccinium myrtillus	25.2	70.0	47.7 ± 6.5	19.6 ± 5.9	41.3 ± 8.0	134.3 ± 11.8	8.1 ± 2.2	201.6
V. vitis-idaea	4.0	8.5	7.1 ± 0.6	5.5 ± 0.8	10.4 ± 1.5	14.7 ± 3.0	3.3 ± 1.3	27.3
V. uliginosum	0.1	5.2	3.0 ± 0.8	17.2 ± 5.2	20.0 ± 6.3	44.9 ± 13.1	3.4 ± 1.5	65.1
Deschampsia flexuosa	0.8	9.1	7.2 ± 1.4	9.1 ± 1.1	35.5 ± 3.8	29.3 ± 4.0	2.2 ± 0.9	45.6
Nardus stricta	0.1	2.3	0.9 ± 0.3	1.1 ± 0.4	7.9 ± 2.5	2.6 ± 0.9	0.3 ± 0.1	4.5
Forbs	0.0	2.8	1.0 ± 0.4	0.7 ± 0.2	4.2 ± 1.5	4.0 ± 1.2	0.8 ± 0.6	5.7
Unrecognizable to sp.						341.5 ± 45.2	185.2 ± 41.6	341.5
Vascular plants total	54.7	142.2	102.0 ± 9.9	89.5 ± 7.3	153.3 ± 16.3	633.7 ± 23.9	205.7 ± 44.4	825.1
Willow thicket								
Carex spp.	0.1	1.6	0.9 ± 0.3	0.4 ± 0.1	4.1 ± 1.4	7.1 ± 1.7	0.7 ± 0.1	8.4
Deschampsia caespitosa	1.1	6.6	4.7 ± 0.8	4.1 ± 0.7	43.2 ± 6.6	12.7 ± 1.3	1.3 ± 0.4	21.5
Festuca rubra	0.5	3.1	1.8 ± 0.4	0.9 ± 0.3	2.9 ± 0.5	3.3 ± 0.5	0.4 ± 0.1	6.0
Other monocotyledons	2.7	7.7	5.7 ± 1.0	15.9 ± 2.3	15.9 ± 2.3	23.0 ± 8.9	1.9 ± 0.5	44.6
Alchemilla vulgaris coll.	0.4	14.8	8.5 ± 2.0	2.8 ± 0.7	10.1 ± 2.6	49.8 ± 8.5	9.1 ± 1.5	61.2
Geranium silvaticum	0.5	13.8	7.0 ± 2.3	2.7 ± 0.6	9.7 ± 1.9	61.8 ± 6.0	7.5 ± 1.4	71.5
Geum rivale	0.6	5.5	3.3 ± 0.8	1.7 ± 1.1	2.0 ± 0.4	15.6 ± 2.6	4.1 ± 1.2	20.6
Ranunculus acris	0.1	3.8	2.0 ± 0.6	0.8 ± 0.4	1.5 ± 0.3	6.6 ± 1.7	0.3 ± 0.1	9.4
Other forbs	0.4	13.0	7.6 ± 2.0	1.6 ± 0.4	31.7 ± 6.1	8.6 ± 1.3	1.0 ± 0.3	17.8
Unrecognizable to sp.						692.2 ± 43.9	20.5 ± 4.2	692.2
Vascular plants total	6.4	69.9	41.5 ± 8.3	30.9 ± 1.8	121.1 ± 10.5	880.7 ± 54.4	46.8 ± 2.8	953.2
Lichen heath								
Salix herbacea	0.0	0.6	0.4 ± 0.2	0.6 ± 0.2	2.3 ± 0.2	2.7 ± 0.9	0.4 ± 0.2	3.7
Empetrum hermaphroditum	9.5	18.5	15.8 ± 1.6	16.5 ± 1.9	67.2 ± 9.6	22.2 ± 4.4	7.7 ± 2.1	54.5
Vaccinium vitis-idaea	5.0	10.2	7.2 ± 1.0	4.0 ± 0.5	38.1 ± 6.7	23.1 ± 2.8	7.1 ± 1.6	34.3
V. uliginosum	0.0	0.9	0.7 ± 0.3	9.5 ± 5.2	8.3 ± 2.6	9.0 ± 1.5	1.5 ± 0.5	19.2
Carex bigelowii	0.2	1.7	1.0 ± 0.2	2.8 ± 0.5	26.2 ± 4.0	13.8 ± 2.2	8.3 ± 2.0	17.6
Festuca ovina	0.8	2.3	1.8 ± 0.2	0.7 ± 0.1	9.2 ± 0.9	4.0 ± 0.9	4.5 ± 3.8	6.5
Unrecognizable to sp.						115.9 ± 14.9	95.8 ± 31.7	115.9
Vascular plants total	15.5	34.2	26.9 ± 1.7	34.1 ± 6.7	151.3 ± 9.3	190.7 ± 11.5	125.3 ± 26.7	251.7

Table 4. Standing crop in g dry weight per m² at Kevo, Finland, for the most important species and groups (understory vegetation only)

Sites and species	Current[a] year leaves	Earlier years leaves	Current[a] year stems	Earlier years stems	Standing dead	Roots	Fruits	Total (alive)
Moist birch forest								
Empetrum hermaphroditum			17.0±4.3	72.5±20.1	37.9± 9.8	165.8±22.3	24.7	280.0
Vaccinium vitis-idaea	9.5±2.4	10.9±4.9	2.0±0.5	7.5± 1.9	21.9±18.3	56.9±17.6	0.3	87.2
V. myrtillus	10.0±5.0	0.0	7.8±2.9	17.1± 5.1	0.2	160.1±48.5	0.1	195.0
V. uliginosum	4.5±2.6	0.0	1.3±0.8	13.1± 5.7	2.0± 0.9	81.3±43.3	0.0	100.1
Deschmapsia flexuosa	2.6±1.6	0.0	0.0	0.0	0.0	1.4	0.0	4.1
Ledum palustre	2.3±1.8	0.4	0.6±0.5	13.8±11.9	0.2	20.4±17.1	0.0	37.3
Other dicotyledons	0.7	0.2	0.3	0.3	0.5	1.0	0.0	2.6
Vascular plants total	29.6	11.6	29.1	124.2	62.7	487.0	25.1	706.5
Main IBP birch forest								
Empetrum hermaphroditum			23.1±5.4	107.8±33.5	46.8±15.1	194.5±36.6	6.5±5.2	331.9
Vaccinium vitis-idaea	13.8±3.8	17.1±4.6	2.0±0.5	8.7± 2.0	1.9± 0.8	115.1±25.4	0.0	156.6
V. myrtillus	1.1±0.7	0.1	1.6±1.2	2.1± 1.2	0.0	4.0± 2.5	0.0	8.9
V. uliginosum	6.0±2.3	0.5±0.4	1.2±0.5	22.7± 8.7	2.6± 1.3	44.1±19.0	0.5±0.3	75.0
Monocotyledons	0.3	0.0	0.3	0.0	1.7	2.5	0.0	3.2
Ledum palustre	4.3±1.1	3.1±1.0	1.0±0.3	13.6± 3.2	2.5± 1.1	154.3±40.1	0.0	176.3
Other dicotyledons	0.2	2.1±2.0	0.0	0.1	0.0	0.1	0.0	2.6
Vascular plants total	25.7	22.9	29.3	155.0	55.4	514.6	7.0	754.5
Oporinia-damaged birch forest								
Empetrum hermaphroditum			8.7±2.6	32.7± 9.8	30.6± 9.5	103.0±39.2	2.6±1.5	147.0
Vaccinium vitis-idaea	5.1±1.8	3.6±1.3	0.7±0.2	2.3± 0.6	1.4± 1.0	31.8± 5.9	0.0	43.4
V. myrtillus	12.1±2.9	0.0	10.0±6.0	30.4± 5.2	4.3± 2.9	242.0±34.7	0.1±0.1	294.7
V. uliginosum	24.0±9.0	0.0	12.8±7.2	49.1±23.2	6.5± 2.6	223.9±80.8	0.6±0.5	310.4
Other dicotyledons	2.9	0.0	1.5	0.3	0.1	2.6	0.0	7.3
Monocotyledons	6.4±5.2	0.0	2.5±2.1	0.0	1.2± 1.2	24.8±23.1	0.0	33.7
Vascular plants total	50.4	3.6	36.3	114.8	44.2	628.0	3.3	836.4

High elevation birch site								
Empetrum hermaphroditum			29.4 ± 9.2	85.4 ± 26.6	58.0 ± 22.6	103.2 ± 32.0	10.4 ± 6.3	228.4
Vaccinium vitis-idaea	4.6 ± 4.2	2.6 ± 1.2	0.6 ± 0.2	3.6 ± 1.5	1.3 ± 0.6	33.6 ± 11.7	0.0	44.9
V. myrtillus	3.4 ± 1.6	0.0	1.7 ± 0.6	12.3 ± 6.2	9.9 ± 7.2	75.6 ± 36.3	0.0	92.9
Monocotyledons	3.5	0.0	0.0	0.0	2.0	0.0	0.0	3.5
Other dicotyledons	1.2	0.2	0.3	1.6	1.0	0.8	0.0	4.5
Vascular plants total	12.7	2.8	32.3	102.9	72.1	213.1	10.4	374.2
IBP Pine forest								
Empetrum hermaphroditum			5.1 ± 2.2	27.9 ± 10.4	9.2 ± 2.9	50.1 ± 15.8	0.2 ± 0.1	83.3
Vaccinium vitis idaea	21.5 ± 5.5	28.6 ± 9.6	4.7 ± 2.5	22.9 ± 6.4	12.5 ± 3.7	229.6 ± 33.0	0.3 ± 0.2	307.7
V. myrtillus	0.0	0.0	0.0	0.1	0.0	1.0	0.0	1.1
Vascular plants total	21.6	28.6	9.6	50.9	21.8	280.8	0.5	392.1
Low alpine heath								
Empetrum hermaphroditum			19.0 ± 3.8	90.7 ± 21.4	48.8 ± 9.1	102.3 ± 36.8	2.5 ± 0.9	214.5
Vaccinium vitis-idaea	6.7 ± 1.3	7.2 ± 2.6	0.5 ± 0.1	4.6 ± 1.1	1.5 ± 0.5	35.6 ± 11.8	0.1	54.8
V. uliginosum	11.2 ± 5.1	0.4 ± 0.4	2.8 ± 1.5	29.1 ± 12.1	6.9 ± 2.3	46.7 ± 21.0	0.0	90.2
Calamagrostis lapponica	1.3 ± 0.9	0.0	0.0	0.0	3.3 ± 2.5	0.0	0.0	1.3
Arctostaphylos alpina	1.1 ± 0.9	0.0	0.1	0.7 ± 0.4	1.4 ± 1.0	10.6 ± 10.3	1.5 ± 1.4	13.9
Vascular plants total	20.4	7.7	22.4	125.2	61.9	195.2	4.0	374.8

[a] Leaves and stem were not separated in *Empetrum*.

possible in the Kevo valley) to about 30 tons ha^{-1} in an area which, disturbed by a heavy storm about six years before the sampling, had approximately 25% of its trees felled. A small amount of birch should be added to the figure.

The birch biomass in the Finnish birch forest sample area is only about $^1/_3$ of the birch biomass in the Norwegian subalpine forest. The basal area of birch stems with a diameter of over 2 cm is only 2.5 m^2 ha^{-1} in the Finnish birch forest sample area. On the basis of gradient analyses of vegetation (KÄRENLAMPI and KAUHANEN, 1972), it is known that basal areas even 3–5 times higher may be found in the Kevo valley. In the Norwegian birch forest the basal area is 8.8 m^2. Thus the most luxuriant birch forests have almost as high maximum tree biomass as found in the Finnish pine forest. The annual production of trees and bushes in the Norwegian birch forest site is almost 2.8 tons ha^{-1} (KJELVIK, 1973). In the Norwegian birch forest there is three to five meters between trees or clusters of trees, which means about 0.2 trees per m^2.

From the Norwegian willow thicket and birch forest a yearly growth of 3.1 kg per kg leaves per year is estimated. Using this value and the measured yearly litter fall at Kevo, a production of 1.4 tons ha^{-1} $year^{-1}$ in the pine forest and 1.8 tons ha^{-1} $year^{-1}$ in the birch forest are seen as an average. The values vary from about 0.4 to 2.3 tons ha^{-1} at the different birch sites with measured litter fall, giving the lowest production at the driest sites. The total macrolitter of birch in Northern Finland varies from 203 to 1111 kg ha^{-1} and year, with an average of 380 kg ha^{-1} at dry sites and 900 kg ha^{-1} in moist birch forest.

Estimates of the bush biomass of the Finnish alpine heath site and the alpine willow thicket site at Hardangervidda, Norway are given in Table 2. The willow thicket has an average of 7 stems per m^2. The Finnish site has a total bush biomass which is nearly 1/10 of the Norwegian willows. The amount of standing dead material is almost as high as the live biomass at the Finnish site. This is due to the damage caused by *Oporinia* larvae in 1965–1966 (KALLIO and LEHTONEN, 1975). The high biomass at the Norwegian site (even the shrubs are less than 1.5 m high) may be explained by a dense shrub layer at the eutrophic site with oxygen-rich, wet soil favorable for growth. This is also the reason for the high willow production, about 2.5 tons ha^{-1} $year^{-1}$ (KJELVIK, 1973), a figure very close to the production of the trees and bushes in the Norwegian subalpine birch forest, although the biomass is much less in the willow thicket.

Ground Vegetation data on vascular plants are shown in Tables 3 and 4. It is seen that the dwarf shrubs *Empetrum hermaphroditum*, *Vaccinium vitis-idaea*, *V. myrtillus* and *V. uliginosum* are the most abundant species in the understory in the oligotrophic sites in northern Finland, and also in the oligotrophic birch forest and lichen heath sites at Hardangervidda, Norway, while the understory vegetation in the eutrophic willow thicket is dominated mainly by forbs. The total live cryptogam biomass in the willow thicket is above 300 g m^{-2}, giving a total live biomass of 1260 g m^{-2}. This is a higher value than for the heath sites and for the understory of the forest sites studied. Again, the reason must be the favorable soil conditions at this site. The high soil water content may cause the high amounts of roots compared to tops at the site. About 90% of the total live vascular plant biomass is roots. The Finnish moist birch and main IBP birch forest

Table 5. Cryptogamic live vegetation at the different sites in $g\,m^{-2}$

	Kevo					Hardangervidda		
	Moist birch forest	Birch forest	High elev. birch	Pine forest	Low alpine heath	Birch forest	Willow thicket	Lichen heath
Elevation meter	131	130	320	115	330	780	1250	1220
Number of samples	8	16	8	16	16	10×8^{b}	10×7^{b}	10×8^{b}
Mosses								
Polytrichum spp.							95.4	
Sphagnum spp.							75.3	
Dicranum spp.	0.3	2.3	8.2	1.4	1.4			
Hylocomium splendens	24.5	2.7	—	—	—			
Pleurozium schreberi	9.1	25.1	16.3	1.3	—		27.4	
Other mosses[a]	5.5	3.0	38.2	15.3	5.9		109.1	
Total (mosses)	39.4	33.2	62.6	18.0	7.2	61.1	307.2	6.7
Lichens								
Cetraria nivalis	—	—	—	—	15.3			23.6
Stereocaulon paschale	—	2.7	43.9	38.8	32.3			
Cladonia alpestris	—	—	8.9	4.7	1.3			
Cladonia mitis	0.2	3.1	8.0	16.5	12.8			176.1
Cladonia rangiferina	3.2	0.9	3.4	11.6	0.2			
Cladonia uncialis	—	0.7	1.7	3.3	4.6			
"Cup lichens"	—	—	0.9	1.5	1.0			
"Horn lichens"	1.8	1.5	2.7	4.2	0.2			
Nephroma arctica	52.6	5.5	2.7	—	0.5			
Peltigera aphthosa	—	4.7	—	1.1	—			
Alectoria ochroleuca	—	—	0.5	—	1.1			
Other macrolichens	—	0.0	1.0	2.7	3.9			182.6
Total (lichens)	58.1	19.2	73.7	84.5	77.2	9.3	0.0	382.3
Total (cryptogams)	97.4	52.4	136.3	102.4	84.5	70.4	307.2	389.0

[a] Includes *Polytrichum* spp. at the Kevo sites.
[b] Ten plots harvested 7 or 8 times during the research period.

sites, together with the Norwegian birch forest and lichen heath, form an intermediate group with 70% roots in the total biomass. The high elevation birch site and the low alpine heath in northern Finland have 55% roots in the total biomass. These two sites are also the driest ones. The biomass of *Empetrum* as percentage of the total biomass increases in the same direction. *Vaccinium vitis-idaea* has the second highest biomass at the drier sites, but in the moist birch forest at Kevo and the birch forest at Hardangervidda the biomass of *Vaccinium myrtillus* is higher. This is certainly also due to the water factor. The greater abundance of *Vaccinium vitis-idaea* generally found at the Kevo sites may be an effect of lower precipitation compared to Hardangervidda. Forbs and grasses are very sparsely represented at the drier sites, but this might also be an effect of low nutrient status at these sites.

Biomasses of the cryptogams are given in Table 5. On the wet Norwegian willow thicket, the moss biomass is higher than at the forest and heath sites. The

Table 6. Summary table of the biomass (live) of primary producers at the sites, dry weights $kg\,ha^{-1}$

	Kevo			Hardangervidda		
	Pine forest	Birch forest	Low alpine heath	Birch forest	Willow thicket	Lichen heath
Pine	29859					
Birch	(400)	11035	1261	34310		
Other bushes		(200)		2165	11276	
Ground vegetation, vascular	3921	7545	3748	8252	9532	2517
Ground vegetation, mosses	180	332	72	611	3072	67
Ground vegetation, lichens	845	192	772	93	0	3823
Total	35205	19304	5853	45431	23880	6407

biomass of cryptogams is nearly 60% of the total biomass at the lichen heath at Hardangervidda. At all the other sites, the percentage of cryptogams is from 5 to 25% of the total understory vegetation. The Norwegian lichen heath has twice as high lichen biomass as any of the Kevo sites, which results from extensive grazing by reindeer in northern Finland. This high biomass also partly explains the higher production at the Norwegian lichen heath site, compared to the Finnish low alpine heath and high elevation birch sites.

The understory vegetation comprises a very high percentage of the whole biomass of primary producers at sites with a fairly moist soil and luxuriant understory vegetation. The same relation is found also in dry ecosystems at higher elevations, where the tree biomass is low. The understory vegetation is nearly 50% of the whole primary producer biomass at the IBP main birch forest sample area at Kevo, while in the pine forest it is only 16.5%. At Hardangervidda the figure is 50% at the willow thicket and 17% in the birch forest, with a denser tree layer than at the Finnish main site.

Table 6 gives a summary of the biomass figures at the different main sites. The biomass of birch in Kevo may be 3–5 times higher at other sites in the Kevo valley, as the biomass is related to basal area. Taking this into consideration, the most luxuriant birch sites in Kevo are very similar to the Norwegian birch site.

The primary production of understory vascular plants at the Norwegian sites (Table 7) is highest at the willow thicket and lowest at the lichen heath. The production at the sites studied seems to be strongly correlated with water and soil nutrient content. Total production follows the same pattern. The understory production at the Finnish sites is highest at the birch forest sites with about $110\,g\,m^{-2}$ per year. About half as high a production is calculated for the high elevation birch site, the pine forest site and the low alpine heath site (KÄRENLAMPI, 1973). The higher production observed at the Hardangervidda sites may partly be caused by the more detailed calculation method used, and partly by the moisture and nutrient conditions. The precipitation is higher at Hardangervidda than at Kevo (SKARTVEIT et al., 1975), and nutrient conditions generally better (HINNERI et al., 1975).

Table 7. Plant production in g dry weight per m^2 per year and day of growing season at Hardangervidda

	Willow thicket		Birch forest		Lichen heath	
	Annual	per day	Annual	per day	Annual	per day
Vascular plants understory						
Green	173	1.7	267	2.1	72	0.6
Non-green above first root	31	0.3	80	0.6	13	0.2
Roots	220	2.2	110	0.8	97	0.9
Subtotal vasc. pl. und. st.	424	4.2	457	3.5	182	1.7
Trees and shrubs	250	2.5	283	2.2		
Cryptogams	210	2.1	40	0.3	88	0.8
Total	884	8.8	780	6.0	270	2.5

Acknowledgements. The technical assistants A. POLDER and E. MIETTINEN LUNDE were responsible for laboratory studies in the Norwegian investigations. The senior author wishes to thank them for their careful work, and also to thank the Norwegian project leader F. E. WIELGOLASKI for permission to use data for which he is directly responsible (from the Norwegian lichen heath and willow thicket sites) in this paper.

References

HINNERI, S., SONESSON, M., VEUM, A. K.: Soils of Fennoscandian IBP tundra ecosystems. In: WIELGOLASKI, F. E. (Ed.): Fennoscandian tundra ecosystems, Part 1: Plants and microorganisms, pp. 31–40. Berlin-Heidelberg-New York: Springer 1975.

KALLIO, P., LEHTONEN, J.: On the ecocatastrophe of birch forests caused by *Oporinia autumnata* (Bkh.) and the problem of reafforestation. In: WIELGOLASKI, F. E. (Ed.): Fennoscandian tundra ecosystems, Part 2: Animals and systems analysis. Berlin-Heidelberg-New York: Springer 1975 in press.

KJELVIK, S.: Brief report on IBP biomass studies in a mountain birch zone at Maurset, Hardangervidda, Norway. In: ROSSWALL, T. (Ed.): Systems Analysis in Northern Coniferous Forests, IBP Workshop, pp. 49–53. Stockholm: Swedish Natural Science Research Council 1972.

KJELVIK, S.: Biomass and production in a willow thicket and a subalpine birch forest, Hardangervidda, Norway. In: BLISS, L. C., WIELGOLASKI, F. E. (Eds.): Primary production and production processes, Tundra Biome, pp. 115–122. Edmonton-Oslo: IBP Tundra Biome Steering Committee 1973.

KÄRENLAMPI, L.: Comparisons between the microclimates of the Kevo ecosystem study sites and the Kevo Meteorological Station. Rep. Kevo Subarctic Res. Stat. **9**, 50–65 (1972).

KÄRENLAMPI, L.: Biomass and estimated yearly net production of the ground vegetation at Kevo. In: BLISS, L. C., WIELGOLASKI, F. E. (Eds.): Primary production and production processes, pp. 111–114. Edmonton-Oslo: IBP Tundra Biome Steering Committee 1973.

KÄRENLAMPI, L., KAUHANEN, H.: A direct gradient analysis of the vegetation of the surroundings of the Kevo Subarctic Station. Rep. Kevo Subarctic Res. Stat. **9**, 82–98 (1972).

SKARTVEIT, A., RYDÉN, B. E., KÄRENLAMPI, L.: Climate and Hydrology of some Fennoscandian tundra ecosystems. In: WIELGOLASKI, F. E. (Ed.): Fennoscandian tundra ecosystems, Part 1: Plants and microorganisms, pp. 41–53. Berlin-Heidelberg-New York: Springer 1975.

SONESSON, M., WIELGOLASKI, F. E., KALLIO, P.: Description of Fennoscandian tundra ecosystems. In: WIELGOLASKI, F. E. (Ed.): Fennoscandian tundra ecosystems, Part 1: Plants and microorganisms, pp. 3–28. Berlin-Heidelberg-New York: Springer 1975.

WIELGOLASKI, F. E.: Vegetation types and primary production in tundra. In: WIELGOLASKI, F.E., ROSSWALL, T. (Eds.): Proceedings IV International meeting on the biological productivity of tundra, Leningrad, October 1971. pp. 9–35. Oslo and Stockholm: IBP Tundra Biome Steering Committee 1972.

WIELGOLASKI, F. E.: Primary productivity of alpine meadow communities. In: WIELGOLASKI, F.E. (Ed.): Fennoscandian tundra ecosystems, Part 1: Plants and microorganisms, pp. 121–128. Berlin-Heidelberg-New York: Springer 1975.

WIELGOLASKI, F. E., KJELVIK, S.: The methodology of net primary production investigations in Norwegian IBP tundra studies. Rep. Kevo Subarctic Res. Stat. **9**, 24–27 (1972).

WIELGOLASKI, F. E., KJELVIK, S.: Production of plants (vascular plants and cryptogams) in alpine tundra, Hardangervidda. In: BLISS, L. C., WIELGOLASKI, F. E. (Eds.): Primary production and production processes, Tundra Biome, pp. 75–86. Edmonton-Oslo: IBP Tundra Biome Steering Committee 1973.

Primary Productivity of Alpine Meadow Communities

F. E. WIELGOLASKI

Introduction and Methods

The plant biomass and the production of tundra areas may vary strongly between plant communities (BLISS, 1962; RODIN and BAZILEVICH, 1967; WIELGOLASKI and ROSSWALL, 1972; BRZOSKA, 1973; BLISS and WIELGOLASKI, 1973; PERKINS et al., 1975), and also between years. The present results from two enclosures of eutrophic alpine meadow communities of the Norwegian IBP tundra studies at Hardangervidda, described elsewhere (SONESSON et al., 1975), are means of the years 1969–1971. The harvesting and sorting procedure is described by WIELGOLASKI and KJELVIK (1972). Three to more than five times lower production values may be found when based on differences between minimum and peak aboveground standing crop values, compared to values estimated from assimilation studies (TIESZEN, 1972; WIELGOLASKI, 1975a). In the present study the primary production estimates are based on a method described by MILNER and HUGHES (1968), and used for various living and dead above- and belowground plant fractions.

Fertilization studies (mixed fertilizer with 11.5% N, 5% P, 14.5% K, 1.2% Mg, 7.5% S and 0.02% B and nitrate fertilizer with 15.5% N in the amounts of 50 and 100 gm^{-2}) were carried out 1972–1973 at a relatively eutrophic dry meadow site with a similar plant species composition as the dry meadow site used in the other investigations, but without *Dryas* at the fertilization plots. HÅLAND (1972) has given a review of earlier research on fertilization of natural mountain vegetation at the southern edge of Hardangervidda. The fertilized plots were harvested by clipping in the fall of 1972, and twice in 1973.

It is very difficult to distinguish between live and dead roots by visual inspection in the peaty soil in the wet meadow. This is therefore studied by $^{14}CO_2$ uptake in the roots by the technique described by BERG et al (1975). Roots were pressed, and after autoradiography for 30 days they were separated quantitatively into live and dead according to the pictures. The same technique was also used for separation in live and dead cryptogams, e.g. for bryophytes in the wet meadow down to about 5 cm in the peat. This is the depth to which single bryophyte plants may be taken out without breaking into small pieces.

Results and Discussion

The average biomass for the three years of study is shown in Table 1, and the annual variation in different plant compartments is given by WIELGOLASKI and KJELVIK (1975). The most important species in the dry meadow are the dwarf shrub *Salix reticulata*, the grass *Deschampsia flexuosa* and the cushion plant *Dryas octopetala*, and in the wet meadow the willow *Salix lapponum* and the sedge *Carex nigra*. Only very few leaves of *Dryas* are green over the winter, in contrast to leaves of *Dryas integrifolia* as observed by SVOBODA (1973). For all species studied, the living root biomass is higher than aboveground live biomass. The shoot-root ratio is found to be lowest for the *Carex* species, especially in the normally waterlogged wet meadow.

Most of the roots are found in the upper 10 cm soil layer, in the wet meadow more than 60%, in the dry meadow somewhat less. By $^{14}CO_2$ uptake in the roots, it is found that only 15–20% of the roots are living below a depth of 10 cm in the wet meadow site, and the variation is relatively small through the summer. In the upper 10 cm, however, more living roots are found, and the percentage increases strongly from early to late summer (40–50% living in mid-July, increasing to more than 90% living in mid-August). In the dry meadow it is easier to separate the roots visually into live and dead at harvesting, and by this method approximately 75% of the roots are found to be living, on an average. As expected, there are more live roots compared to dead in the well-aerated dry meadow than in the waterlogged wet meadow. The live to dead ratio of the wet meadow aboveground parts (0.46) is somewhat higher than found by MUC (1973) for a similar community type in high arctic Canada, possibly because of the higher summer temperatures and more rapid decomposition in low-mid alpine regions of southern Norway.

By the $^{14}CO_2$ it is found that only approximately 4–5% of the total material of bryophytes is dead in the upper 5–6 cm in the wet meadow. The amounts of dead bryophytes are expected to be higher at the dry meadow, but only visual inspections are carried out at this site, and the values given for dead bryophytes at this site in Table 1 are probably too high. Only very small amounts, about 2%, of lichens are found to be dead in the $^{14}CO_2$ observations, and the fraction of dead lichens is therefore neglected in the dry meadow.

From the average minimum and maximum values of green material given in Table 1 it is found that the green vascular plant material in the dry meadow may increase by about 85 gm^{-2} during the summer, and about 80 gm^{-2} in the wet meadow. These values are, as expected, less than half the values given in Table 2 for the same plant compartment estimated from the more detailed procedure shown in Fig. 1. Primary production estimation of green material based on minimum and maximum standing crop presupposes that no biomass is lost by death, decomposition, mechanical damage or intake by animals before the time of peak biomass, further that no translocation of minerals and organic compounds takes place after they are incorporated in the green material, and finally that production of green materials stops at the time of peak aboveground biomass. None of these assumptions are realistic, and cause underestimates of production by the maximum-minimum method. COUPLAND (1973) found 6–7 times higher primary pro-

Table 1. Average standing crop in g dry weight per m² at meadow sites, Hardangervidda, Norway, for the most important plant species and groups ± standard error n = 10

Sites and species	Above 1. root					Roots	
	Green			Non-green	Dead	Live	Dead
	Min	Mean	Max				
Dry meadow							
Vaccinium vitis-idaea	1.5	2.2 ± 0.2	2.7	1.2 ± 0.1	3.0 ± 0.9	4.2 ± 0.6	1.0 ± 0.2
Salix herbacea	0	3.1 ± 0.7	6.1	3.8 ± 0.5	14.2 ± 1.4	10.0 ± 1.6	0.8 ± 0.2
Salix reticulata	0.1	5.4 ± 1.1	8.9	13.4 ± 1.0	21.9 ± 2.1	33.2 ± 4.0	3.2 ± 0.7
Carex bigelowii	0.5	2.6 ± 0.3	3.5	3.0 ± 0.4	12.9 ± 1.7	20.1 ± 3.4	2.2 ± 0.5
Deschampsia flexuosa	0.5	6.2 ± 1.3	11.3	3.4 ± 1.0	13.9 ± 2.0	11.6 ± 2.3	1.0 ± 0.1
Other monocotyledons	0.8	5.9 ± 1.1	9.2	3.1 ± 0.6	12.4 ± 1.6	10.0 ± 0.9	1.9 ± 0.7
Dryas octopetala	0.7	13.3 ± 2.4	23.0	22.7 ± 2.7	133.4 ± 22.5	58.5 ± 11.5	7.6 ± 2.8
Polygonum viviparum	0.0	3.9 ± 1.0	6.8	7.1 ± 2.5	21.8 ± 2.1	24.7 ± 2.5	2.0 ± 0.3
Other forbs	5.6	17.1 ± 2.2	24.3	12.2 ± 2.1	24.6 ± 2.9	42.3 ± 3.6	4.0 ± 1.1
Unrecognizable to sp.						330.5 ± 18.2	160.3 ± 18.0
Subtotal vascular plants	9.7	59.7 ± 7.1	95.8	69.9 ± 7.3	258.1 ± 25.8	545.1 ± 31.9	184.0 ± 19.1
Bryophytes		16.1 ± 1.1		15.3 ± 2.0	40.9 ± 9.6		
Lichens		19.0 ± 1.7					
Total		95.0 ± 6.2		85.2 ± 8.8	299.0 ± 23.1	545.1 ± 31.9	184.0 ± 19.1
Wet meadow							
Salix lapponum	1.3	7.8 ± 2.6	17.3	11.7 ± 1.2	19.2 ± 3.5	119.5 ± 15.2	32.9 ± 8.4
Salix herbacea	0.1	5.3 ± 1.2	7.7	4.7 ± 0.7	5.6 ± 0.8	17.0 ± 2.1	1.6 ± 0.8
Carex nigra	2.7	25.6 ± 5.1	38.6	37.0 ± 3.8	198.9 ± 29.6	330.0 ± 34.6	49.0 ± 12.5
Other monocotyledons	0.5	10.3 ± 2.2	15.1	3.7 ± 0.7	17.0 ± 2.9	22.5 ± 6.2	3.6 ± 1.1
Forbs	0.1	5.9 ± 1.5	8.8	3.1 ± 0.7	7.4 ± 1.1	10.6 ± 2.1	1.0 ± 0.4
Unrecognizable to sp.						816.0 ± 76.2	1257.0 ± 130.7
Subtotal vascular plants	4.7	54.9 ± 11.5	87.5	60.2 ± 3.4	248.1 ± 33.4	1315.6 ± 120.1	1345.1 ± 128.2
Bryophytes		20.9 ± 4.5		154.5 ± 17.2	8.1 ± 0.9		
Total		75.8 ± 15.2		214.7 ± 16.2	256.2 ± 31.2	1315.6 ± 120.1	1345.1 ± 128.2

Table 2. Plant production in g dry weight per m^2 per year and day of growing season (105 days in dry meadow, 100 days in wet meadow), based on the principles used for calculations shown in Fig. 1

	Dry meadow		Wet meadow	
	Annual	Per day of grow. seas.	Annual	Per day of grow. seas.
Vascular plants				
Green	190	1.8	186	1.8
Non-green ab. 1 root	51	0.5	66	0.7
Roots	245	2.3	410	4.1
Subtotal vascular plants	486	4.6	660	6.6
Cryptogams	48	0.5	173	1.7
Total	534	5.1	833	8.3

duction of tops in a grassland community by correcting some of these errors than using the average yield of hay in the area.

Changes in the amount of plant material in any compartment between two harvestings are caused either by direct production of new material, by transfer from one compartment to another, or by decomposition of dead material. Consumption by animals is neglected in the present study. To reduce the statistical errors in the observations, the production calculations are not carried out on a plant species level, but only on the compartments of the groups vascular plants and cryptogams, respectively. The decay rates at the two meadow sites studied are assumed to be about 10% per month in the vegetation period of approx. $3^1/_2$ month and 10% for the whole non-vegetation period. These estimates are of the same order as observed values of newly dead material (ROSSWALL et al., 1975), except for material in the waterlogged part of the soil in the wet meadow. It is found, however, that variation in the decomposition rate by 5% between two harvestings only affects the production estimates slightly. In many cases, a greater reduction of dead compartments than 10% per month are found from the harvestings. In this case the actual observed values are used. When, however, dead material in a compartment increases more than live material of the compartment decreases in the same period, it is impossible to calculate the decay rate from the harvestings. The consumption by vertebrates is expected to be low in the studies because of fences around the sites, and is therefore neglected, although a few small rodents and birds occur inside the fences. The effect of herbivorous invertebrates is also expected to be relatively small. As seen in the flow diagram example from the dry meadow in Fig. 1, about 1/3 of the annual vascular plant primary production may take place after the time of peak aboveground biomass at the beginning of August. This increment is therefore important to include, in order to obtain a true picture of the annual production.

In the example in Fig. 1, dead roots increased from about 160 to 177 gm^{-2} during the period July 1st to August 4th. Ten percent decomposition of dead roots in the period is 17 gm^{-2}. This means that the increase in dead material has been 177–160 + 17 = 34 gm^{-2}. Living roots were found to increase from about 464 to about 578 gm^{-2}, or about 114 gm^{-2} in the same period. The production of roots must have been 114 + 34 = 148 gm^{-2}. As there has been no increase in

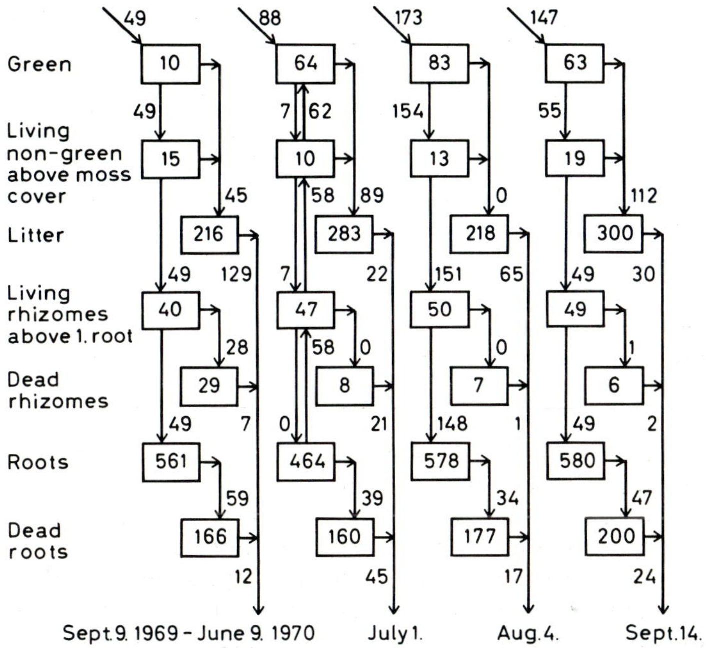

Fig. 1. Biomass in g dry weight per m^2 for various components of vascular plants at the dry meadow site, Hardangervidda, Norway. Explanations in text

litter and dead rhizomes in this period, but a reduction of more than 10%, the total production between the two harvestings is only the sum of the root production and the sum of increments in the compartments living rhizomes above the first root, living non-green above the moss cover and green, making a total primary production estimate for the period of about 173 gm^{-2}. In the period July 1st to August 4th most of this production is caused by root production (148 gm^{-2}), which is found because of the low living root biomass observed in early July. This low value could partly be caused by errors in the observations, but may also be a result of root death and respiration and translocation of material to the new green parts in spring. As a rough estimate from $^{14}CO_2$ studies (BERG et al., 1975), 50–70% of the observed weight decrease in living roots from June 9th to July 1st is assumed to be caused by translocations to the tops, and 60% is therefore used in Fig. 1. This means that 60% of the 561–464 = 97 gm^{-2} root biomass reduction in the period, i.e. 58 gm^{-2} has been transported to the tops, and thus only 39 gm^{-2} of the decrease is caused by death. The amount of dead roots has, however, also decreased, and the total decomposition of roots in this period will be (166–160) + 39 = 45 gm^{-2}. The total primary production in the period is: 10% decomposition of litter = 22 gm^{-2} + (283–216) + (47–40) + (64–10) − 62 = 88 gm^{-2} (see Fig. 1). If the period is more than 1 month, as e.g. from Aug. 4th to September 14th, more than the 10% decomposition is used.

A flow diagram like the one presented in Fig. 1 will, of course, include several errors (standard errors of the estimates are of the order of 10–20%), but is still

Table 3. Dry weight of tops (clipped at soil surface) of vascular plants in $g\,m^{-2}$ in dry meadow vegetation after different fertilizer treatments. Harvestings July 27th and September 13th, 1973

Treatment		July 27th 1973		September 13th 1973	
		$g\,m^{-2}$	% dead	$g\,m^{-2}$	% dead
Control	(n=10)	93± 5	39	109± 9	69
50 $g\,m^{-2}$ N fertilizer	(n= 5)	159±30	16	182±25	70
100 $g\,m^{-2}$ N fertilizer	(n= 5)	137±13	22	147± 6	70
50 $g\,m^{-2}$ mixed fertilizer	(n= 5)	216±62	11	267±11	56
100 $g\,m^{-2}$ mixed fertilizer	(n= 5)	432±62	13	491±23	23

expected to give a much better picture of the real primary production than just the minimum and maximum biomass values. It is also possible from the flow diagram to calculate death and decomposition values. Because of the approximately steady state at the sites, these values are of the same order as the production estimates (WIELGOLASKI and KJELVIK, 1973).

The primary production of roots is found to be considerably higher in the wet than in the dry meadow (Table 2), although the aboveground vascular plant production on a ground area basis is almost similar at the two sites. This may be seen as a consequence of the high root biomass in the wet meadow. If the roots in the wet meadow (mostly of *Carex nigra*) are assumed to remain functional for three years, as found for *Molinia caerulea* by SHEIKH and RUTTER (1969), it means that 1316 gm^{-2}:3 = 438.7 gm^{-2} dry matter of roots has to be produced each year. This is only slightly higher than the values estimated (Table 2), and the difference may even be explained by the slow building up of the peat in the mire, about 1 cm per 100 years according to radiocarbon dating.

The production of cryptogams on a ground area basis is found to be 3–4 times higher in the wet than in the dry meadow. This may be partly explained by the low percentage of dead bryophytes found in the wet meadow, but is of course also a result of the much higher percentage cover of the bottom layer in the wet meadow (average about 90%) than in the dry meadow (average about 45% and often much lower), according to LYE (1972).

The harvesting of the plants in the fertilizer study in the fall 1972, two months after fertilization, showed only a weak tendency to increased yield with the largest amount of mixed fertilizer. The plants, however, remained green later into the fall at the fertilized plots, especially with the highest amount of mixed fertilizer. This effect is reported also by LILAND (1970). Results from the harvestings in 1973 are given in Table 3 for the tops. The lowest amount of mixed fertilizer increased the yield by 158 gm^{-2}, and the highest amount of the same fertilizer gave an additional increase of 224 gm^{-2}. The N fertilizer also shows a small yield increase, but the strongest N fertilization may be overoptimal, as the yield shows a tendency to decrease. A lower percentage of dead material was found, particularly by the strongest mixed fertilizing treatment, compared to the control plots.

Harvesting of roots in autumn 1973 showed the average biomass 343 gm^{-2} at the control plots and 400 gm^{-2} at the mixed fertilizer plots, but the variation was

Table 4. Dry weights 1973 in gm^{-2} of tops of various species after treatment with 100 gm^{-2} mixed fertilizer and control (no fertilizer) ± standard error

	100 $g\,m^{-2}$ mixed fertilizer		Control	
	July 27.	Sept. 13.	July 27.	Sept. 13.
Salix herbacea	2± 1	0	9±3	7±2
Salix reticulata	0	0	1±0.7	4±2
Anthoxanthum alpinum	39±11	38 ±11	6±2	5±2
Carex bigelowii	38±11	41 ± 4	17±3	18±6
Deschampsia flexuosa	18±10	72 ±11	5±1	13±7
Poa alpina	28± 8	36 ±11	7±2	6±2
Phleum commutatum	24	4	0	2
Festuca rubra	89±25	143 ±31	5±2	11±4
Other monocotyledons	40± 6	81 ±11	12±3	10±3
Astragalus alpinus	4± 2	0.8	1	1±0.8
Polygonum viviparum	35± 7	11 ± 2	12±2	9±1
Ranunculus acris	21± 7	5 ± 2	2±1	1±0.3
Thalictrum alpinum	10± 5	0.4	2±1	2±0.6
Alchemilla vulgaris coll.	37±19	28 ±10	3±2	2±1
Saussurea alpina	12± 4	14 ± 5	3±1	1±0.2
Leontodon autumnalis	0	1	0	0
Other forbs	26±10	8 ± 3	6±1	9±1
Not sorted to species	8± 5	5 ±1	3±1	9±2

high between the replicates. The root values are somewhat lower than for the intensively studied dry meadow, although the dry weights of tops at the peak of the growing season on the control fertilizer plots are of the same order as in the other dry meadow studies.

The species composition of the plots (Table 4) changed with the highest amount of mixed fertilizer. The amount of grasses, e.g. *Anthoxanthum alpinum* and *Festuca rubra*, and the forb *Alchemilla vulgaris* increased, while the amount of *Salix herbacea* and *S. reticulata* decreased. In autumn 1973 76% of the yield on the plots fertilized with 100 gm^{-2} mixed fertilizer was grasses, in contrast to 43% when unfertilized, showing that the grasses increase while the dwarf shrubs and forbs decrease on a relative scale. This effect of fertilization is also reported by SELSJORD (1960) and by ANDERSSON (1969). A similar effect is reported after sheep grazing (WIELGOLASKI, 1975b).

Acknowledgements. Research fellow S. KJELVIK carried out part of the calculations, while research fellow A. K. VEUM was co-responsible for the fertilizer studies in the first year. The author wishes to thank them both.

References

ANDERSSON, S.: Gjödsling av fjällbete—lönsamt eller ej? Rennäringsnytt **2**, 8–9 (1969).

BERG, A., KJELVIK, S., WIELGOLASKI, F. E.: Distribution of ^{14}C photosynthates in Norwegian alpine plants. In: WIELGOLASKI, F. E. (Ed.): Fennoscandian tundra ecosystems. Part 1: Plants and microorganisms, pp. 208–215. Berlin-Heidelberg-New York: Springer 1975.

BLISS, L. C.: Net primary production of tundra ecosystems. In: LIETH, H. (Ed.): Die Stoffproduktion der Pflanzendecke, pp. 35–46. Stuttgart: Fischer 1962.

BLISS, L. C., WIELGOLASKI, F. E.: Primary production and production processes, Tundra Biome. Edmonton-Oslo: IBP Tundra Biome Steering Committee 1973.

BRZOSKA, W.: Stoffproduktion und Energiehaushalt von Nivalpflanzen. In: ELLENBERG, H. (Ed.): Ökosystemforschung, pp. 225–233. Berlin-Heidelberg-New York: Springer 1973.

Coupland, R. T.: Producers: III. Rates of dry matter production and of nutrient and energy flow through shoots. Matador Project Tech. Rep. **33**. Saskatoon: CCIBP 1973.

HÅLAND, Å.: Gjødsling til naturleg fjellvegetasjon i sørvest-Norge. Statens forsøksgard Særheim. Meld. **53**, 1972.

LILAND, P. J.: Orienterende gjødslingsforsøk på fjellbeite. Norden **2**, 65–66 (1970).

LYE, K. A.: Vegetation of selected localities for IBP on investigation in Hardangervidda, Southern Norway. IBP i Norden **8**, 101–111 (1972).

MILNER, C., HUGHES, R. E.: Methods for the measurements of the primary production of grassland. IBP Handbook No. **6**. Oxford: Blackwell 1968.

MUC, M.: Primary production of plant communities of the Truelove Lowland, Devon Island, Canada.—Sedge meadows. In: BLISS, L. C., WIELGOLASKI, F. E. (Eds.): Primary production and production processes, Tundra Biome, pp. 3–14. Edmonton-Oslo: IBP Tundra Biome Steering Committee 1973.

PERKINS, D. F., JONES, V., MILLAR, R.: Primary production and mineral nutrient studies in the grassland ecosystem. In: HEAL, O. W., PERKINS, D. F. (Eds.): The ecology of some British moors and montane grasslands. Berlin-Heidelberg-New York: Springer Verlag 1975 (in press).

RODIN, L. E., BAZILEVICH, N. I.: Production and mineral cycling in terrestrial vegetation. Edinburgh-London: Oliver and Boyd 1967.

ROSSWALL, T., VEUM, A. K., KÄRENLAMPI, L.: Plant litter decomposition at Fennoscandian tundra sites. In: WIELGOLASKI, F. E. (Ed.): Fennoscandian tundra ecosystems. Part 1: Plants and microorganisms, pp. 268–278. Berlin-Heidelberg-New York: Springer 1975.

SELSJORD, L.: Beitedyrking på Langsetra i Alvdal. Forskn. Fors. Landbr. **11**, 277–289 (1960).

SHEIKH, K. H., RUTTER, A. J.: The responses of *Molinia caerulea* and *Erica tetralix* to soil aeration and related factors. I. Root distribution in relation to soil porosity. J. Ecol. **57**, 713–726 (1969).

SONESSON, M., WIELGOLASKI, F. E., KALLIO, P.: Description of Fennoscandian tundra ecosystems. In: WIELGOLASKI, F. E. (Ed.): Fennoscandian tundra ecosystems. Part 1: Plants and microorganisms, pp. 3–28. Berlin-Heidelberg-New York: Springer 1975.

SVOBODA, J.: Primary production of plant communities of the Truelove Lowland, Devon Island, Canada.—Beach ridges. In: BLISS, L. C., WIELGOLASKI, F. E. (Eds.): Primary production and production processes, Tundra Biome pp. 15–26. Edmonton-Oslo: Tundra Biome Steering Committee 1973.

TIESZEN, L. L.: Photosynthesis in relation to primary production. In: WIELGOLASKI, F. E., ROSSWALL, T. (Eds.): Proceedings IV International Tundra Biome meeting of IBP on the biological productivity of tundra, Leningrad, pp. 52–62. Oslo-Stockholm: IBP Tundra Biome Steering Committee 1972.

WIELGOLASKI, F. E.: Primary production of tundra. In: COOPER, J. P. (Ed.): Photosynthesis and productivity in different environments, pp. 75–106. Cambridge: University Press 1975a.

WIELGOLASKI, F. E.: Comparison of plant structure on grazed and ungrazed tundra meadows. In: WIELGOLASKI, F. E. (Ed.): Fennoscandian tundra ecosystems. Part 1: Plants and microorganisms, pp. 86–93. Berlin-Heidelberg-New York: Springer 1975b.

WIELGOLASKI, F. E., KJELVIK, S.: The methodology of net primary production investigations in Norwegian IBP tundra studies. Rep. Kevo Subarctic Res. Stat. **9**, 24–27 (1972).

WIELGOLASKI, F. E., KJELVIK, S.: Production of plants (vascular plants and cryptogams) in alpine tundra, Hardangervidda. In: BLISS, L. C., WIELGOLASKI, F. E. (Eds.): Primary production and production processes, Tundra Biome, pp. 75–86. Edmonton-Oslo: IBP Tundra Biome Steering Committee 1973.

WIELGOLASKI, F. E., KJELVIK, S.: Plant biomass at the Norwegian IBP sites at Hardangervidda 1969–1972. In: VIK, R. (Ed.): IBP in Norway. Methods and results. Sections PT-UM Grazing project, Hardangervidda, Botanical investigation, pp. 1–88. Oslo: Norwegian national IBP committee 1975.

WIELGOLASKI, F. E., ROSSWALL, T.: Proceedings IV International Tundra Biome meeting of IBP on the biological productivity of tundra, Leningrad. Oslo-Stockholm: IBP Tundra Biome Steering Committee 1972.

Growth in Populations of Andromeda polifolia on a Subarctic Mire

J. G. K. FLOWER-ELLIS

On the various microsites at Stordalen, *Andromeda polifolia* L. is ubiquitous, forming populations in which density, age structure and size of individuals vary greatly. Estimates of the standing crop and production of *Andromeda* for the mire as a whole are in consequence uncertain.

Variation may be reduced by making individuals in microsite populations defined by density the basis for production estimates. The residual variation within the microsite, associated mainly with age differences is, however, still great. It may be further reduced by expressing the performance of individuals over the life cycle characteristic of the microsite population. Even this fails to remove more than part of the variation. Some of that remaining may give rise to fluctuations in dry weight, when the course of biomass change during the growing season is expressed in terms of an individual of known age. Such fluctuations may have to be ascribed to random sources.

However, if at the same time the development of single organs, i.e. leaves, shoots, flowers and fruits, is studied, it may be possible to identify the seat, if not the source, of the dry weight fluctuations. In *Andromeda*, about 64% of the aboveground biomass is leaf (FLOWER-ELLIS, 1973). There are generally three, for short periods four, active leaf generations. Older leaves senesce and are shed continuously, although spring and autumn peaks in leaf litterfall are discernible. The annual loss of leaves is estimated to be equivalent to about 30% of the aboveground standing crop of *Andromeda* (FLOWER-ELLIS, 1973), hence it is of importance in maintaining turnover in the communities of which *Andromeda* is a constituent.

Leaves increase in weight from their first to their final growing season, reaching about 154% of the weight of the fully expanded current leaf. Since much of this weight increment occurs after leaves have attained their final size, it is of interest to trace its course and establish, if possible, its relation to environmental factors. A first hypothesis might be that leaf weight increases continuously throughout the growing season. To confirm this it is necessary to express weight changes on a rectified basis, so that the random variations, e.g. in leaf length, encountered in sampling, are substantially excluded. The next step is to relate the course of leaf weight increment to the biomass changes at the level of the individual.

Table 1. Separation categories for individuals of *Andromeda polifolia*. The symbol x indicates that the character was always recorded, (x) that it was recorded for subsamples. F = flower and fruits, L = leaves, S = shoots

Fraction	Current (C)			2nd season (C+1)		3rd season (C+2)		Older (≥C+3)		Attached dead tot.		
Character	F	L	S	L	S	L	S	L	S	F	L	S
Weight	x	x	x	x	x	x	x	(x)	x	(x)	(x)	x
Number	x	x	x	x	x	x	x	x	x	x	x	x
Length	x		(x)	x	(x)	x	(x)	(x)	(x)		(x)	(x)
Age			(x)		(x)		(x)		x			(x)

Methods and Material

The material of *Andromeda polifolia* L. presented here was collected from typical examples of the following microsites, distributed over the whole area of the Stordalen mire (SONESSON et al., 1973). The community types as classified by MADSEN and WIDELL (1974) are given in brackets:

1. Elevated areas, lichen-dominated: exposed summit (1 a)
2. Elevated areas, lichen-dominated: sheltered slope (1 b)
3. Pure stands in small depressions (4)

The aim of sampling was to cover the range of ages present on the microsite; 25 individuals were therefore collected over the size range of the population within every microsite. Duplicate samples were collected on most occasions. The individuals collected were not drawn from quadrats of specific area, since the range of population density for the microsites could be obtained from the random samples of the main sampling series. Sampling frequency depended on the growth phase (see Fig. 1); for the present purpose, eight sampling occasions (275 individuals), covering the period of development of the current leaves, have been selected.

After collection, samples were immediately frozen (−35° C) and also kept at low temperature during fractionation to reduce respiration losses. Individuals were separated (FLOWER-ELLIS, 1972) under 10 X magnification into the fractions shown in Table 1, oven-dried (70° C for 48 hrs) and weighed to 0.1 mg. Leaves were measured to 0.2 mm, shoot lengths to 0.5 mm, at 10–15 X magnification. Leaf area was measured on dried leaves with a photocell device.

Results and Discussion

The regression of leaf weight on length was calculated for all leaf generations for which material was adequate, on every sampling occasion. It has previously been shown that the form of the relationship is the same, irrespective of microsite, within any leaf generation (FLOWER-ELLIS, 1973). When differences in slope of the regression are found in populations from different microsites sampled on a single occasion, they may be ascribed to differences in phenological development. They occur mainly when leaf expansion is still going on. The regression lines in Fig. 1

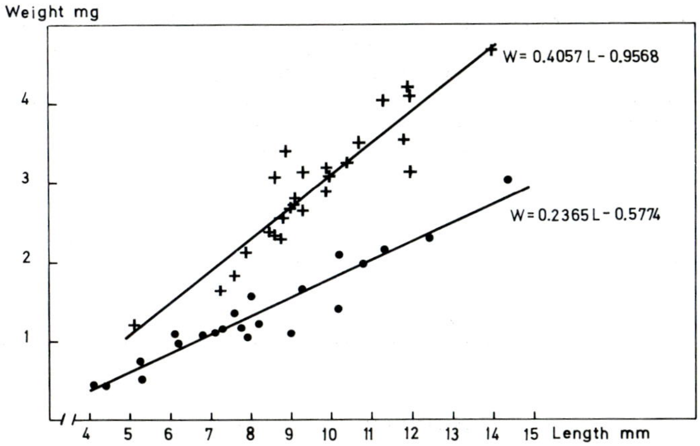

Fig. 1. Regression of leaf weight on leaf length for current leaves of *A. polifolia* from lichen-dominated microsites at Stordalen on two occasions in 1973. Leaf expansion was complete at the latest measurement (+), 5th September, probably complete at earliest measurement (●), 3rd July. Regression line fitted by method of least squares

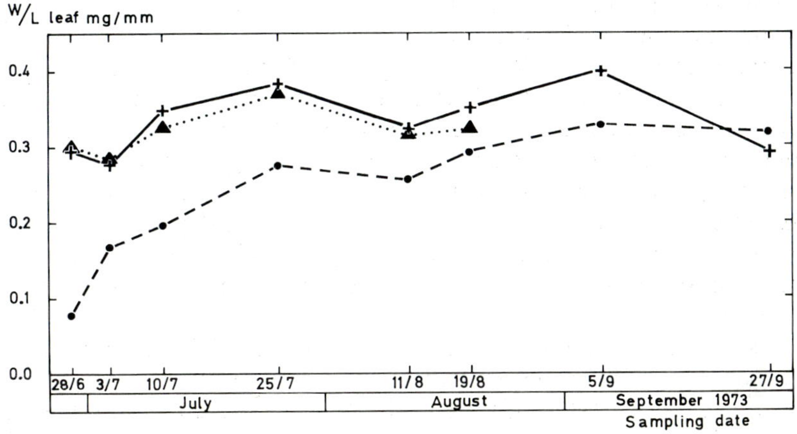

Fig. 2. Changes in mean weight per unit length of the leaf of mean length (8.53 mm) for C, C+1, and C+2 leaves (explained in text) between 28th June and 27th September 1973, Stordalen. C leaves (●), C+1 (+), C+2 (▲)

represent two sampling occasions for the lichen-dominated sites, between 3rd July and 5th September 1973, i.e. after leaf expansion on these sites was substantially complete. The increase in the slope of the regression therefore indicates an increase in leaf weight per unit length. This is shown for the entire growing season in Fig. 2, as weight per unit length for the leaf of mean length 8.53 mm, for all sampling occasions, microsites and leaf generations ($N = 4823$). This lay within the modal class limits of leaf length distribution found in random samples from the main sampling series. Leaf length is normally distributed.

Table 2. Leaf weight changes, mg $day^{-1} \times 10^3$, for C, C+1, and C+2 leaves (explained in text) of *A. polifolia* at Stordalen, 1973. The minus sign denotes a decrease in weight. Based on the leaf of mean length (8.53 mm), with projected (single surface) area 11.8 mm^2 (area = 1.7158 length −2.75). The values refer approximately to mg day^{-1} dm^{-2}. Lack of material is indicated by n.d.

Period days	28/6–3/7 5	3/7–10/7 7	10/7–25/7 15	25/7–11/8 17	11/8–19/8 8	19/8–5/9 17	5/9–27/9 22
C	93	35	44	−9	39	19	−5
C+1	−31	89	20	−30	29	24	−41
C+2	−29	53	26	−28	11	n.d.	n.d.
Sum	33	177	90	−67	79	43	−46

The weight changes in the current (C), second (C+1), and third (C+2) seasons' leaves shown in Fig. 2 had a similar course, in which three periods are worthy of note: (1) 28th June–3rd July, (2) 25th July–19th August, (3) 5th–27th September. In Table 2, the weight changes are expressed as mg day^{-1}; period (1) was a period of net weight increase, although both C+1 and C+2 leaves decreased in weight. The extension of the C leaves was still in progress. The depression in period (2) is characterised by a net weight decrease for all three leaf generations (least in C), followed by an increase in which the daily weight increment of C and C+1 leaves exceeded that of the C+2 leaves. After 19th August, these last were too few for further weight changes to be calculated, although the main sample series will provide material for this purpose. During period (3) both C and C+1 leaves decreased in weight, the former insignificantly, the latter by ca. 27%. This decrease is consistent with what is known of the movements of assimilates labelled with ^{14}C at this time (JOHANSSON, 1974). The rates of change calculated as in Table 2 agree well with the values for CO_2 fixation presented for C and C+1 leaves for the period in question (JOHANSSON and LINDER, 1975).

The weight changes presented here reinforce previous indications that *Andromeda* leaves store reserve substances (JOHANNSON, 1974; cf. also BLISS, 1971); the presence of starch in the leaves in autumn is reported by LIDFORSS (1907). It is also evident that they must be taken into account when budgets for assimilated $^{14}CO_2$ are being calculated (JOHANSSON, 1974).

The total dry weight (including attached dead) and biomass of the mean individual from the different sampling occasions varied greatly (between 33.7 and 223.9 mg), largely as a result of the age-class composition of the samples. In consequence, the mean dry weight of the fractions varies too much for direct comparison. Weight changes are therefore expressed in Fig. 3 in percent of the total dry weight of the mean individual from every sampling. The means are based on not less than 25 and on two occasions 50, individuals.

The close agreement between the relative proportions of the fractions and the course of weight changes in the leaves (Fig. 2) is clearly brought out. Considering the dominance of the leaves in the total biomass, it is not surprising that small weight changes should produce such a marked effect on the weight of the individual. The relative participation of the various leaf generations in the weight decrease of period (2) is the product of the absolute decrease per leaf and the number

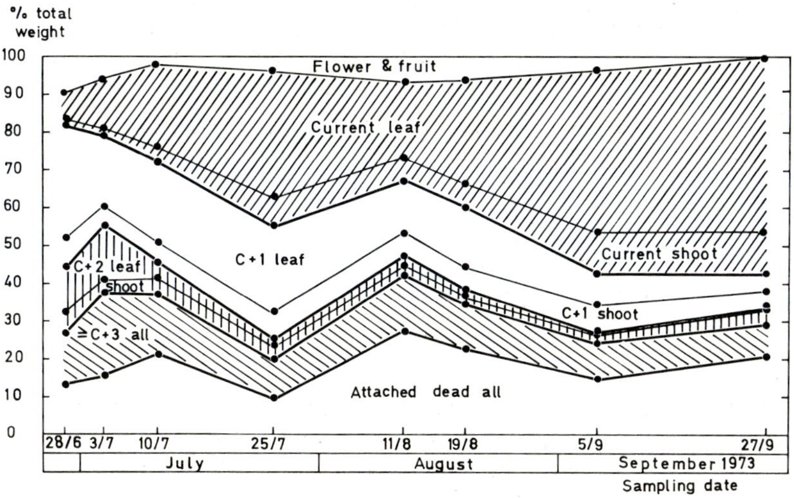

Fig. 3. Weight changes in individuals of *A. polifolia* between 28th June and 27th September 1973, Stordalen. Expressed as percent of the total weight, including attached dead, of the mean individual from every sampling occasion, by the fractions shown in Table 1. The thick lines separate C, C+1, and C+2, etc., fractions, (explained in text), the fine lines, leaf and shoot of the same year

of leaves involved. Figure 3 also brings out the decline in leaf numbers of the C+1 and C+2 generations during the growing season, earlier inferred from the relationship between the numbers of leaves present in the three generations at a point in time (FLOWER-ELLIS, 1973). Comparison of the rate of leaf loss in the present material with that inferred for the random samples from 1971 also shows agreement; by the end of their second growing season, about 30% of the C+1 leaves remain. At the end of their third season, about seven per cent of the original leaves (now C+2) remain, while at the end of the fourth season about five percent of the original number (now C+3) remain. The depletion curve probably reaches zero in the fifth growing season. The data at present available indicate that the greatest rate of loss of the C+2 leaves is in early summer, whereas C+1 leaves appear to be lost at a constant rate throughout the season. Further information is needed to confirm this.

Variations in the other fractions shown in Fig. 3 may be explained as follows: For the flowers and fruits, the decrease in the proportion up to 10th July is probably related to the fall of the corolla, the gradual increase to 11th August to the ripening of the capsule (on that date 53% of the fruits were fully formed). From 11th August to 19th August there was little change, but after that date an increasing proportion of ripe capsules opened, and many were empty. At the end of the season, 44% of the capsules had still not opened and about seven percent were not ripe. The variations in the mean weight of the flowers and fruits correspond closely to the changes shown in Fig. 3. The attached dead fraction varied considerably; here comparison is more difficult, since three components are involved. The attached dead leaves, expressed as a percentage of all leaves, showed fluctuations which paralleled those of the whole attached dead fraction. Many leaves appeared to have been killed by frost; it is characteristic of such leaves and shoots in *Andromeda* that they remain attached, whereas senescent leaves absciss and fall readily. Since the input of attached dead fractions, other than leaves, to litter is important, the dynamics of shoot and flower mortality require further investigation.

This study indicates that the approach adopted in 1971 for estimating production and litterfall in *Andromeda* and other species with evergreen leaves may also be useful for exploring the relationship between assimilation and growth. It has been shown here that leaf weight increases even after leaf expansion has ceased, and that it may also decrease in the middle of the growing season. This implies that the origin of the weight increment of the current leaves, at least to the extent of the decrease in weight, is the storage of mobilisable resources. The parallel behaviour of the C+1 and C+2 leaves confirms this. The marked decline in weight of the C+1 leaves at the end of their second growing season gives a measure of the withdrawal of these reserves. Their apportioning between other organs is still largely obscure (but cf. JOHANSSON, 1974). However, some of the weight increment may depend on the accumulation of other metabolic products, e.g. tannins, known to be present in quantity in the leaves (HEGI, 1927). Secondary changes in leaf structure may also be involved. Until the relative importance of the various components of leaf weight increment has been established, attempts to relate it directly to photosynthesis must be made with caution, and must include other components of the budget, such as respiration and translocation.

References

BLISS, L. C.: Arctic and alpine plant life cycles. Ann. Rev. Ecol. and Systemat. **2**, 405–438 (1971).

FLOWER-ELLIS, J. G. K.: Primary production studies, Stordalen 1971. Progress report 1971. Swedish IBP Tundra Biome Project Tech. Rep. **9**, 9–17 (1972).

FLOWER-ELLIS, J. G. K.: Growth and morphology in the evergreen dwarf shrubs *Empetrum hermaphroditum* and *Andromeda polifolia* at Stordalen. In: BLISS, L. C., WIELGOLASKI, F. E. (Eds.): Primary production and production processes, pp. 123–136. Edmonton-Oslo: IBP Tundra Biome Steering Committee 1973.

HEGI, G.: Illustrierte Flora von Mittel-Europa. Dicotyledones V (3), pp. 1651–1654. München: J. F. Lehmanns Verlag 1927.

JOHANSSON, L. G.: The distribution and fate of ^{14}C photoassimilated by plants on a subarctic mire at Stordalen. Progress report 1973. Swedish IBP Tundra Biome Project Tech. Rep. **16**, 165–172 (1974).

JOHANSSON, L. G., LINDER, S.: The seasonal pattern of photosynthesis of some vascular plants on a subarctic mire. In: WIELGOLASKI, F. E. (Ed.): Fennoscandian tundra ecosystems, Part 1: Plants and microorganisms, pp. 194–200. Berlin-Heidelberg-New York: Springer 1975.

LIDFORSS, B.: Die wintergrüne Flora. Lunds Univ. Årsskrift N.F. **2**, 1–78 (1907).

MADSEN, I.-L., WIDELL, S.: A vegetation map of the Stordalen site. Progress report 1973. Swedish IBP Tundra Biome Tech. Rep. **16**, 3–15 (1974).

SONESSON, M., FLOWER-ELLIS, J. G. K., JONSSON, S., ROSSWALL, T., RYDÉN, B. E.: Word model for Abisko (Stordalen), Sweden. April 1973. Progress report 1972. Swedish IBP Tundra Biome Tech. Rep. **14**, 4–13 (1973).

3.3. Plant Processes

Weight Increase of Some Lichens as Related to Carbon Dioxide Exchange and Thallus Moisture

L. KÄRENLAMPI, J. TAMMISOLA, and H. HURME

The relationship between growth of lichens and environmental factors has been studied within the IBP Tundra Biome by KÄRENLAMPI (1971), and the relationship between CO_2 exchange and moisture content of the lichen thallus is fairly well known (reviewed e.g. by KALLIO and KÄRENLAMPI, 1975). In the present study a new moisture-time variable is introduced to predict growth, and the correlation between weight increase and photosynthetic capacity is reported.

Materials and Methods

Tall individuals of *Cetraria nivalis* and *Cladonia alpestris* were collected from the surroundings of the Kevo Research Station in northern Finland. They were cut into sections of definite lengths and kept in perforated plastic boxes. The growth data in Tables 1 and 2 are based on direct weighings of the lichens grown in the boxes. The method is described by KÄRENLAMPI (1971).

The data of the CO_2 exchange (Table 1) were calculated by computer using a multiple regression model described by KÄRENLAMPI and TAMMISOLA (1970) and KALLIO and KÄRENLAMPI (1975). The model predicts the CO_2 exchange of lichens using the following environmental variables: illumination, temperature, moisture and transformations and combinations of these factors.

The moisture-time variables (frequency in number of hours of thallus moisture content within certain limits) in Table 2 were computed by multiple regressions. These predict the moisture of the lichens when certain parameters, e.g. precipitation and radiation as provided by the meterological station, are known. The moisture-time variables are integrated at three-hour intervals and changed into relative values.

Results and Discussion

Almost 40% of the growth for whole plants of *Cetraria nivalis* during the season (Table 1) occurred in the top 1 cm compartments. The sections with a higher proportion of basal parts grew much more slowly, e.g. the 5-cm long thalli grow about 8% annually. The predicted values of CO_2 exchange (Table 1) corre-

Table 1. Percentage annual weight growth of *Cetraria nivalis* in various plant segments at different distances from the plant top or base, and an example of the photosynthetic capacity

Part of the lichen	Growth in per cent dry weight[a]	CO_2 exchange[b] in mg CO_2 g^{-1} h^{-1}
1 cm long parts from top	39.0	0.573
2 cm long parts from top	34.8	0.414
3 cm long parts from top	22.2	0.228
4 cm long parts from top	14.0	0.133
5 cm long parts from top	8.0	0.083
4 cm long parts from base	8.3	0.062
3 cm long parts from base	7.2	0.068

[a] Means of two boxes; period May 30–Sept. 22, 1971.
[b] Predicted with the model for the following conditions: thallus moisture content 125%; temperature 6° C; illumination 7430 lux.

Table 2. Cumulative relative values for the dry weight growth of 5 cm long sections of *Cladonia alpestris* and for some moisture-time variables

Date[a]	Weight growth[b]	Moisture-time		
		100% limit	50% limit	0% limit
June 11.	0.00	0.00	0.00	0.00
June 30.	0.05	0.05	0.11	0.17
July 22.	0.22	0.23	0.25	0.36
July 30.	0.47	0.34	0.33	0.43
Aug. 19.	0.52	0.50	0.50	0.62
Nov. 1.	1.00	1.00	1.00	1.00

[a] Summer 1969.
[b] Means of six boxes.

late well with the weight increase ($r = 0.984$, $n = 7$). Correlations between the weight increase and CO_2 exchange were calculated for several different combinations of the environmental factors (illumination, moisture and temperature). Correlation coefficients were high for the combinations commonly found in nature.

The extent to which there is correlation between weight increase and CO_2 exchange seems to suggest that yearly production could be estimated when photosynthetic capacity has been measured. There are, however, differences between microhabitats, annual climatic conditions, as well as between lichen species. With these differences the estimates of production based solely on the photosynthetic capacity cannot be considered general until more data are available.

The cumulative relative values of the moisture-time variable at 100% moisture (dry weight) limit are closest to the cumulative relative values of plant dry weights (Table 2) for *Cladonia alpestris*. The moisture-time variable at the 50% limit is inferior in the dry, first half of the summer. During a period with high precipitation at the end of July the weight increase of lichens is clearly higher than expected on the basis of any of the moisture-time variables given in Table 2. The lichen moisture was mostly close to the optimum during that period, about 150% of dry weight. There may also be some other explanation for the strong growth. It is difficult to conceive of 25% of annual growth taking place during one week only, although the summer of 1969 was very dry throughout, except for the last week of July.

Generally, however, it seems possible to predict production of lichens on the basis of very simple moisture-time variables. The moisture of lichens can be fairly easily calculated by short-interval or continuous recording of precipitation, radiation and other relevant climatological parameters. The values obtained can then be used again for calculating the CO_2 exchange of the lichen, the next step being the conversion of the value to production as weight increase or calories.

References

KALLIO, P., KÄRENLAMPI, L.: Photosynthetic activity in mosses and lichens. In: COOPER, J. P. (Ed.): Photosynthesis and productivity in different environments. Cambridge: University Press 1975 (in press).

KÄRENLAMPI, L.: Studies on the relative growth rate of some fructicose lichens. Rep. Kevo Subarctic Res. Stat. **7**, 33–39 (1971).

KÄRENLAMPI, L., TAMMISOLA, J.: Preliminary report on models of reindeer lichen CO_2 exchange, environment, and productivity. IBP i Norden **5**, 23–25 (1970).

CO_2 Exchange and Growth of Rhacomitrium lanuginosum and Dicranum elongatum

P. KALLIO and S. HEINONEN

Dicranum elongatum Schleich. ex Schwaegr. and *Rhacomitrium lanuginosum* (Hedw.) Brid. are very common and often abundant mosses in the Arctic and, to a lesser degree, in the subarctic, where they prefer acid substrates (HERZOG, 1926; STEERE, 1954; TALLIS, 1959; HOLMEN, 1960; BRASSARD, 1971). *Rhacomitrium* occurs most abundantly in an oceanic climate and at high altitudes and latitudes, and is widely distributed all over the globe (TALLIS, 1958, 1964; LYE, 1966). The distribution in Finnish Subarctic is mostly restricted to less continental mountains in the northernmost corner, but even here the low alpine belt represents a subxeric type of fell vegetation (KÄRENLAMPI and KAUHANEN, 1972).

Dicranum elongatum also demands humid habitats, but is not so dependent on oceanity. It grows in a peaty soil on heaths and moors and is very typical on abrupt slopes on mountains. Here it forms tough cushions, often hanging on shadowy slopes. The tufts gather and retain rain water for some time, which enables them to maintain their moisture unchanged better than poikilohydric organisms normally do. In winter the habitats are mostly without permanent snow cover, and so photosynthetic activity is possible in spring even though the ground elsewhere is still covered by a thick snow layer.

The present paper records a study of photosynthetic and respiratory activities of these mosses, and their responses to the ecological factors which organisms in arctic and subarctic conditions have to withstand. Some comparative experiments were made with forest mosses from S. Finland, e.g. *Hylocomium proliferum*. In addition, the growth of *Dicranum elongatum* at its normal growing sites in the subarctic was compared with results under laboratory conditions.

Material and Methods

The *Dicranum* material was collected in Finnish Lapland from a slope on the shore of Lake Kevojärvi, only 3–13 m above the water level. Some of the *Dicranum* carpets on rocky ground are wet during most of the growth period, but particularly in this respect wide differences exist.

The *Rhacomitrium* material is from different parts of the globe. One sample was taken near the Kevo IBP low alpine heath site in northern Finland (69°45′ N), another from southern Finland near Turku (60°54′ N). The northernmost strain was obtained from Spitsbergen (78°55′ N), others coming from western Norway (60°23′ N), Western Ireland (54°12′ N), eastern Scotland (57°10′ N), Wales (53°14′ N), the Austrian Alps (47°17′ N), and the southernmost one from South Georgia (53°30′ S). These samples were transported by air to our laboratory in Turku, where experiments were started as soon as possible, or otherwise the material was put into a culture cabin at a temperature of approximately 0° C and a rhythmic light of 4000 lux.

The CO_2 exchange has been measured under laboratory conditions only. The measuring technique used is the same infrared gas analyzer method (URAS) as is already described in KALLIO and HEINONEN (1971).

For the experiments carried out at the Kevo Research Station, the samples were immediately placed in assimilation chambers, while the material to be analyzed in the laboratory in Turku was transferred there by air. It was in many cases preserved in Turku for some time before handling at a temperature near 0° C and in rhythmic light conditions. For the measurement of the CO_2 exchange, only the top segments (green part) of 5 mm *(Dicranum)* or 10 mm *(Rhacomitrium)* were used, but the moss sward was kept as natural as possible. Moisture varied between 200–300% from the dry weight.

To measure growth, changes of length were investigated in field conditions, and changes of dry weight were measured in continuous cultivation under optimal laboratory conditions.

The method used for estimating length-growth and production of *Dicranum* in the field corresponds, slightly modified, to the "cranced wire" method described by CLYMO (1970) and used by SONESSON (1973): instead of wires a plastic plate of 2 × 1.5 cm with a ruled scale is used having a stalk with tiny teeth which fix the plate in the moss sward so that, e.g. frost action cannot move it. The plastic plates were inserted into the bryophyte carpet at the beginning of the growth period in the first year of the study (1972), so that the lowermost line on the plate was approximately at the same level as the tips of *Dicranum* stems. The plates were inserted in six plots on the slope, and each plot had 10–24 plates, depending on the size of the tuft. During the first study summer the growth was controlled once in a month only, but in 1973 this was done at intervals of two weeks during the whole growth period until mid-September. The ground was then already frozen.

Another method was also used to estimate growth in the field. Fine-meshed nylon nets (in size 10 × 10 cm with 1.5 × 1.5 mm and 3 × 3 mm mesh openings) were laid on the surface of the moss carpet so that they touched the tips of the *Dicranum* shoots, which grew through the net. These nets were in the field from the beginning of summer 1972. Length-growth can be measured directly from the part of the shoot above the net. The annual shoot production of *Dicranum* was calculated on an area basis by a corer with a cross-section area of 1.12 cm^2. Stem apices 3.5 mm long representing a mean annual length-growth were cut, dried and weighed. To find out which part of the *Dicranum* stem is still alive or how long the ability to regenerate continues in peat, the topmost 5 mm (green part) of the shoot (of some samples) was cut away, and the rest put into an open, wooden box; of other samples, 10 mm, 15 mm and so on were cut off to 7.5 cm. The box contain-

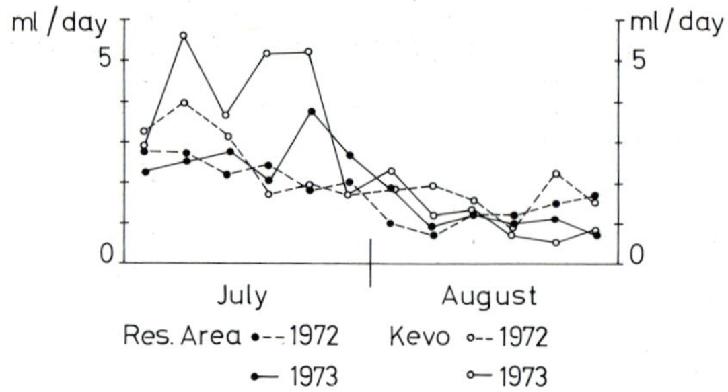

Fig. 1. Five-day means of the daily Piche evaporation in 1972 and in 1973, in the research area and near the Meterological Station (Kevo)

ing these samples was placed back in the natural habitat of *Dicranum* and left there for one year.

The evaporation was measured on the research plot by a PICHE evaporometer. These results were compared with those of the dry plain at the Kevo Meteorological Station (Fig. 1).

Growth was estimated by weighing in the laboratory during 1972–1973. A sample of top segments of *Dicranum*, 5 mm in length and 65 cm^2 in size, the density being the same as in nature, was kept in an assimilation chamber for five days at constant optimal conditions of 10° C and in rhythmic light with a daily dark period of 8 hrs, the light intensity being 10000 lux for 16 hrs. Growth rate can be measured on the basis of the increase in dry weight during the experimental time. The values obtained were compared with the mean CO_2 exchange.

Results

Rhacomitrium lanuginosum. In a previous publication, KALLIO and HEINONEN (1973) studied ecological problems of *Rhacomitrium lanuginosum*, elucidating its distributional geography and adaptation by comparing the Kevo strain with different provenances of this species. These provenances were from different parts, and different climatic conditions of the globe.

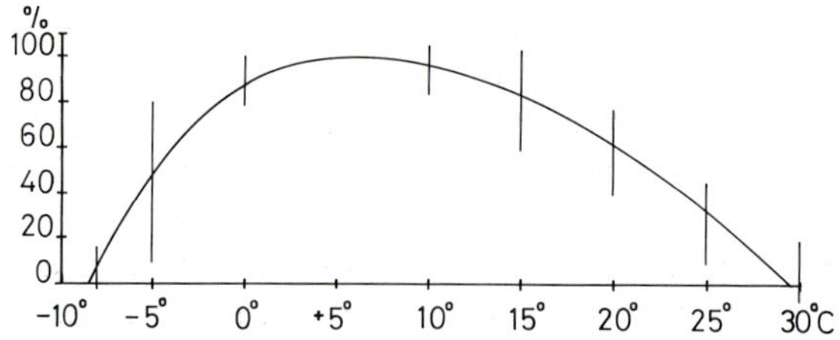

Fig. 2. Mean temperature response curve for apparent photosynthesis of 9 different *Rhacomitrium* strains. The photosynthetic activity at +5° C is indicated as 100, the other values representing percentages from it

Table 1. Mean photosynthesis of different *Rhacomitrium* strains under the optimal temperature of +5° C and 10000 lux

Locality	mg CO_2 g^{-1} h^{-1}	mg CO_2 m^{-2} h^{-1}
Spitsbergen	0.147	120
Northern Finland	0.187	58
Southern Finland	0.608	150
Western Norway	0.239	148
Western Ireland	0.284	202
The Austrian Alps	0.457	128
South Georgia	0.123	76

The response of apparent photosynthesis to temperature in the different *Rhacomitrium lanuginosum* strains is summerized in Fig. 2, which shows that the temperature range for the positive CO_2 balance is rather wide, mostly near 40°. These experiments were carried out unter moisture conditions of 200–300% d. w. Responses of the different strains to temperature seem to be similar, although there is a great deal of variation in the photosynthetic activity. The optimum temperature in all provenances is ca. 5° C, but there is little variation between the temperatures 0° and 10°, and the curve is rather flat. Apparent photosynthesis continues down to the temperature −8° C, the maximum temperature for the positive CO_2 balance being above 30° C. Quantitative differences between the strains seemed to be considerable, as indicated by the values in Table 1.

Examples of the dependence of apparent photosynthesis on light intensities are given in Fig. 3. The compensation point varies between 200–400 lux and 6000 lux in a temperature range from −5° to +15° C. In light intensities of over 10000 lux the curve rises only very slowly, and in high light intensities—over 20000 lux—the curve is already descendant. Moisture was kept optimal during the experiment.

The effect of continuous light to apparent photosynthesis was also studied with *Rhacomitrium* strains and with some forest mosses from S. Finland (KALLIO and VALANNE, 1975). By a continuous light of 13000 lux, and +5° C, apparent photosynthesis often begins to decrease in a few days, while the same light energy given in a daily photoperiod maintains the activity.

The effect of a short, low temperature is shown in Fig. 4. The technique of freezing experiments has been described in KALLIO and HEINONEN (1971). The recovery after freezing (−30° C overnight, 8–10 hrs.) in continuous light, during the first three hours after the temperature has reached 0° C, is about 60% of the value before freezing. When freezing was done in the dark, the recovery of apparent photosynthesis was total in most cases in twelve hours.

Strains from Kevo and southern Finland were transplanted in summer 1971 to the same plot on a rocky hill slope near the Kevo Research Station. The strains were compared in 1973, two years after they were transplanted. Apparent photosynthesis curves of these samples are seen in Fig. 5, and they show that after two years' acclimatization there are still differences in the apparent photosynthesis of the two strains, but these are clearly smaller than at the beginning of the transplantation period. The chlorophyll contents of the two strains show rather clear equalization, although the material is too scanty for detailed analysis (Table 2).

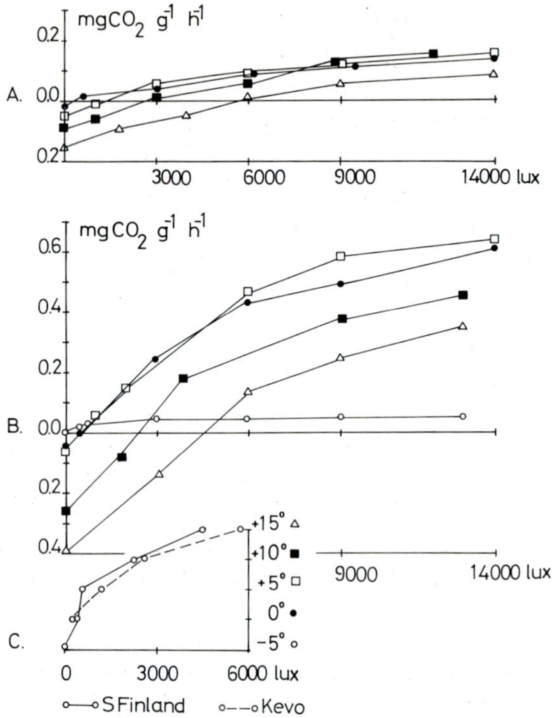

Fig. 3A–C. Dependence of apparent photosynthesis on light intensities in *Rhacomitrium lanuginosum*. (A) Kevo (northern Finland) strain. (B) Southern Finland strain. (C) The compensation points in different temperatures. Symbols: ○ = −5° C, ● = 0° C, □ = +5° C, ■ = 10° C, △ = 15° C

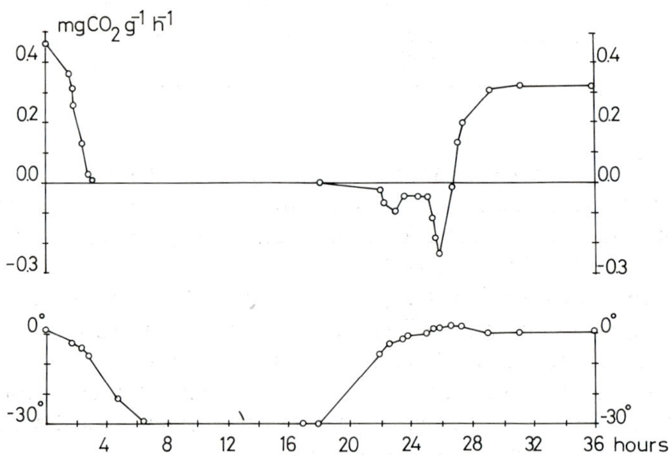

Fig. 4. Effect of short-term low temperature on apparent photosynthesis of *Rhacomitrium lanuginosum* in continuous light (South Georgia strain, Oct. 1972). Temperature changes indicated by the lower curve

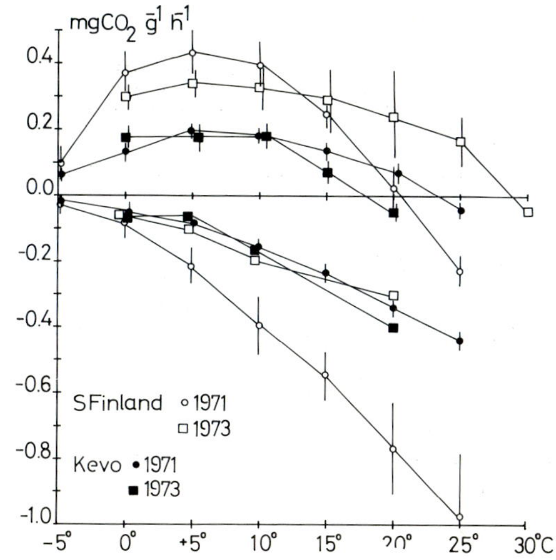

Fig. 5. Apparent photosynthesis and dark respiration of *Rhacomitrium lanuginosum*, Kevo and southern Finland strains, before and after transplantation at Kevo in 1971 and in 1973

Table 2. Chlorophyll contents of Kevo (northern Finland) and Yläne (southern Finland) strains in 1971 and in 1973, after they had been two years at Kevo

Rhacomitrium strain	Chlorophyll content mg g^{-1} dw.		Date of measurement
	a	b	
Kevo strain	0.421	0.357	Oct. 1971
	0.610	0.366	Oct. 1973
Yläne strain	1.259	0.650	Oct. 1971
	0.801	0.383	Oct. 1973

The chlorophyll content had decreased in the southern Finland strain, while it increased in the Kevo strain during the two years. Apparently, a reaction to the long day is seen in the southern Finland strain. This problem is more thoroughly studied by KALLIO and VALANNE (1975).

Dicranum elongatum. The temperature curves of the apparent photosynthesis of *Dicranum elongatum*, seen in Fig. 6, are very similar to those of *Rhacomitrium*. The optimum temperature is 5° C, but no great differences are discernible in the photosynthetic activity between 5° and 10°, and photosynthesis continues down to a temperature of ca. −8° C. The maximum temperature for the positive apparent photosynthesis seems to be around 30° C.

The response to freezing is similar to that of *Rhacomitrium* (Fig. 7). The stress effect of the low temperature is seen in both cases as a respiration peak (CO_2 surge) immediately after the freezing at temperatures between −10° and 0° C.

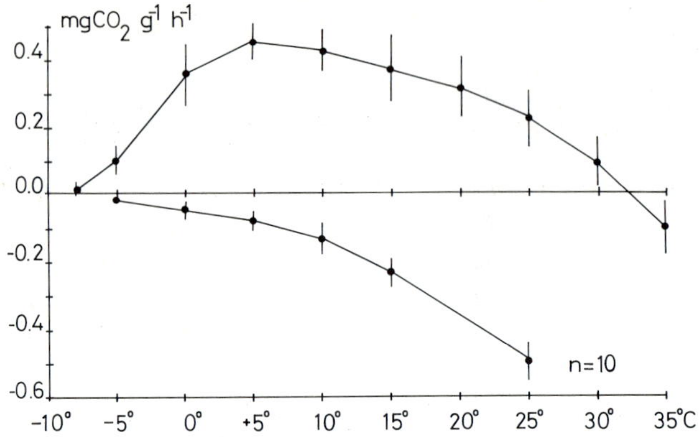

Fig. 6. *Dicranum elongatum.* Dependence of apparent photosynthesis and dark respiration on temperature (measurements made in August 1971–1973)

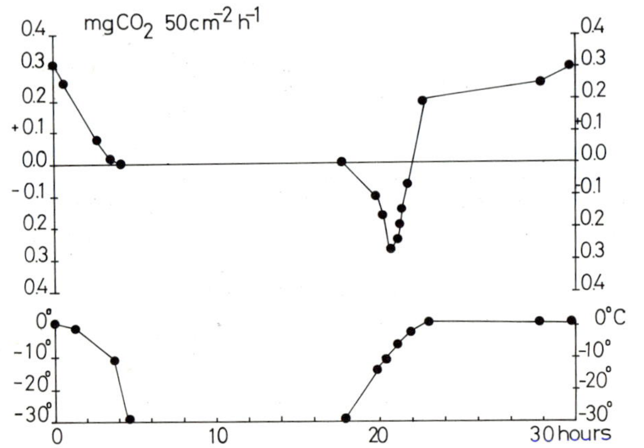

Fig. 7. Response to freezing in *Dicranum elongatum* (May 1970) on apparent photosynthesis. The lower curve indicates the changes in temperature during the experiment

The effect of light intensity to apparent photosynthesis in *Dicranum* is seen in Fig. 8. The response to light is almost the same as in *Rhacomitrium*, although a slightly less heliophilous feature may be seen. Compensation points for *Dicranum* vary between 300 lux and 3000 lux, depending on the temperature. The effect of long days and continuous light is less clear in *Dicranum* than in *Rhacomitrium* (KALLIO and VALANNE, 1975).

The plate method gave a growth range from 0 to 3.9 mm (mean 1.9 mm) year^{-1}. Differences in moisture are probably the main reason for the differences in growth range. Fig. 9 shows the growth in different plots during one year. The length-growth of *Dicranum* measured by the nylon nets was from 3.2 mm $\pm$ 0.49 (larger mesh openings) to 3.8 mm $\pm$ 0.56 (smaller mesh openings) per year, which is a somewhat higher value than that obtained with the plate method.

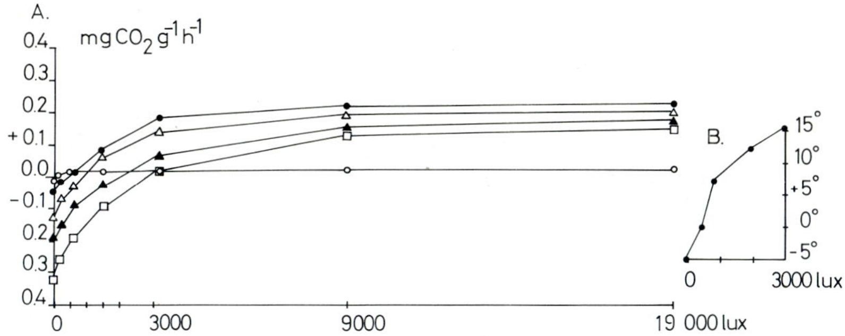

Fig. 8A and B. (A) Dependence of apparent photosynthesis on light at different temperature in *Dicranum elongatum* (February 1970). (B) The compensation points as dependent on temperature. For legend to the symbols, see Fig. 3

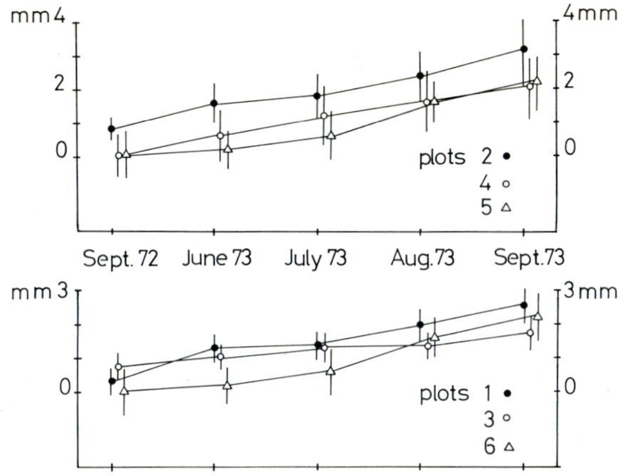

Fig. 9. Growth of *Dicranum elongatum* from Sept. 1972 to Sept. 1973, measured with the plastic plates in six plots within the research area

The average weight of the stem apices of *Dicranum* was found to be 0.027 ± 0.004 g cm^{-2} ($n = 28$), and hence the annual shoot production is 270 g m^{-2}. According to the average growth range measured with the plates, 1.9 mm $year^{-1}$, the net production will be smaller, only 148 g m^{-2} $year^{-1}$. Because the density of the *Dicranum* swards varies—155 ± 31 shoots cm^{-2}—the variation in production is high.

The apparent photosynthesis was 2.09 g CO_2 m^{-2} day^{-1} ($n = 13$) in *Dicranum* by continuous cultivation for five days in assimilation chambers at favorable conditions. The dry weight increase in the same period was 4.07 g m^{-2} day^{-1}.

In *Dicranum elongatum*, the growth of the current year is not clearly demarcated from the "dead" brown peaty part. Therefore the method based on morphology, which has been used advantageously in e.g. *Polytrichum alpestre* (LONG-

TON, 1970), is not used in *Dicranum*. It was found that the ability to regenerate continues until 8 cm down the stem. In the lowest parts of the stem, the new shoot had grown 2 mm, and the growth rate increased toward the top, to ca. 3–4 mm in the topmost centimeters of the *Dicranum* stem. This corresponds to the maximum normal growth in this habitat. Although these observations need confirmations in laboratory tests, they agree well with results from $^{14}CO_2$ studies in Norwegian alpine bryophytes (WIELGOLASKI, 1975; cf. also BAKER, 1972).

Discussion

The photosynthetic capacity of *Rhacomitrium lanuginosum* and *Dicranum elongatum* is low. It represents the lowest values in mosses (STÅLFELT, 1937), and can be compared with the average for the ground lichens (HADLEY and BLISS, 1964; LANGE, 1965; KALLIO and HEINONEN, 1971; KALLIO and KÄRENLAMPI, 1975). TALLIS (1959) also reports a very low production for *Rhacomitrium lanuginosum* in his studies.

It seems that arctic and subarctic mosses show a lower photosynthetic capacity than mosses in temperate zones, e.g. southern Finland (KALLIO and HEINONEN, 1973). On the other hand, rather high photosynthetic values are known from the Antarctic in some moss genera (RASTORFER, 1972).

As regards the temperature response, mosses with wide distribution have developed "local provenance ecotypes" with only slight differences. They seem to be "preadapted" to a wide range of temperatures for positive apparent photosynthesis, the optimum being in all provenances almost the same—i.e. near $+5°$ C. A similar reaction to low temperatures during periods of inactivity (e.g. night or winter temperatures) in quite different climatic zones is another typical feature.

The change of activity in the southern Finland strain of *Rhacomitrium lanuginosum*, which took place during two years at subarctic conditions, is an example of acclimatization known in literature both for mosses and lichens (STÅLFELT, 1960; TAMM, 1964; BÜTTNER, 1971; KERSHAW and ROUSE, 1971; ADAMS, 1971; SKRE, 1975), but it also shows a difference in the capacity of the two strains after a period of two years (when all the photosynthetically active parts of the moss have been renewed). It may be that acclimatization is partly based on changes in the chlorophyll content.

The highest annual shoot production values in the present studies are below 300 g m^{-2}, and of the same order as of BAKER (1972) in *Chorisodontium aciphyllum* and lower than in the *Pohlia* studies in South Georgia by CLARKE et al. (1971). The latter investigation reports annual shoot production (ASP) values varying between 380 g m^{-2} and 1055 g m^{-2}. The shoot-growth (lenght increase) of *Dicranum* at Kevo is almost the same as at Stordalen (Abisko, Sweden), where SONESSON (1973) reports values between 0 and 2 mm per year, the maximum being 4 mm.

Kevo represents a dry ecosystem in the subarctic. Low moisture content is the most important factor that limits growth in lichens (KÄRENLAMPI, 1970), and this observation also applies to mosses in our experiments. The capacity of the antarctic and Kevo mosses is probably not very different, but the ecological conditions deviate. A capacity of shoot production of 4.07 g m^{-2} day^{-1} was found in optimum conditions in assimilation chambers at Kevo. This corresponds to a yearly production of 570 g m^{-2}, when the growth season is taken as 140 days in Kevo (KÄRENLAMPI, 1970). An annual shoot production of ca. 280 g is the maximum measured in the field, i.e. about half the possible capacity in shoot production.

The *Dicranum elongatum* tufts in northern Finland are often over 10 cm thick. Only the topmost millimeters are green and photosynthetically active. There are, however, living cells in the brown part of the moss carpet. The energy used in this layer affects the shoot production.

It seems that even the subarctic and arctic mosses, growing in conditions where the sun does not set at all for 2–3 months, are not totally adapted to the continuous light, but there is a difference in comparison with strains from southern Finland (KALLIO and VALANNE, 1975).

Acknowledgements. We are very indebted to all the persons who have helped us in getting the *Rhacomitrium* material, especially Mrs. P. KARLSTRÖM, and the technical assistance of Mr. M. KARUNEN M.Sc. and Mrs. A. ALONEN is gratefully acknowledged.

References

ADAMS, M. S.: Temperature response of carbon dioxide exchange of *Cladonia rangiferina* from the Wisconsin pine barrens, and comparison with an alpine population. Midland Naturalist **86**, 224–227 (1971).

BAKER, J. H.: The rate of production and decomposition of *Chorisodontium aciphyllum* (Hook. f. & Wils). Broth. Br. Antarct. Surv. Bull. **27**, 123–129 (1972).

BRASSARD, G. R.: The mosses of Northern Ellesmere Island, Arctic Canada. I. Ecology and Phytogeography, with an analysis for the Queen Elizabeth Islands. Bryologist **74**, 233–281 (1971).

BÜTTNER, R.: Untersuchungen zur Ökologie und Physiologie des Gasstoffwechsels bei einigen Strauchflechten. Flora **160**, 72–99 (1971).

CLARKE, G. C. S., GREENE, S. W., GREENE, D. M.: Productivity of bryophytes in polar regions. Ann. Botany. **35**, 99–108 (1971).

CLYMO, R. S.: The growth of *Sphagnum:* methods of measurement. J. Ecol. **58**, 13–49 (1970).

HADLEY, E. B., BLISS, L. C.: Energy relationships of alpine plants on Mt. Washington, New Hampshire. Ecol. Monogr. **34**, 331–357 (1964).

HERZOG, T.: Geographie der Moose. Jena 1926.

HOLMEN, K.: The mosses of Peary land, North Greenland. Medd. Grønland **163**, 1–96 (1960).

KALLIO, P., HEINONEN, S.: Influence of short-term low temperature on net photosynthesis in some subarctic lichens. Rep. Kevo Subarctic Res. Stat. **8**, 63–72 (1971).

KALLIO, P., HEINONEN, S.: Ecology of *Rhacomitrium lanuginosum* (Hedw.) Brid. Rep. Kevo Subarctic Res. Stat. **10**, 43–54 (1973).

KALLIO, P., KÄRENLAMPI, L.: Photosynthetic activity in mosses and lichens. In: COOPER, J. P. (Ed.): Photosynthesis and productivity in different environments. Cambridge: University Press 1975 (In press).

KALLIO, P., VALANNE, N.: On the effect of continuous light on photosynthesis in mosses. In: WIELGOLASKI, F. E. (Ed.): Fennoscandian tundra ecosystems, Part 1: Plants and microorganisms, pp. 149–162. Berlin-Heidelberg-New York: Springer 1975.

KÄRENLAMPI, L.: Morphological analysis of the growth and productivity of the lichen *Cladonia alpestris*. Rep. Kevo Subarctic Res. Stat. **7**, 9–15 (1970).

KÄRENLAMPI, L.: Comparisons between the microclimates of the Kevo ecosystem study sites and the Kevo Meteorological Station. Rep. Kevo Subarctic Res. Stat. **9**, 50–65 (1972).

KÄRENLAMPI, L., KAUHANEN, H.: A direct gradient analysis of the vegetation of the surroundings of the Kevo Subarctic Station. Rep. Kevo Subarctic Res. Stat. **9**, 82–98 (1972).

KERSHAW, K. A., ROUSE, W. R.: Studies on lichen-dominated systems. I. The water relations of *Cladonia alpestris* in spruce-lichen woodland in northern Ontario. Can. J. Botany. **49**, 1389–1399 (1971).

LANGE, O.: Der CO_2-Gaswechsel von Flechten bei tiefen Temperaturen. Planta **64**, 1–19 (1965).

LONGTON, R. E.: Growth and productivity of the moss *Polytrichum alpestre* Hoppe in Antarctic regions. Antarctic Ecology **2**, 818–837 (1970).

LONGTON, R. E.: Reproduction of Antarctic mosses in the genera *Polytrichum* and *Psilopilum* with particular reference to temperature. Br. Antarctic Surv. Bull. **27**, 51–96 (1972).

LYE, K. A.: A quantitative and qualitative investigation of oceanic bryophyte communities and their relation to the environment. Nytt Magasin Botanik **13**, 87–133 (1966).

RASTORFER, J. R.: Comparative physiology of four west Antarctic mosses. Antarctic Res. Series **20**, 143–161 (1972).

SKRE, O.: CO_2 exchange in Norwegian tundra plants studied by infrared gas analyzer technique. In: WIELGOLASKI, F. E. (Ed.): Fennoscandian tundra ecosystems, Part 1: Plants and microorganisms, pp. 168–183. Berlin-Heidelberg-New York: Springer 1975.

SONESSON, M.: Studies in production and turnover of bryophytes at Stordalen 1972. Progress Report 1972. Swedish IBP Tundra Biome Project Tech. Rep. **14**, 66–75 (1973).

STÅLFELT, M. G.: Der Gasaustausch der Moose. Planta **27**, 30–60 (1937).

STÅLFELT, M. G.: Flechten und Moose. In: RUHLAND, W. (Ed.): Handbuch der Pflanzenphysiologie, Vol. V:2, 364–375. Berlin-Heidelberg-New York: Springer 1960.

STEERE, W. C.: Bryophytes. Botan. Rev. **20**, 425–450 (1954).

TALLIS, J. H.: Studies in the biology and ecology of *Rhacomitrium lanuginosum* Brid. I. Distribution and ecology. J. Ecol. **46**, 271–288 (1958).

TALLIS, J. H.: Studies in the biology and ecology of *Rhacomitrium lanuginosum*. II. Growth, reproduction and physiology. J. Ecol. **47**, 325–350 (1959).

TALLIS, J. H.: Growth studies on *Rhacomitrium lanuginosum*. Bryologist **67**, 417–422 (1964).

TAMM, C. O.: Growth of *Hylocomium splendens* in relation to tree canopy. Bryologist **67**, 423–427 (1964).

WIELGOLASKI, F. E.: Primary productivity of alpine meadow communitites. In: WIELGOLASKI, F. E. (Ed.): Fennoscandian tundra ecosystems, Part 1: Plants and microorganisms, pp. 121–128. Berlin-Heidelberg-New York: Springer 1975.

On the Effect of Continuous Light on Photosynthesis in Mosses

P. KALLIO and N. VALANNE

STÅLFELT (1937) observed the detrimental effect of a high light intensity on forest mosses, and TALLIS (1959) a rapid chlorophyll degradation in *Rhacomitrium lanuginosum*. RASTORFER (1970) was probably the first to study the chlorophyll content and the photosynthetic activity of mosses in polar regions with reference to dependency on the light intensity. The brown color and the low chlorophyll content in antarctic mosses changed within a few days to "normal" when the mosses were cultivated under rhythmic or low light intensity conditions. Sometimes the chlorophyll *a* seems to be more sensitive than *b* to adverse light conditions. Temperature also plays a role in these changes: a low temperature and high light intensity seem to have a particular effect (RASTORFER and HIGINBOTHAM, 1968; cf. also STÅLFELT, 1937). In his studies, however, RASTORFER did not particularly stress the problem of "day length" as being an additional factor besides light intensity. The significance of time factor, combined with high light intensities, has been stressed e.g. by GODISH and SCHEIN (1966). In their studies of mosses, KALLIO and HEINONEN (1973) observed the effect of continuous light, but did not analyse it in more detail. The effect of continuous light or long day on chloroplast structure has not been studied in mosses.

Our point of view in this study is geobotanical and ecological. In the polar regions (e.g. Spitzbergen), the thermal summer is largely a season of continuous day, i.e. the sun is above the horizon. Are the mosses of this area, therefore, so adapted to the light conditions that they can take advantage of the long light period, or are they able to compete also in non-optimum conditions as regards photosynthesis? In subarctic Finland, e.g. Kevo area, where the Finnish IBP tundra project is carried out (SONESSON et al., 1975), the thermal summer is longer than the period of everlasting days. At these latitudes there is also an autumn season with a clearly rhythmic light. As the summer activity is normally very low in forest mosses (KALLIO and HEINONEN, 1973), their growth-rhythm may be connected with an adaptation to light-rhythm.

Since photosynthesis must be closely correlated to chlorophyll content and the structure of the chloroplast, these were studied in addition to the photosynthetic activity. The ratio of chlorophyll *a* to *b*, which seems to be a particular feature in the polar regions (RASTORFER, 1972), may be of special value.

Material and Methods

The mosses studied were *Pleurozium schreberi* (Brid.) Mitt., *Hylocomium splendens* (Hedw.) BSG, *Hypnum cupressiforme* Hedw., *Rhacomitrium lanuginosum* (Hedw.) Brid. and *Dicranum elongatum* Schleich. The specimens were collected at the Kevo Subarctic Research Station (69°45′ N), and around Turku in SW Finland (60° N). The *Dicranum elongatum* specimens are only from Kevo, and *Hypnum cupressiforme* only from Turku. Some experiments were made using the *Dicranum* and *Rhacomitrium* material, also from Spitzbergen. All the experiments were carried out in the laboratory of the Botanical Department, University of Turku. In some experiments, the mosses were transferred immediately from natural conditions to the assimilation chambers. In most cases, however, they were cultivated for some time in the culture boxes in low rhythmic light and at a temperature mostly between 5 and 10° C. Under these conditions, the activity did not usually show any great changes for a few weeks.

There is generally a short period of acclimatization to assimilation cuvette conditions. This usually lasts 1–3 days. A longer preculture would be of importance to experiments of this kind. On the other hand, however, this would eliminate the seasonal differences, which are essential in ecological work.

Photosynthesis experiments are based on infrared gas exchange measurements (URAS), and are described in KALLIO and HEINONEN (1971). The continuous light period usually lasted 5–7 days, and the control species were under rhythmic light conditions at the same time. The light intensity in these experiments was usually between 12000 and 13000 lux. In some experiments, the light intensity under continuous light conditions was half of that under the rhythmic light, so that the light energy per day was the same in both cases. The moisture was kept at between 200 and 500% of dry weight. The results are indicated in $mg \cdot 100\ cm^{-2} \cdot h^{-1}$ or $mg \cdot g^{-1} \cdot h^{-1}$ dry weight. Relative values are also used in the diagrams.

For the electron microscopy, the material was fixed in glutaraldehyde-osmium (VALANNE, 1971). The sections were stained with Reynold's lead citrate, and examined with an AEI EM 6 B electron microscope or Siemens Elmiscope I. The chlorophylls were extracted with 80% acetone, and determined spectrophotometrically according to ARNON (1949).

Results

Figure 1 indicates the decrease of apparent photosynthesis in the southern Finland strains and Fig. 2 in the Kevo (northern Finland) strains under continuous light for *Pleurozium schreberi*. The effect is clear after a period of one week, and there are no essential differences in the response to light between these two strains. Under the rhythmic conditions, no decrease was observed during the same period. On the contrary, the curve is frequently ascendent due to acclimatization of the moss to the "cuvette conditions". Also, the effects of continuous light intensities of 6000 lux and 12000 lux are compared with the rhythmic light of 12000 lux in Fig. 1. Continuous light of even 6000 lux reduces the photosynthetic

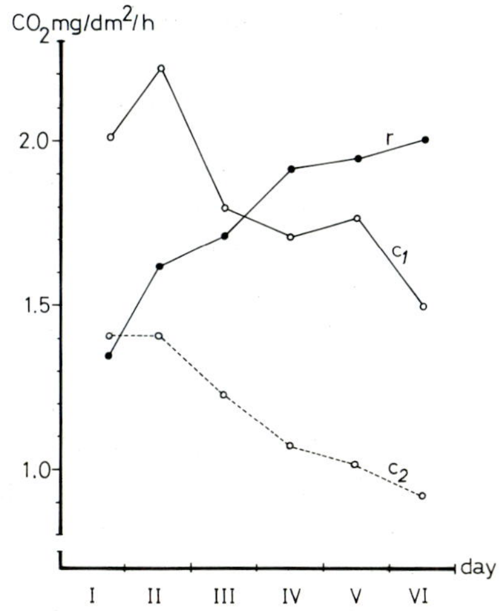

Fig. 1. *Pleurozium schreberi*, southern Finland strain. Effect of continuous light of 12000 lux (c_1) and 6000 lux (c_2), compared with the effect of rhythmic light (*r*), on apparent photosynthesis. Temperature 10° C. February 1974

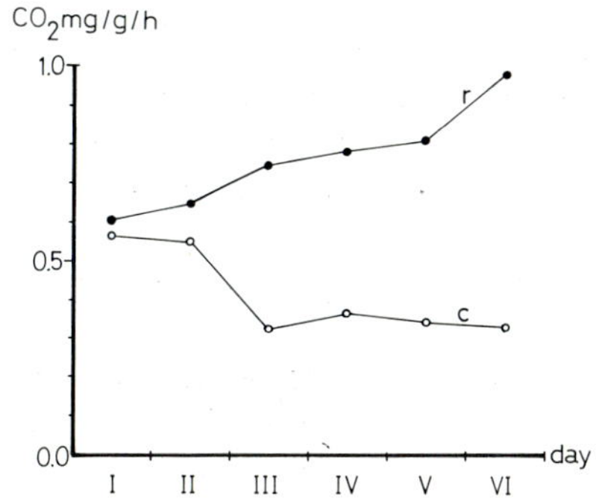

Fig. 2. *Pleurozium schreberi*, Kevo strain. Effect of continuous light (*c*) of 12000 lux on apparent photosynthesis, as compared with rhythmic light (*r*) 12 + 12 hrs of 12000 lux. Temperature 10° C. August 1973

activity. The effect on chlorophyll is seen in Table 1: The total content decreases in both experiments with continuous light, but less in that with the lower light intensity. The ratio of *a* to *b* shows a slight increase in both cases, but relatively more in the lower light intensity. The very light green colour is seen clearly after continuous light. The same change takes place also in all the other experiments. Plate I indicates that the effect of continuous light on chloroplasts is to some extent dependent on the material used. Large and numerous microbodies in *Pleurozium schreberi* occur typically in continuous light (Plate II A), as compared with rhythmic light conditions.

Table 1. Chlorophyll contents of mosses after different light treatments

Species, place and date of collection, eventual storage	Date of chlorophyll assays	Treatment in light	Total chlorophyll $\mu g\, g^{-1}$ d.w.	Ratio of *a* to *b*
Pleurozium schreberi				
(Lieto) 4. 11. 1973	12. 11. 1973	rhythmic light 12000 lux	2320	2.0
	12. 11. 1973	continuous 12000 lux	1623	2.2
after storage in rhythmic light 2000 lux, 5° C	28. 1. 1974	start (control)	4925	1.8
	28. 1. 1974	continuous 12000 lux	3440	2.0
Turku (Katariinan laaksö) 12. 2. 1974	20. 2. 1974	rhythmic 12000 lux	5660	1.9
	20. 2. 1974	continuous 12000 lux	3800	2.0
		continuous 6000 lux	5350	2.1
Hylocomium splendens				
Turku (Katariinan-laakso) 18. 2. 1974	25. 2. 1974	rhythmic 12000 lux	7170	1.3
		continuous 12000 lux	4730	1.4
		continuous 6000 lux	5440	1.7
Rhacomitrium lanuginosum				
Kevo-Utsjoki Oct. 1972	28. 4. 1973	start (control)	1380	1.7
After storage in the field at Ruissalo	28. 4. 1973	rhythmic 13000 lux	1200	1.8
After storage in the field at Ruissalo	28. 4. 1973	continuous 13000 lux	1040	1.8
(Lieto) 4. 11. 1973	24. 1. 1974	start (control)	2095	1.8
after storage in rhythmic light 2000 lux, + 5° C	24. 1. 1974	continuous 6000 lux	1520	2.2

Figure 3 shows the effect of continuous light on *Hylocomium splendens*, apparent photosynthesis in the southern Finland strain and Fig. 4 in the Kevo material. The effect of continuous light of different intensities is compared with rhythmic light in Fig. 3. The effect of no rhythm on this species is very similar to that seen for *Pleurozium*. The color difference is seen, and the chlorophyll contents shown in Table 1 indicate that the total chlorophyll content decreases in continuous light compared to rhythmic light conditions. The ratio of *a* to *b* chlorophyll also increases in this case, more in the low light intensity. In continuous light, the stroma lamellae are poorly developed as compared with those of the chloroplasts in rhythmic light (Plate III A and B), while the grana lamellae show less difference.

The effect of continuous light on apparent photosynthesis of *Hypnum cupressiforme* is seen in Fig. 5. This is a marked decrease in apparent photosynthesis in

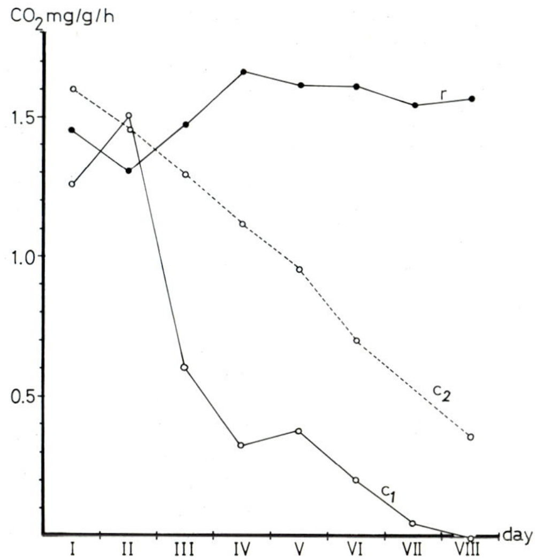

Fig. 3. *Hylocomium splendens* from southern Finland. Effect of continuous light (c_1) 12000 lux and (c_2) 6000 lux, compared with rhythmic light 12+12 hrs of 12000 lux (r). Temperature 10° C. February 1974

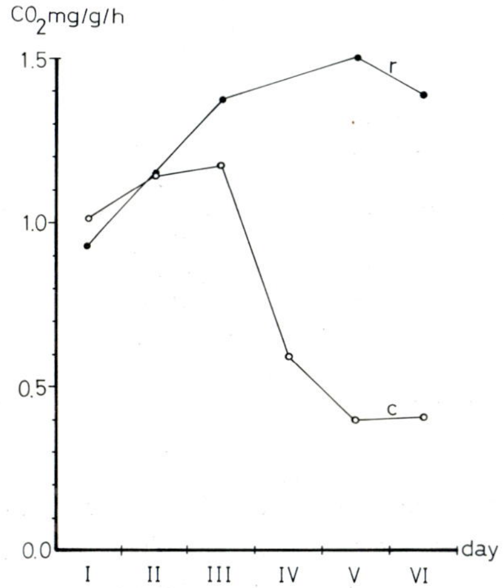

Fig. 4. *Hylocomium splendens*, Kevo strain. Apparent photosynthesis in rhythmic (r) light 12+12 hrs, 12000 lux and continuous light, 12000 lux (c). Temperature 10° C. July 1973, after a preculture of 8 days in low light (2000 lux) conditions

continuous light compared with rhythmic light. No chlorophyll measurements and ultrastructure observations were made.

In some experiments, the apparent photosynthesis in *Rhacomitrium lanuginosum* of the southern Finland strain under continuous light decreases from the

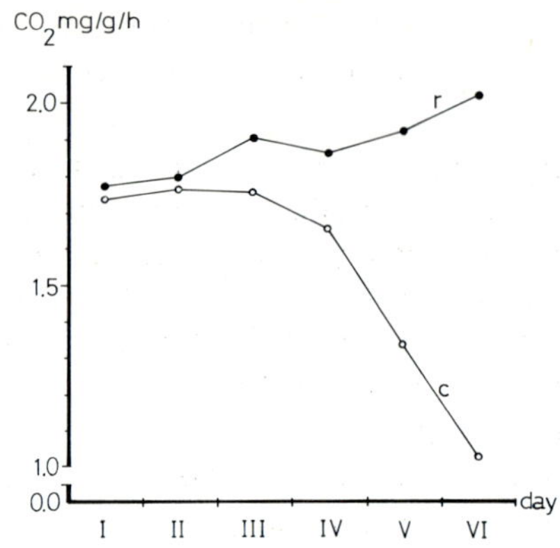

Fig. 5. *Hypnum cupressiforme*, southern Finland strain. Apparent photosynthesis in rhythmic (*r*) and continuous light, 12000 lux (*c*) at 15° C. September 1973

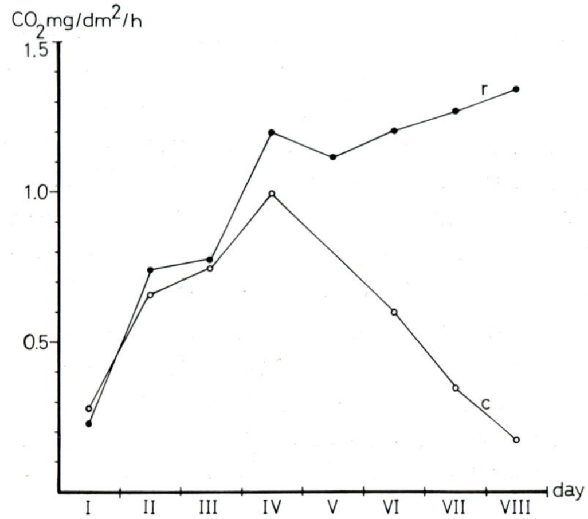

Fig. 6. *Rhacomitrium lanuginosum*, Scotland strain. Apparent photosynthesis in rhythmic (*r*) and continuous (*c*) light (12000 lux) at the temperature of 5° C. After a cultivation of two months in low rhythmic light conditions and a temperature of 0–5° C. Acclimatization after the preculture in low light intensity is seen clearly

relative value of 100 to 20–30 within 6 days, while the curve in rhythmic light descends only slightly, or even ascends. The result, particularly in this species, however, is very variable, depending on season and pretreatment conditions (NYGAARD, 1975). In some experiments, no differences are seen when rhythmic and continuous light conditions are compared. In Table 1 the chlorophyll measurements represent experiments where a difference was observed. The chlorophyll content has decreased, and the ratio of *a* to *b* chlorophyll has increased. No significant difference is seen in the chloroplast structure under either of the experimental conditions. Abundant plastoglobuli (Plate III D) are, however, a typical feature in continuous light, thus making a distinction between the two light conditions.

In Kevo and Spitzbergen strains, wide variation is also observed, but mostly there are only slight differences when effects of continuous and rhythmic lights are compared. In one Spitzbergen strain precultivated under low rhythmic light conditions for one year, no detrimental effect of continuous light was observed in numerous treatments. Even a strain from Scotland was exposed to these experiments. Similar differences as those found for *Pleurozium*, *Hylocomium*, and *Hypnum* where seen in this strain (Fig. 6).

Continuous light seems to have no detrimental effect on *Dicranum elongatum*, either in the Kevo or in the Spitzbergen strain; the Kevo strain was kept for two weeks in continuous light, without any decrease of activity.

Discussion

Mosses are well adapted to severe environmental conditions (AHMADJIAN, 1967; 1970; GANNUTZ, 1970; RASTORFER, 1970; SCHOFIELD and AHMADJIAN, 1972; KALLIO and HEINONEN, 1973, 1975). The temperature between 5° C and 15° C is the normal range of maximum apparent photosynthesis in most northern mosses, but activity continues at temperatures below 0°. The northern species are well adapted to the short time "active" temperatures between freezing periods. They are also very flexible, which is seen in the "preadaptation" to different thermal environments (KALLIO and HEINONEN, 1973, 1975).

Two of the species in this study, *Rhacomitrium lanuginosum* and *Dicranum elongatum*, are widely distributed in the Arctic. *Rhacomitrium lanuginosum*, has a very wide distribution, particularly in the areas with oceanic climate. *Pleurozium schreberi* and *Hypnum cupressiforme* are typical forest mosses which are not distributed north of the forest zone, while *Hylocomium proliferum* also occurs in the Arctic.

All three boreal forest species studied show some detrimental effect of continuous light on photosynthesis. The laboratory experiments indicate than an effect of no rhythm is, in fact, involved (not only the high light energy per day). This is in agreement with moss studies by STÅLFELT (1960), GANNUTZ (1970), CLARKE and GREENE (1971) and VERSEGHY (1972), although they have not analyzed in more detail the day length effect.

The open tundra at 80° N, e.g. in Spitzbergen, which has continuous light during the whole thermal growth period, may be an environment to which many species cannot adapt owing to light conditions. Local species also show some nonadaptative features, such as the yellowish or brownish-green color (RASTORFER, 1970). The activity of *Pleurozium schreberi* is low in summer at Kevo (KALLIO and HEINONEN, 1973), the active season being the rhythmic light season—as is the case for many mosses, also in the more southern areas (STÅLFELT, 1973; ROMOSE, 1940; TAMM, 1953). Further investigation is needed to find whether the arctic strain of *Hylocomium splendens* var. *alaskanum* is a special ecotype of the light climate, thus being different from those e.g. in southern Finland and Kevo.

The widely distributed *Rhacomitrium lanuginosum*, which is similar in Ireland, Bergen, Kevo and Spitzbergen as regards adaptation to temperature (KALLIO and HEINONEN, 1975), is differentiated in special light climatic ecotypes: The Spitzbergen strain is highly, if not totally adapted to continuous light conditions, but not so the southern Finland strain. There is, however, an ecological factor to be accounted for before light conditions as an ecological variable in the arctic are totally understood: It is possible that the light intensity rhythm in the arctic is sufficient to cause the plants to function differently than e.g. in continuous day or night in the laboratory experiments. It seems, however, that nonadaptation is involved to a certain extent, the opposite of the homeostatic response of mosses to the use of the long light period as compensation for the short growth period. Even the low production, however, is enough under these conditions to guarantee survival in competition with other organisms.

The results of the chloroplast and chlorophyll assays obtained in this study with mosses are in accordance with those obtained with higher plants (BRONCHART, 1967; SIRONVAL and ENGLERT, 1967; COSTES et al., 1973). The decrease of total chlorophyll in continuous light, particularly at high intensities, is observed in all studies. The positive correlation between chlorophyll content and photosynthesis is also a general "rule". The results are also in agreement with the effects of different light rhythms in *Acetabularia* (VANDEN DRIESSCHE, 1970), *Micrasterias* (KALLIO, 1953) and some other freshwater algae, (GOLDMAN et al., 1963) as well as with results in symbiotic lichen algae (HARRIS and KERSHOW, 1972).

The increase in the number and size of microbodies in *Pleurozium schreberi* after exposure to continuous light suggests that photorespiration may be involved in the decrease of apparent photosynthesis (TOLBERT, 1971; RASTORFER, 1970; MACHE and LOISEAUX, 1973). However, the decreased apparent photosynthesis under long day conditions may also be due to other, enzymatic regulations of CO_2-fixation (BRULFERT et al., 1972; QUEIROZ, 1972; FAIR et al., 1973). Therefore, decreased chlorophyll content must not be taken as the only reason for lowered CO_2 uptake in continuous light.

The ultrastructure of the chloroplasts shows varying reactions to continuous light. One reason for this may lie in the different initial conditions (e.g. seasonal). Nevertheless, there was a general trend for stroma lamellae to disappear in continuous light (particularly in high light intensitiy). This agrees with the results obtained under long day conditions (PARK and DRURY, 1967). On the other hand, many intervening factors, as for example pollution (e.g. MASUCH et al., 1973) and darkness (VALANNE, 1971), cause chloroplasts to be very similar, with poorly developed fret membranes. Particularly in *Rhacomitrium lanuginosum*, which is a very resistant species with high lipid content, numerous plastoglobuli are typical under continuous light conditions.

The difference in the sensitivity of northern species to light damage effect may be dependent on some accessory pigments, which have often been understood to be operants in acclimatization and adaptation (RASTORFER, 1970; VERSEGHY, 1972). Therefore, one way of understanding adaptation to continuous light is further study of the carotenoid physiology and some other pigments.

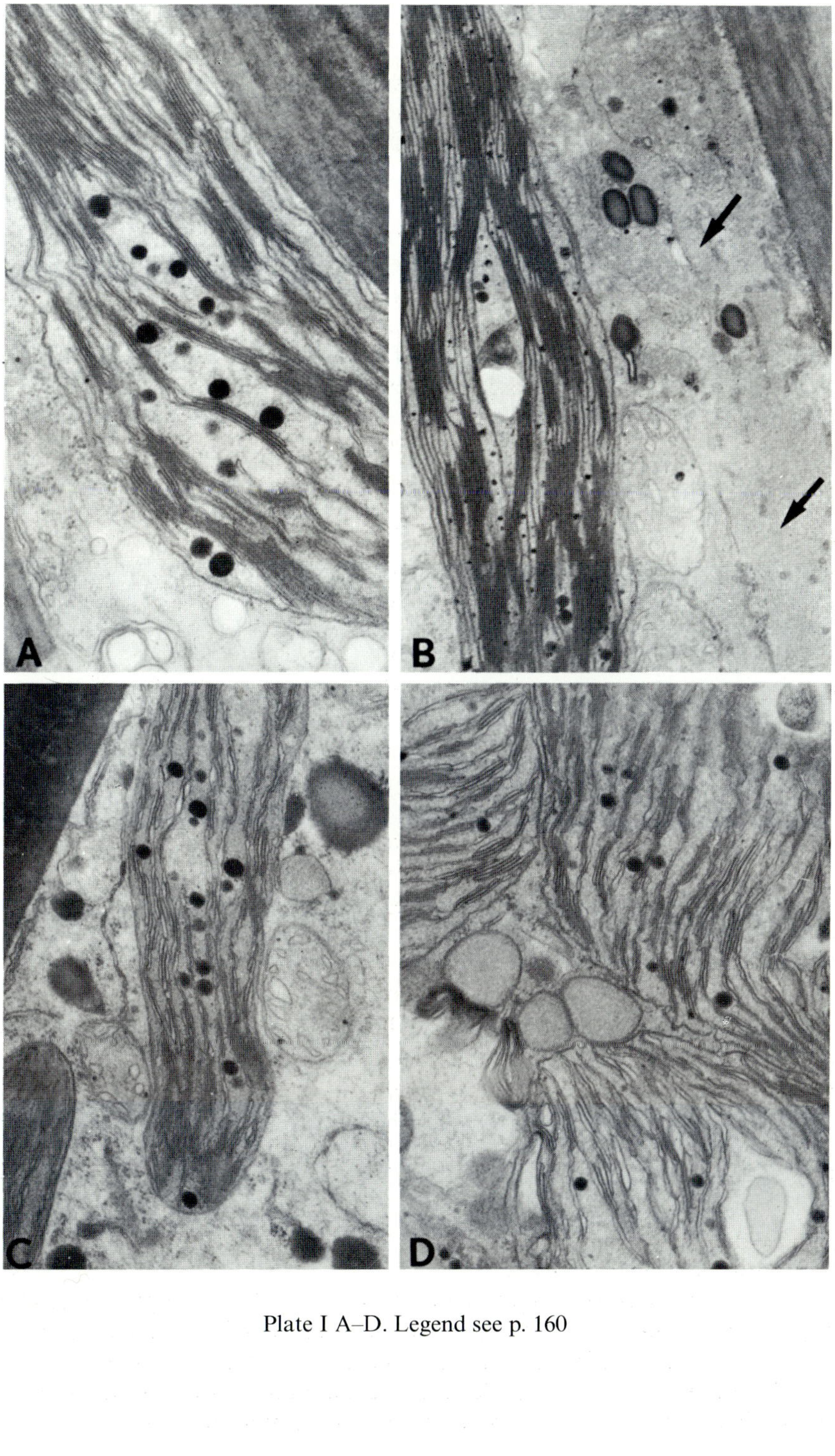

Plate I A–D. Legend see p. 160

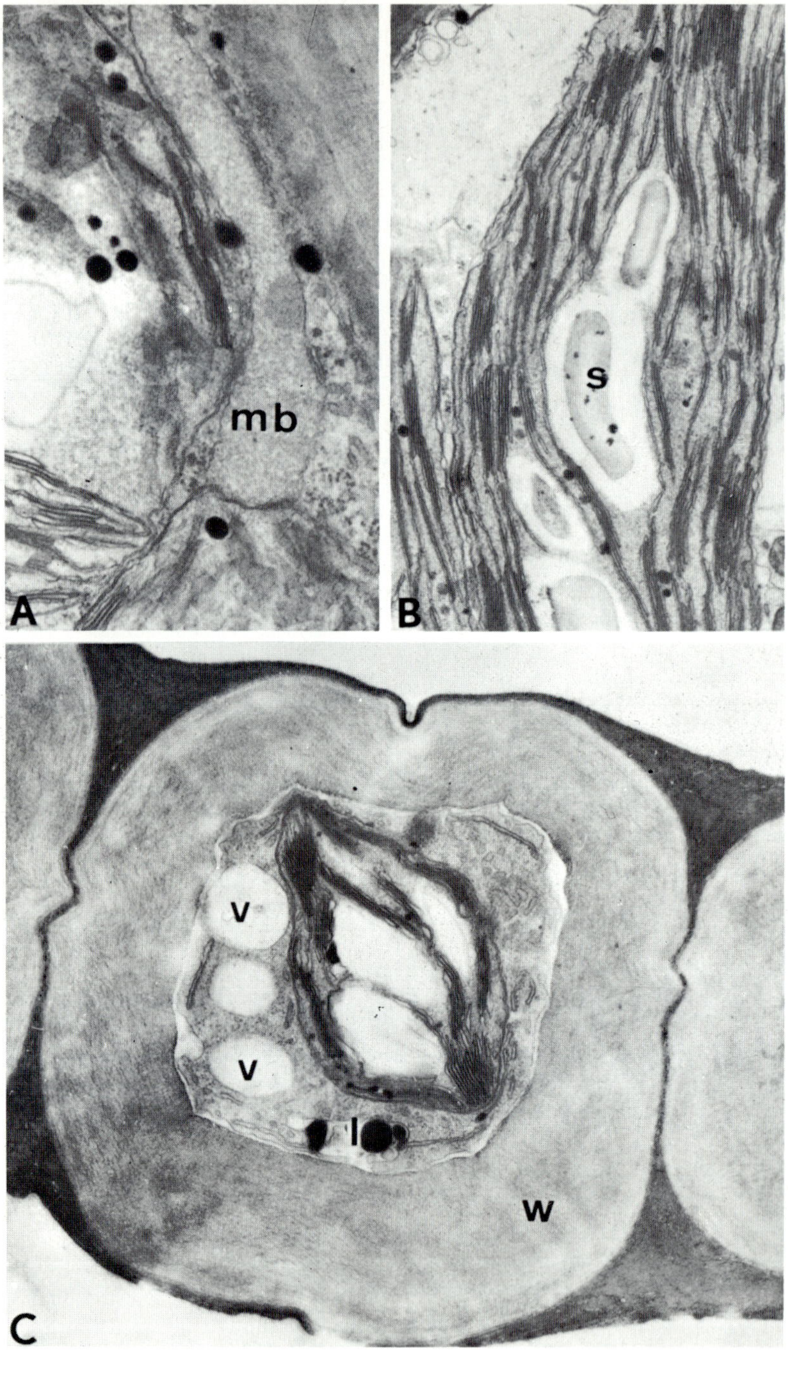

Plate II A–C. Legend see p. 160

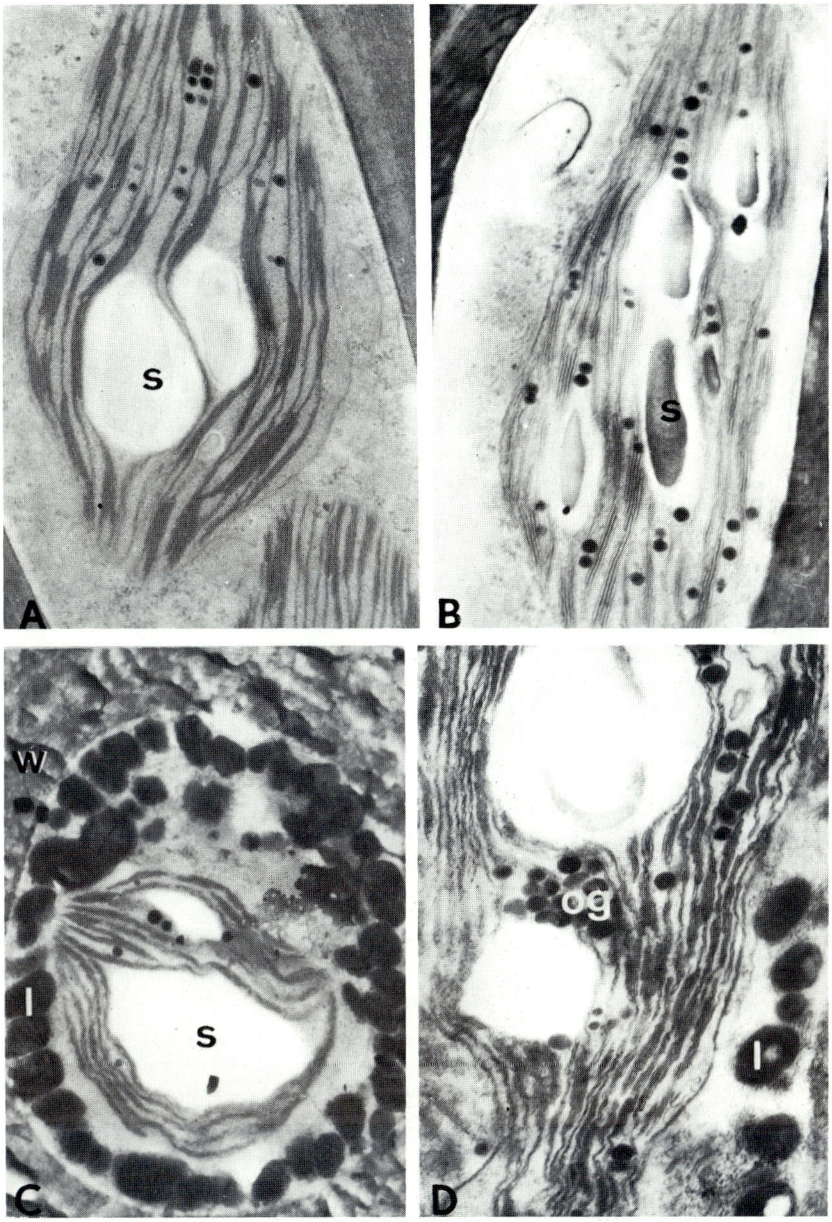

Plate III A–D. Legend see p. 160

Legend to plates pp. 157–159

Plate IA—D. Ultrastructure of the leaf cells of *Pleurozium schreberi*. (A) A chloroplast of material cultured for 3 months in rhythmic light of low intensity (2000 lux, 5° C). Well-defined grana and stroma lamellae. × 23000. (B) The same material after 6 days in continuous light (12000 lux, 10° C, no apparent photosynthesis). Both grana and stroma lamellae are well developed, but loculi of the grana are particularly dense. In the cytoplasm, large areas (arrows) with microbodylike content. × 23000. (C) Cytoplasm of material collected from the field in winter, and cultured for 6 days in rhythmic light (12000 lux). A typical chloroplast of high light intensity, with reduced grana and poorly developed stroma lamellae. The shape of the chloroplast is long and narrow. × 23000. (D) The former material after 6 days in continuous light of high intensity (12000 lux). The stroma lamellae are reduced, but grana are well developed. The shape of the chloroplasts is rounded. × 20000

Plate II A–C: *Pleurozium schreberi* continuous light. (A) A large microbody (mb), typical of cells in continuous light, of high intensity (12000 lux). × 26000. (B) Chloroplast of the same material as in Plate IC and D, but cultured for 6 days in continuous light of lower intensity (6000 lux, 10° C). Welldefined stroma and grana lamellae. *s* = starch. × 20000. (C) A transverse section of the leaf of *Hylocomium splendens*. The material was collected in summer and cultured for 6 days in continuous light (12000 lux, 10° C). Thick-walled (w) cells with very few lipid particles in the cytoplasm. The grana are well developed. *v* = vacuoles, *l* = lipids. × 18000

Plate III A–D: *Hylocomium splendens*. (A) Chloroplast of material collected in spring and cultured for 6 days in rhythmic light (13000 lux, 10° C). Well-defined grana and stroma lamellae. × 23000. (B) The same material after 6 days in continuous light (13000 lux). The stroma lamellae are almost deficient. × 23000. (C) and (D): *Rhacomitrium lanuginosum*, continuous light (12000 lux). (C) A transverse section of the cell showing numerous lipid particles inside the plasmalemma, a typical feature of this species. × 15000. (D) Part of chloroplast with aggregations of plastoglobuli (og) due to the continuous light. × 23000

Acknowledgements. This work is part of the Finnish IBP/PT/Tundra work at Kevo Subarctic Research Station of the University of Turku. The authors are indebted to Mrs. A. ALONEN, Miss K. HILJANEN, Miss H. HURME, Miss H. OJANEN, Mrs. U. M. SUORANTA, and Mr. M. KARUNEN, M.Sc., for technical help.

References

AHMADJIAN, V.: The lichen symbiosis. Waltham, Mass.: Blaisdell 1967.

AHMADJIAN, V.: Adaptation of antarctic terrestrial plants. In: HOLDGATE, M. W. (Ed.): Antarctic ecology, Vol. 2, pp. 801–811. New York: Academic Press 1970.

ARNON, D. I.: Copper enzymes in isolated chloroplasts. Polyphenoloxidase in *Beta vulgaris*. Plant Physiol. **24**, 1–15 (1949).

BRONCHART, R.: Effets de la durée des jours sur la structure du chloroplaste d'épinard. In: SIRONVAL, C. (Ed.): Le chloroplaste, pp. 325–327. Paris: Masson & Cie 1967.

BRULFERT, J., GUERRIER, D., QUEIROZ, O.: Photopériodisme et activité enzymatique: présence en jours longs d'un système inhibiteur d'une voie métabolique active en jours courts. Compt. Rend. **274**, 2671–2674 (1972).

CLARKE, G. C., GREENE, S. W.: Reproductive performance of two species of *Pohlia* from temperate and sub-Antarctic stations under controlled environmental conditions. Trans. Brit. Bryological Soc. **6**, 278–295 (1971).

COSTES, C., BAZIER, R., EVRARD, C.: Recherches sur la stabilité des chloroplastes sous la lumière artificielle: effects de l'éclairement sur les pigments et sur les lipides. Physiol. Végét. **11**, 301–326 (1973).

FAIR, P., TEW, J., CRESSWELL, C. F.: Enzyme activities associated with carbon dioxide exchange in illuminated leaves of *Hordeum vulgare* L. I. Effects of light period, leaf age, and position, on carbon dioxide compensation point (*r*). Ann. Botany **37**, 831–844 (1973).

GANNUTZ, T. P.: Photosynthesis and respiration of plants in the Antarctic Peninsula area. Antarctic J.U.S. **5**, 49–52 (1970).

GANNUTZ, T. P.: Ecodynamics of lichen communities in Antarctica. Res. Antarctic **1971**, 213–226.

GODISCH, T., SCHEIN, R. D.: The ecodynamics of lichens assimilation and respiration. Plant Physiol. Suppl. **37** (1966).

GOLDMAN, C. R., MASON, D. T., WOOD, B. J. B.: Light injury and inhibition in antarctic freshwater phytoplankton. Limnol. Oceanogr. **8**, 313–322 (1963).

HARRIS, G. P., KERSHOW, K. A.: Thallus growth and the distribution of stored metabolites in the phycobionts of the lichens *Parmelia sulcata* and *Parmelia physodes*. Can. J. Botany **49**, 1367–1372 (1972).

KALLIO, P.: The effect of continued illumination on the desmids. Arch. Soc. "Vanamo" **8**, 58–72 (1953).

KALLIO, P., HEINONEN, S.: Influence of short-term low temperature on net photosynthesis in some subarctic lichens. Rep. Kevo Subarctic Res. Stat. **8**, 63–72 (1971).

KALLIO, P., HEINONEN, S.: Ecology of *Rhacomitrium lanuginosum* (Hedw.) Brid. Rep. Kevo Subarctic Res. Stat. **10**, 43–54 (1973).

KALLIO, P., HEINONEN, S.: CO_2 exchange and growth of *Rhacomitrium lanuginosum* and *Dicranum elongatum*. In: WIELGOLASKI, F. E. (Ed.): Fennoscandian tundra ecosystems, Part 1: Plants and microorganisms, pp. 138–148. Berlin-Heidelberg-New York: Springer 1975.

MACHE, R., LOISEAUX, S.: Light saturation of growth and photosynthesis of the shade plant *Marchantia polymorpha*. J. Cell Sci. **12**, 391–401 (1973).

MASUCH, G., WEINERT, H., GUDERIAN, R.: Wirkungen von Chlorwasserstoff und Schwefeldioxid auf die Ultrastruktur der Chloroplasten von *Spinacia oleracea* L. In: Proc. of the Third International Clean Air Congress, **A**, pp. 160–163. Düsseldorf 1973.

NYGAARD, R. T.: Acclimatization effect in photosynthesis and respiration. In: WIELGOLASKI, F. E. (Ed.): Fennoscandian tundra ecosystems, Part 1: Plants and microorganisms, pp. 163–167. Berlin-Heidelberg-New York: Springer 1975.

PARK, R. B., DRURY, S.: The effects of daylength on quantasome structure and chloroplast photochemistry in *Spinacia oleracea* L. Var. Early hybrid 7. In: SIRONVAL, C. (Ed.): Le chloroplaste, pp. 328–334. Paris: Masson & Cie 1967.

QUEIROZ, O.: Circadian rhythms in enzyme activity and phytochrome. In: MITRAKOS, K., SHROPSHIRE, W., JR. (Eds.): Phytochrome, pp. 295–316. London-New York: Academic Press 1972.

RASTORFER, J. R.: Effects of light intensity and temperature on photosynthesis and respiration on two east antarctic mosses, *Bryum argenteum* and *Bryum antarcticum*. Bryologist **73**, 544–556 (1970).

RASTORFER, J. R.: Comparative physiology of four west antarctic mosses. Antarctic Res. Ser. **20**, 143–161 (1972).

RASTORFER, J. R., HIGINBOTHAM, N.: Rates of photosynthesis and respiration of the moss *Bryum sandbergii* as influenced by light intensity and temperature. Am. J. Botany **55**, 1225–1229 (1968).

ROMOSE, V.: Ökologische Untersuchungen über *Homalothecium sericeum*, seine Wachstumsperioden und seine Stoffproduktion. Dansk. Bot. Arkiv **10**, 1–134 (1940).

SCHOFIELD, E., AHMADJIAN, V.: Field oservations and laboratory studies of some antarctic cold desert cryptogams. Antarctic Terrestrial Biology **20**, 97–142 (1972).

SIRONVAL, C., ENGLERT, E.: Action de la durée des jours sur la croissance et le contenu des chloroplastes d'épinard. In: SIRONVAL, C. (Ed.): Le chloroplaste, pp. 335–348. Paris: Masson & Cie 1967.

SONESSON, M., WIELGOLASKI, F., KALLIO, P.: Introduction to Fennoscandian tundra sites. In: WIELGOLASKI, F. E. (Ed.): Fennoscandian tundra ecosystems, Part 1: Plants and microorganisms, pp. 3–28. Berlin-Heidelberg-New York: Springer 1975.

STÅLFELT, M. G.: Der Gasaustausch der Moose. Planta **27**, 30–60 (1937).

STÅLFELT, M. G.: Flechten und Moose. In: RUHLAND, W. (Ed.): Handbuch der Pflanzenphysiol., Vol. V-2, pp. 364–375. Berlin-Heidelberg-New York: Springer 1960.
TALLIS, J. H.: Studies in the biology and ecology of *Rhacomitrium lanuginosum* Brid. II. Growth reproduction and physiology. J. Ecol. **47**, 325–350 (1959).
TAMM, C. O.: Growth, yield and nutrion in carpets of a forest moss (*Hylocomium splendens*). Meddn. St. Skogsforskn. Inst. **43**, 1–140 (1953).
TOLBERT, N. E.: Microbodies-peroxisomes and glyoxysomes. Ann. Rev. Plant Physiol. **22**, 45–74 (1971).
VALANNE, N.: The effects of prolonged darkness and light on the fine structure of *Ceratodon purpureus*. Can. J. Botany **49**, 547–554 (1971).
VANDEN DRIESSCHE, T.: Temporal regulation in *Acetabularia*. In: BRACHET, J., BONOTTO, S. (Eds.): Biology of *Acetabularia*, pp. 53–73. New York-London: Acad. Press 1970.
VERSEGHY, K.: Saisonale Veränderung des Chlorophyllgehaltes einiger xerothermen Flechtenarten. Botan. Közlem. **59**, 109–117 (1972).

Acclimatization Effect in Photosynthesis and Respiration

R. T. NYGAARD

MOONEY et al. (1964) found that plant growth at high temperatures (HT) increased apparent photosynthesis (AP) by 2–3 times in several alpine species from California, at the same time as dark respiration (DR) was reduced by 10–20%. All plants were grown at an artificial 20 hrs photoperiod and 16–26° C. The increase in AP was correlated to higher chlorophyll content. Similar results were obtained by PEARSON and HUNT (1972) on alfalfa and MOONEY and SHROPSHIRE (1967) on *Polygonum* and *Encelia*. The dark respiration at 15° C in alfalfa was slightly higher in low temperature (LT) plants than in HT plants, but Q_{10} was higher after HT treatment. On the other hand, BILLINGS and GODFREY (1968) found that AP, as well as dark respiration, was higher after growing alpine and arctic populations of *Oxyria digyna* under LT then under HT treatment. The increase in DR was large enough to bring about a 10° C decrease in the optimum temperature of AP from 30° to 20° C. Acclimatization effects are also reported by BÜTTNER (1971) and STÅLFELT (1960) in cryptogams. Temperature pre-treatment is thus important to ecophysiology of plants. When photosynthesis and respiration of vegetation from different sampling sites are measured, the temperature history of the communities often varies and should preferably be known.

Methods

Vaccinium myrtillus, *Vaccinium vitis-idaea* and *Deschampsia flexuosa* were used in the experiments. Whole mats of vegetation were taken out in the field during summer and autumn, and placed in two Vötsch climatic chambers for temperature-acclimatization in one to four days. Material pre-treated at low temperatures had 8–12 hrs periods in the dark at +2–3° C and 12–16 hrs periods in the light at +10° C. Material pre-treated at high temperatures had 8–12 hrs periods in the dark at +15° C and 12–16 hrs periods in the light at +23–25° C. The light source was ordinary tubes (Philips TL 113), giving an intensity at the plant level of 8–9000 lux. The relative humidity was about 80%.

Gas-exchange measurements were carried out just after acclimatization with an infrared gasanalyzer, URAS 2, from Hartmann and Braun (sensitivity range 0–0.05 Vol-% CO_2). Plant material was cut and placed in plastic containers with water, and inserted into assimilation chambers. The chambers were lowered into a thermostatic water bath. The same type of light source was used as for acclimatization, and the intensity on the plant level was 13500 lux (standard luxmeter Dr. B. Lange). The humidity approached 100% in the plant chambers, and the speed of the air-stream was 60 liters per hour.

Results and Discussion

The experiment with *V. myrtillus* carried out in August (Table 1) shows that even 24 hrs temperature pre-treatment is enough to give an acclimatization effect on the apparent photosynthesis. A higher photosynthesis activity is shown by high-temperature acclimatized material (HT) than material acclimatized at low temperatures (LT). In acclimatization experiments lasting 4 days and carried out in early July, the photosynthesis activity of the HT-material is three to four times higher than the activity of the LT-material. Dark respiration is highest in low temperature pre-treated material. The real gross assimilation cannot be calculated, as the photorespiration activity is unknown.

The apparent photosynthesis in *V. vitis-idaea* and *D. flexuosa*, pre-treated at high temperatures, is also found to be higher than for low-temperature pre-treated plants (Fig. 1). The photosynthesis optimum for HT-material is found for both species at higher temperatures (about 15° C) than for the LT-material (7–8° C). The results for the two species agree well, but the difference in photosynthesis after HT and LT acclimatization respectively, seems to be greatest for the monocotyledonous species studied. The acclimatization effect for *V. vitis-idaea* was smaller in experiments carried out in the summer than in the autumn studies. The reason may be the different physiological status of the plants at these times.

Table 1. Vegetation mat with *Vaccinium myrtillus*, temperature acclimatized for 96 and 24 hrs. Dark respiration and apparent photosynthesis in mg CO_2 per g dry weight per hour. "Low" temperature pre-treated (LT), 12 hours at +3° C and 12 hrs at 10° C. "High" temperature pre-treated (HT), 12 hrs at 15° C and 12 hrs at 25° C. Exposed to artificial light for 12 hours, light intensity 8000–9000 lux, during acclimatization. Gas exchange measured on cut plants at 12° C, photosynthesis measured at light intensity 13500 lux

	Period	Dark respiration		Apparent photosynthesis		Apparent photosynthesis
		HT	LT	HT	LT	HT/LT
Long period acclimatization (96 hrs)	Early July	0.69	1.15	4.03	1.47	2.8
	Early July	0.62	0.91	3.33	0.77	4.3
	Late Aug.	0.41	0.61	2.50	1.85	1.4
	Late Aug.	0.54	0.71	2.88	1.92	1.5
Short period acclimatization (24 hrs)	Late Aug.	0.60	0.66	2.21	1.73	1.3

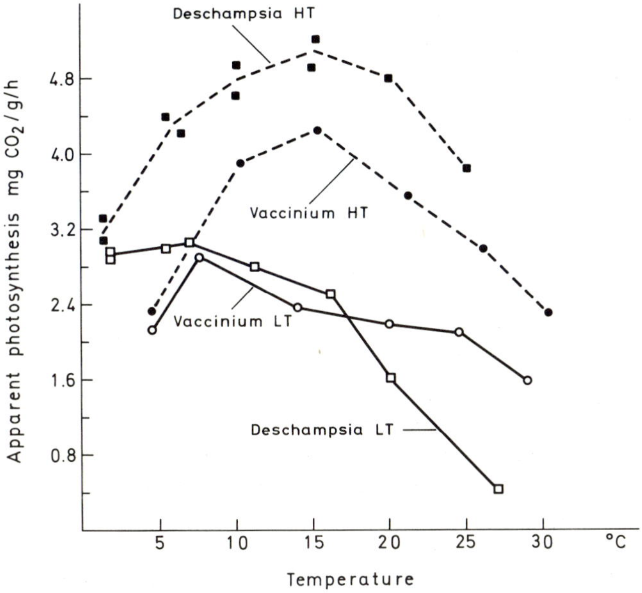

Fig. 1. Apparent photosynthesis of *Vaccinium vitis-idaea* and *Deschampsia flexuosa* at different temperatures. The first species studied after acclimatization for 96 hrs at "low" temperature (LT) +2° C in the dark for 8-hrs periods and 8° C in 1000 lux light intensity for 16 hrs periods and at "high" temperatures (HT) 15° C in the dark for 8 hrs periods and 23° C in 1000 lux light intensity for 16 hrs periods. The second species studied after acclimatization for 72 hrs at LT +2° C in the dark for 12 hrs periods and 10° C in 9300 lux light intensity for 12-hrs periods, and at HT 15° C in the dark for 12-hrs periods and 25° C in 9300 lux light intensity for 12-hrs periods. Relative humidity during acclimatization approx. 80%. Photosynthesis at relative humidity approx. 100%, and light intensity 13500 lux

On analyzing the CO_2-compensation points (Ccp) as a function of the temperature (July) for *V. vitis-idaea*, the curve for LT-material showed somewhat higher values than the HT-material. These results agree with those from the experiment showing the effect of temperature acclimatization in photosynthesis in *V. vitis-idaea*. The differences between the Ccp-curves in Fig. 2 are, however, too small to be significant, while the differences between "untreated" and temperature-treated materials are distinct and may indicate an effect of temperature acclimatization on the photorespiration and/or photosynthesis activity. The increasing Ccp-values with increasing temperature could have been caused by an increasing photorespiration and/or decreasing photosynthesis activity. The experimental time for the Ccp-measurements was about 48 hrs. A possible acclimatization effect may then have been weakened in the course of the time taken for the experiment.

Figure 3 shows dark respiration as a function of temperature in *D. flexuosa*, measured directly after a period with temperature acclimatization. The respiration experiments show, as for *V. myrtillus*, a somewhat higher activity for LT pre-treated than for HT pre-treated plants. When the respiration studies are carried out after completed photosynthesis measurements (which may take 24 hrs), the acclimatization effect seems to disappear. The higher dark respiration activity in

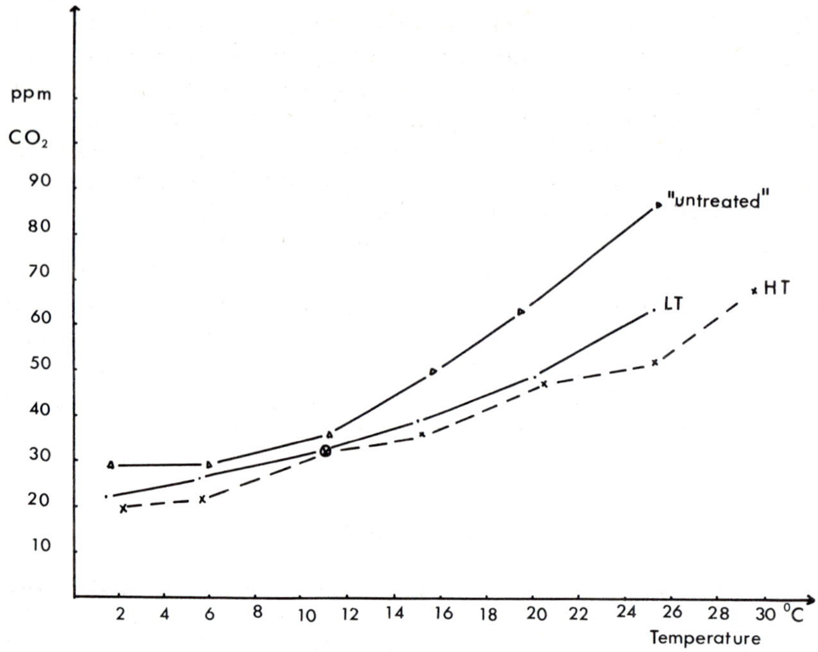

Fig. 2. CO_2-compensation point measurements of *Vaccinium vitis-idaea* (July 1972). Acclimatization as mentioned for *Deschampsia flexuosa* (Fig. 1). *LT*, "low" temperature; *HT*, "high" temperature

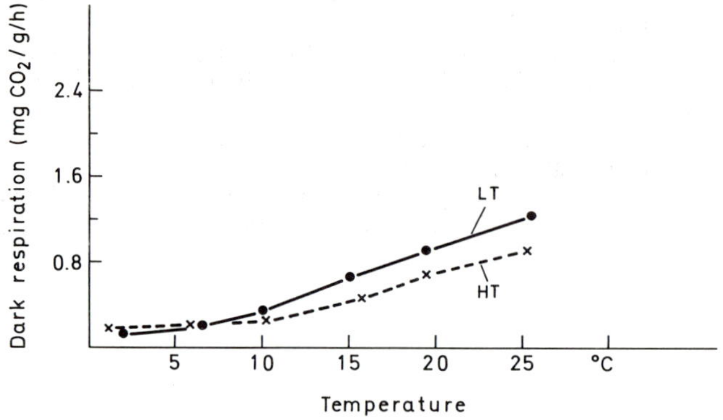

Fig. 3. Dark respiration of *Deschampsia flexuosa* at different temperatures. Acclimatization as mentioned for the same species in Fig. 1

LT pre-treated material cannot explain the difference in the photosynthesis activity between HT- and LT-material of *D. flexuosa*. FORRESTER et al. (1966) conclude from their experiments with soybean leaves that photorespiration replaces dark respiration in light. Further, HOLMGREN and JARVIS (1967) say that the influence of temperature is different on photorespiration and dark respiration.

The conclusion from the present experiments is that the temperatures to which the plants are exposed before gas-exchange measurements are very important for the degree of photosynthesis or respiration activity. Thus, for some species at Hardangervidda, a higher respiration activity was shown at the sites in the low-medium alpine zone 200–300 m above the tree line (1200–1300 m) than for the same species in the subalpine birch forest (800 m). The reason for the difference may be a different temperature acclimatization of the plant material before the respiration was measured by the infrared gas analyzer in a field laboratory.

Other experiments have shown that an acclimatization effect on the rate of apparent photosynthesis may also be seen in *Phleum pratense* and in shoots of *Picea abies* (I. P. BJØRNSETH, pers. comm.). Several of the species mentioned differ greatly, both morphologically and systematically. It is possible, therefore, that a temperature acclimatization effect on the photosynthesis and respiration of higher plants is a general physical/chemical phenomenon.

References

BILLINGS, W. D., GODFREY, P. J.: Acclimation effects on metabolic rates of arctic and alpine *Oxyria* populations subjected to temperature stress. Bull. Ecol. Soc. Am. **49**, 68–69 (1968).

BÜTTNER, R.: Untersuchungen zur Ökologie und Physiologie des Gasstoffwechsels bei einen Strauchflechten. Flora **160**, 72–99 (1971).

FORRESTER, M. L., KROTKOV, G., NELSON, C. D.: Effects of oxygen on photosynthesis, photorespiration and respiration in detached leaves. Plant Physiol. **41**, 422–427 (1966).

HOLMGREN, P., JARVIS, P. G.: Carbon dioxide efflux from leaves in light and darkness. Phys. Plant. **20**, 1045–1051 (1967).

MOONEY, H. A., SHROPSHIRE, F.: Population variability in temperature related photosynthetic acclimation. Oecologia **2**, 1–13 (1967).

MOONEY, H. A., WRIGHT, R. D., STRAIN, B. R.: The gas exchange capacity of plants in relation to vegetation zonation in the White Mountains of California. Am. Midland Naturalist **72**, 281–297 (1964).

PEARSON, C. J., HUNT, L. A.: Effects of pretreatment temperature on carbon dioxide exchange in alfalfa. Can. J. Botany **50**, 1925–1930 (1972).

STÅLFELT, M. G.: Flechten und Moose. In: RUHLAND, W. (Ed.): Handbuch der Pflanzenphysiologie, Vol. 2, pp. 364–375. Berlin-Heidelberg-New York: Springer 1960.

CO_2 Exchange in Norwegian Tundra Plants Studied by Infrared Gas Analyzer Technique

O. Skre

The aim of the present investigation is to estimate the primary production in terms of net or apparent photosynthesis (AP) at the six IBP tundra sites at Hardangervidda described by Sonesson et al. (1975). Instead of the snowbed at Finse a locality near the lichen heath at Stigstuv was chosen. The climatic conditions at the willow thicket were close to those measured at Stigstuv (Skartveit et al., 1975).

Estimation of plant photosynthetic production may be carried out in different ways, as described by Sestak et al. (1971). In IBP studies field methods are often used (Kjelvik et al., 1975; Johansson and Linder, 1975; Mayo et al., 1972; Tieszen, 1972) because they are the most reliable to the aim of the project, i.e. estimation of primary production in natural environments. However, the great distance between the field sites imposed the use of our infrared gas analyzer in field. For the present study an arrangement was chosen in which measurements could be done on intact leaves and branches in a climatically controlled greenhouse or phytotrone.

The pretreatment of plants may be of great importance for the CO_2 exchange (Nygaard, 1975). Light conditions may cause anatomical and physiological differences adjusting the leaves to low or high light intensity levels (Björkman and Holmgren, 1963). These adjustments may influence light as well as temperature reactions. Billings and Godfrey (1968) found that the effect of a preceeding high or low temperature was greater than the genotypic difference between *Oxyria digyna* populations. Surveys of the temperature related photosynthetic reactions are given by e.g. Mooney and Shropshire (1967).

It has been shown recently that respiration in light may be as high as 2–5 times the dark respiration. The low Q_{10} of apparent photosynthesis may be explained by the Q_{10} of photorespiration being higher than that of photosynthesis (Zelitch, 1966). A rough estimate of the photorespiration (PR) is found by circulating the air through the plant chambers in a closed system and measuring the CO_2 compensation value (Ccp). In plants possessing PR Ccp may be as high as 50 p.p.m. Reviews are given by Jackson and Volk (1970) and Berner (1972).

In addition to the photosynthesis and respiration of leaves and stems, root respiration is of great interest. Roots may compose 80–90% of the live biomass in tundra (Wielgolaski, 1972), and most of this is located in the upper 10 cm of the soil BILLINGS et al., 1973; WIELGOLASKI and KJELVIK, pers. comm., 1974). The total carbon losses from tundra soils (SVENSSON et al., 1975) originate from live roots and microorganisms.

Material and Methods

Vegetation mats of the size 40 × 80 cm were cut out to 15 cm depth in field and transferred to isopor boxes of equal size, four parallells from each site. The samples were allowed to equilibrate for two weeks after cutting and then transported to the phytotrone where they were placed in a daylight cell at 15/9° C (15/6° C for the last sample) and with high temperature periods of 6 hrs. These temperatures were as close to those naturally occurring at Hardangervidda as possible. However, the light quality may be different as a result of the great altitudinal difference between Oslo and Hardangervidda. The sampling was repeated three times during the 1973 season with roughly 5 weeks intervals. The dates of transport were July 1 (N_1), August 4 (N_2) and September 9 (N_3). In the daylight cell air humidity was chosen at 80% and the soil was kept at field capacity by means of a tensiometer. From time to time sample boxes were taken out for 48 hrs acclimation at 10000 lux, 18 hrs photoperiod and high (18/12° C) or low (9/6° C) day/night temperature treatment (HT and LT) before CO_2 exchange measurements. For dates see Table 1. The relatively low HT treatments was chosen because meteorological data (SKARTVEIT et al., 1975) seem to indicate that air temperatures above 20° C are rather infrequent at Hardangervidda. Small differences (usually less than 3° C) are also found between leaf and air temperatures in tundra, as reported by BLISS (1966).

A plant sampling (L) was also done in August the previous year and kept at "lowland" conditions during the winter season. On March 20 all boxes were moved to a daylight greenhouse with day temperatures ranging from 15° C at the end of March to 30° C in mid-June and night temperatures from 10 to 18° C. Air humidity ranged from 70 to 40% and soil moisture was kept at field capacity. This vegetation was used as an object of gas exchange measurements on single species after flowering in June and early July.

On April 1 some of these plants were carefully transferred from the sampling boxes to plastic pots filled with sand and modified Hoagland solution (ARNON, 1938; HEWITT, 1966) in ratio 5:1. The plants were kept at a constant photoperiod of 18 hrs and 80% humidity. The day and night temperatures were 15° and 9° C and the light intensity at pot level 20000 lux. For light quality see LINDER (1972). After growing in pots for 3–5 months some plants (S) were taken out for measurements of CO_2 exchange and harvested. Other plants (W) were transferred to Hoagland solution and equilibrated for 1 week before root respiration measurements and harvesting.

For the experiments five square 16 × 16 × 36 cm (Volume 9/1) cuvettes made of 4 mm plexiglass were constructed (Fig. 1). Single vascular plants and transections of plants could be included through a 4 mm slit in the bottom of the cuvette before sealing the "door". An air stream of $1\ l\ min^{-1}$ and 350 p.p.m. CO_2 was forced through the cuvette. A rotating fan was mounted near the air intake to

Fig. 1. Cuvette with mounted vegetation

prevent overheating of leaves and CO_2 depletion at leaf surface (Björkman and Holmgren, 1963). The wind speed was about 1 m sec^{-1} at the vegetation surface. All photosynthetic studies were carried out in artificial light of the same quality as described by Linder (1972). Temperature differences within the cuvettes did not exceed 1° C. The air temperature ranging from 6 to 25° C can be taken as leaf temperature at the actual air humidity (75–85%) and wind speed. By means of a cooling device lowering of temperatures could be done quickly. CO_2 exchange was measured in darkness and at 20000 lux at 6°, 10°, 15°, 20°, and 25° C; at 6° and 20° also at light intensities of 500, 2000, 5000, 10000 and 25000 lux.

In experiments on *Alchemilla* the equilibration time for apparent photosynthesis in the present study was found to be 20 min after change from darkness to 20000 lux, and 60 min are needed to establish a constant respiration level (Fig. 2). The equilibration times were chosen at 5 min for small changes in light intensity, 20 min for changes from darkness to maximum light intensity and *vice versa* and 30 min for each 5° C change in temperature. Vascular plants were supplied with water and cryptogams carefully moistened before each experiment to maintain full turgor. The humidity of incoming air was kept at 60–70% increasing to 85–90% in outgoing air due to transpiration. A complete experiment lasted for 6 hrs and within this period usually no drift in AP was found. The cuvettes were allowed to stand overnight in a closed system at 15° C and 10000 lux in order to find the CO_2 compensation value (Ccp). Then the plant tissue inside the cuvettes was harvested, dried at 80° C and separated into species and green, non-green and standing dead tissue to calculate the CO_2 exchange per g biomass. The number and size of leaves, stems and shoots were also recorded for each species. All photosynthetic measurements were referred to the amounts of green tissue (leaves

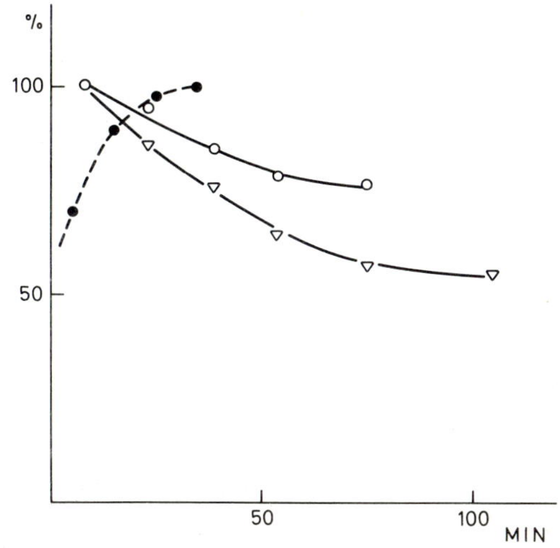

Fig. 2. Times of equilibration after mounting plants for root respiration measurements (open circles), top dark respiration measurements (open triangles) and apparent photosynthesis after change from dark to light (filled circles). All in percentage of maximum values

and green stems) while dark respiration was calculated in relation to the entire standing crop. The nitrogen content (SKRE et al., 1975) was used as a reference in order to correct for different metabolic capacity. The number of parallells was 8 in the case of vegetation transections and 4 in the case of single species. The standard deviation between parallell readings was usually of the magnitude $\pm 10\%$.

Root respiration was studied on plants grown in sand and transferred to 2 l suction bottles filled with Hoagland solution. After equilibration for one week the bottles were sealed, air was bubbled through the solution and sucked out at a rate of 1 l min^{-1}. Temperature control was achieved by emerging the bottles in a thermostate water bath. The roots were allowed to stay for 55 min at constant temperature (10° C) to equilibrate (Fig. 2), then the temperature was increased in steps and new readings were taken at 15°, 20°, and 25° C. At last a CO_2 exchange measurement of the top at 15° C and 20000 lux and in darkness was carried out. The total experiment lasted for 3 hrs. As reference a suction bottle filled with nutrient solution was used.

For biomass data and further details of CO_2 exchange measurements and the methods used, see SKRE (1975). 1000 lux correspond to 0.013 cal cm^{-2} min^{-1}.

Results and Discussion

CO_2 Exchange at 15° C. To compare different values of CO_2 exchange a certain pretreatment and stage of development must be selected. Maximum rates of photosynthesis are usually achieved by high temperature (HT) pretreatment (NYGAARD, 1975) and when plants are in the flowering stage (HADLEY and BLISS, 1964). Consequently, the HT treatment (18/12° C) and stage N_1 or S were chosen

Table 1. CO_2 exchange data at 15° C for plants transported from Hardangervidda on phytotron chambers before measurements: F = field site, D = date of experiment, + and AP=apparent photosynthesis at 20000 lux and 15° C. E_n=compensation light in 1000 lux AP + DR. DR = dark respiration at 15° C. AP and DR are

F	Plants	N_1-plants									
		D	AP	E_n 6°	E_n 20°	E_x	GP Q_{10}	AP t_{opt}	DR	DR Q_{10}	Ccp
DM	Vascular	+ July	11.3	0.6	2.0	30	1.4	13	3.7	2.2	48
		− 22	11.6	0.7	2.4		1.1	13	3.2		64
WT	Vascular	+ July	15.9	0.6	2.0	20	1.3	13	3.2	2.2	54
		− 15	11.0	0.8	3.2		1.6	18	3.0		61
	Moss	+	2.5	0.4	1.0	15	1.3	14	1.0	1.8	
		−	0.6	1.0	6.0		1.2	12	0.8		
WM	Vascular	+ July	13.6	0.8	3.8	22	1.3	22	4.8	2.1	
		− 20	15.2	0.8	3.6		1.7	17	3.8		
	Moss	+	4.0	0.3	3.0	22	1.4	14	2.1	2.2	
BF	Vascular	+ July	7.9	0.6	1.4	23	1.3	23	1.9	1.9	
		− 18	5.7	0.8	3.0		1.5	20	2.6	2.1	
	Moss	+	1.1	0.2	1.5	10	1.2	14	1.2	1.9	
LH	Vascular	+ July	2.5	0.8	3.2	20	1.2	14	0.8	1.8	
		− 27	1.8	1.4	3.4		1.3	16	0.8		
	Lichen	+	0.3	0.5	6.0	35	1.0	13	0.2	2.6	
		−	0.1	4.5	[b]		1.2	7	0.7	2.3	
SB	*Salix*	+ Aug.	11.3	0.4	1.4	26	1.3	15	1.6	2.0	
	herbacea	− 15	11.5	0.4	4.5		1.5	16	1.7		
BF	*Vacc. myrt.*	+ July	9.5	0.3	1.0	20	1.3	27	2.1	1.6	
	leaves	− 18	8.0	1.2	3.5		2.0	23	2.9	2.1	
WM	*Carex*	+ July	12.5	0.3	1.8	22	1.3	22	4.5	1.8	66
	nigra	− 20	16.5	0.4	3.5		2.0	17	3.6		67

[a] Green stems. [b] No positive AP

as standard conditions. The HT treatment is also equivalent to a 15° mean temperature, where apparent photosynthesis (AP) usually is highest.

At these conditions AP at 20000 lux ranged from 16 mg CO_2 $g^{-1}h^{-1}$ in transections of vegetation from the willow thicket to 2.5 mg CO_2 $g^{-1}h^{-1}$ at the lichen heath (Table 1). The dark respiration (DR) measured at 15° C ranged from nearly 5 mg CO_2 $g^{-1}h^{-1}$ at the wet meadow to less than 1 mg CO_2 $g^{-1}h^{-1}$ at the lichen heath.

The apparent photosynthesis in HT sand culture plants of *Festuca rubra* and *Deschampsia flexuosa* grown at 15/9° C and 18 hrs photoperiod in artificial light decreased from the first (S) to the second harvest (W) which was done one month later, while *Ranunculus acris*, *Geum rivale*, *Alchemilla glabra* and *Salix lapponum* mostly maintained their photosynthetic capacity when measured at 20000 lux and 15° C (Table 2).

CO_2 exchange measurements of the L plants grown at high temperature showed a small decrease in AP caused by ageing in *Festuca rubra*, *Deschampsia flexuosa* and *Salix lapponum* relative to the *S* plants (Tables 2 and 3). The dark respiration (DR) at 15° C was usually highest in the L plants, probably as a result of ageing. The values of DR on single species in the present study corresponded well with those reported by SCOTT and BILLINGS (1964), HADLEY and BLISS (1964)

July 1st (N_1), August 4th (N_2) and September 9th (N_3) and placed in cooled daylight – indicate high (18/12° C) and low (9/6° C) day/night temperature pretreatment, respectively. at 6° and 20° C. E_x = saturation light in 1000 lux at 15° C. GP = "gross" photosynthesis, measured in mg CO_2 g^{-1} h^{-1}. See Figs. 3a–b for field site symbols.

N_2-plants									N_3-plants								
D	AP	E_n 6°	E_n 20°	E_x	GP Q_{10}	AP t_{opt}	DR	DR Q_{10}	D	AP	E_n 6°	E_n 20°	E_x	GP Q_{10}	AP t_{opt}	DR	DR Q_{10}
Aug.	9.2	2.0	3.5	22	1.2	13	3.2	1.6	Oct.	7.0	0.8	8.2	20	1.4	20	3.7	1.5
24	8.5	2.0	3.3		1.4	15	2.4		8	6.6	1.2	10.5		1.5	20	4.4	
Aug.	9.2	0.5	1.8	32	1.3	17	1.3	1.7	Sept.	8.4	0.0	2.2	24	1.2	15	3.1	1.7
14	7.4	0.6	2.0		1.5	18	1.4		25	5.6	0.5	2.0		1.2	17	1.8	
	4.2	1.0	20.0	30	1.3	19	1.1	3.0		0.7	2.2	7.5	15	1.3	20	0.9	1.5
	3.5	1.2	[b]		1.5	15											
Aug.	8.2	0.5	3.0	26	1.5	15	1.0	1.9	Oct.	5.9	0.2	0.4	25	1.6	20	1.4	2.2
28	9.8	1.5	4.0		1.6	14	1.1		1	5.5	0.3	0.5		1.5	17	1.3	
Aug.	4.3	0.4	1.8	28	1.5	20	0.8	2.4	Sept.	2.7	0.4	3.5	30	1.3	21	0.7	2.0
20	4.1	0.5	6.0		1.7	18	1.2		22	2.2	0.7	5.5		1.5	19	0.9	
Sept.	2.4	0.6	1.3	27	1.1	17	0.3	1.7	Oct.	1.8	0.5	1.2	20	1.5	20	0.3	2.4
1	2.5	2.2	4.5		1.3	13	0.5		14	1.5	2.0	2.8		2.0	19	0.6	
	0.2	2.0	10.0	30	1.5	8	0.4	2.8		0.2	0.5	8.5	25	1.1	10	0.4	2.4
	0.1	1.5	[b]			5	0.4			0.2	0.5	4.5		1.2	7	0.3	
Sept. 18	9.5	0.2	2.0	25	1.3	16	2.6	2.5									
Aug.	11.4	0.3	1.2	20	1.7	21	0.8	1.9	Sept.	1.1[a]	3.0	10.0	35	1.7	22	0.8	1.8
20	11.4	1.1	3.5		1.9	19	1.6		25	0.5	5.0	18.0		1.2	22	0.8	
Aug.	8.4	1.0	3.3	28	1.5	18	0.9	2.5	Oct.	7.7	0.3	1.8	26	1.3	21	1.6	2.1
28	8.4	2.2	4.5		1.5	17	1.5		1	8.3	0.2	2.6		1.5	18	1.5	

and MOORE et al. (1973). In *Salix lapponum* and *Dryas octopetala* grown at high temperatures, considerably higher values were found than reported by NYGAARD (pers. comm., 1972) probably again as a result of ageing.

SCOTT and BILLINGS (1964), HADLEY and BLISS (1964) and MOORE et al. (1973) found about half as high maximum AP values in American *Geum* species as in the present *Geum rivale* (Tables 2 and 3). The maximum AP values for *Ranunculus glacialis* are found to be lower in the Alps (MOSER, 1973) than for the present *R. acris*, while DOUGLAS JOHNSON (pers. comm., 1973) found somewhat higher AP in *R. nivalis* from the Rocky Mountains. AP values of *Carex stans* (MAYO et al., 1972) and *C. aquatilis* (TIESZEN, 1972) correspond well with the present values of *C. nigra* (Table 1). The maximum photosynthesis in different species of *Salix* from tundra areas varies strongly, e.g. in Alaskan *S. pulchra* 12 mg CO_2 $g^{-1}h^{-1}$ (TIESZEN, 1972), in *S. arctica* from western Taimyr 25 mg CO_2 $g^{-1}h^{-1}$ (SHVETSOVA and VOZNESENSKII, 1970), in *S. glauca* from Wrangel Island 6 mg CO_2 (GERASIMENKO and ZALENSKY, 1973), in *S. lapponum* in the present study 17.8 mg CO_2 and in *S. herbacea* 11.3 mg CO_2 $g^{-1}h^{-1}$ (Table 1). The environmental conditons, however, varied in the different studies. The present low value of *Empetrum hermaphroditum* correspond well with the results of TIESZEN (1972) when stem biomass is accounted for, but are higher than the maximum values found by HADLEY and

Table 2. CO_2 exchange data for plants grown in sand culture for 4–6 months at 15°/9° C day/night temperatures and 20000 lux (S), and for S plants after 1 week in nutrient solution (W), RR = root respiration. See Table 1 and Fig. 3 for other symbols

			S-plants								W-plants			
			AP	E_n 6°	20°	E_x	GP Q_{10}	AP t_{opt}	DR	DR Q_{10}	AP	DR	RR	RR Q_{10}
BF	*Deschampsia*	+	12.1	0.2	1.2	22	1.2	14	1.0	1.8			0.5	1.7
	flexuosa	–	9.9	0.8	1.8		1.2	17	1.9		2.5	0.5	0.8	
WM	*Festuca*	+	15.9	0.3	1.2	24	1.2	22	1.7	1.7				
	rubra	–	12.9	0.4	1.2		1.6	20	1.6		5.9	1.0	1.1	2.5
WT	*Geum*	+	10.4	0.5	1.2	20	1.1	13	1.9	2.2				
	rivale	–	6.3	0.9	5.0		1.6	18	3.8		8.0	2.1	0.9	2.0
DM	*Ranunculus*	+	13.6	0.4	2.7	20	1.2	13	2.3	2.0				
	acris	–	9.0	0.4	3.7		1.1	17	2.9		6.5	1.5	2.1	1.4
DM	*Alchemilla*	+	6.2	0.3	2.5	20	0.8	13	2.9	1.8				
	glabra	–	5.1	1.1	5.0		1.3	15	2.6	2.2	6.1	3.3	1.6	1.5
WM	*Salix*	+	17.8	0.4	4.7	30	1.5	14	3.3					
	lapponum	–	9.7	0.9	4.5		1.4	16	3.5		6.7	1.0	1.5	2.0
	Thalictrum	–				25					6.5	2.1	3.8	1.6
	Saussurea	–									12.5	9.8[a]	2.7	1.6
	Salix herb.	–									5.9	1.5	0.6	1.8
	Vacc. v-i.	–									3.9	3.2	2.1	2.0
	Carex nig.	–									1.8	1.1	1.1	2.5

[a] Wounding effect.

BLISS (1964) and JOHANSSON and LINDER (1975), perhaps as a result of higher air humidity. Rather high values are found in *Vaccinium vitis-idaea* and *Poa alpina* (Table 3) as compared to those reported by HADLEY and BLISS (1964) and SCOTT and BILLINGS (1964), but the values for *V. vitis-idaea* and *V. myrtillus* agree well with the AP values of detached shoots of these species found by NYGAARD (1975) when woody tissue is accounted for. The maximum AP in *Dryas octopetala* in the present experiment correspond well with Canadian high arctic studies in a closely related species (MAYO et al., 1973) and also with studies in western Taimyr (SHVETSOVA and VOZNESENSKII, 1970).

The rates of root respiration at 15° C of water culture (W) plants pretreated at LT (9/6° C) range from about 0.5 mg CO_2 $g^{-1}h^{-1}$ to nearly 4.0 mg (Table 2). High values were found for the forbs *Thalictrum alpinum*, *Ranunculus acris*, and *Saussurea alpina* while the tuberous roots of *Geum rivale* and *Alchemilla glabra* had low repiration rates. Low values were also found in monocotyledons, and these rates are in good agreement with the findings of BILLINGS et al. (1973) on arctic monocotyledons. Woody plants showed in the present experiment low *(Salix herbacea)* to intermediate root respiration (Table 2). The values of root respiration were partly lower and partly higher than the corresponding leaf and stem dark respiration.

In cryptogams the present values of AP at 15° C and 20000 lux after HT pretreatment seem to correspond well to earlier readings. In agreement with BÜTTNER (1971) somewhat higher values (0.35 and 0.28 mg CO_2 $g^{-1}h^{-1}$) were found in *Cetraria nivalis* than in *Cladonia* species from the lichen heath, while the

Table 3. CO_2 exchange data for overwintered plants grown at "lowland" conditions (daylight, 25–30° C day temperature, 10–18° C night temperatures). Plants are in the post-flowering stage (L). See Table 1 and Fig. 3 for symbols

			D	AP	E_n 6°	20°	E_x	GP Q_{10}	AP t_{opt}	DR	DR Q_{10}	Ccp
WM	*Carex*	+	June	15.4	0.6	4.0	30	1.6	17	4.6	2.0	
	nigra	−	5	8.6	0.4	5.5		1.2	19	3.4		
DM	*Anthoxanthum*	+	June	12.4	0.3	0.8	20	1.5	20	2.0	2.2	
	alpinum	−	7	11.0	0.3	1.8		1.4	22	1.5		
DM	*Poa*	+	July	12.1	0.8	0.4		1.4	17	1.5	2.0	
	alpina	−	6	10.5	1.4	3.5		1.5	15	4.2		
DM	*Carex*	+	July	12.7	0.6	1.5		1.4	20	2.1	1.7	
	bigelowii	−	5	8.8	1.2	4.6		1.3	15	4.4		
BF	*Vaccinium*	+	June	8.5	0.2	0.4		1.2	22	1.2	1.8	
	myrtillus	−	8	5.7	1.0	1.6		1.6	23	1.1		
BF	*Vaccinium*	+	June	5.0	2.0	3.0		1.3	15	1.4	1.8	
	vitis-idaea	−	23	6.7	1.8	3.0		1.7	18	1.0		
LH	*Empetrum*	+	July	4.8	0.8	4.5		1.1	8	1.1	1.9	
	hermaphroditum	−	2	3.7	2.8	12.0		1.7	12	1.1		
SB	*Salix herbacea*	+	June	15.1	1.1	4.0	20	1.7	19	3.6	2.4	
		−	27	7.1	3.5	10.0		2.2	22	2.2		
DM	*Dryas*	+	July	8.2	1.5	4.0	30	1.4	15	2.8	1.8	
	octopetala	−	5	6.8	2.5	3.5		1.4	15	2.3		
BF	*Deschampsia*	+	June	7.3	0.3	6.0	25	1.4	17	3.4	2.2	40
	flexuosa	−	3	6.8	2.1	7.2		1.5	18	4.2		50
WM	*Festuca*	+	June	11.1	1.2	3.1	30	1.0	9	2.9	1.9	35
	rubra	−	10	8.6	1.0	7.8		1.6	12	4.3		55
WT	*Geum*	+	June	9.3	0.3	4.5		1.4	12	2.7	2.5	48
	rivale	−	4	6.9	0.2	13.0		1.7	15	4.3		68
DM	*Ranunculus*	+	June	17.4	0.9	3.5	20	1.3	15	2.6	2.8	45
	acris	−	25	14.7	0.6	2.5		1.2	20	2.3		75
DM	*Alchemilla*	+	June	10.4	0.9	4.0	28	1.3	13	1.6	3.2	52
	glabra	−	24	11.1	0.6	3.5		1.8	15	2.9		75
WM	*Salix*	+	June	10.0	0.2	2.2		1.4	20	3.8	1.7	
	lapponum	−	1	9.8	0.4	6.4		1.9	20	3.7		

dark respiration was lower (0.15 and 0.25 mg CO_2 $g^{-1}h^{-1}$). The AP values of lichens fit well with the results of KALLIO and KÄRENLAMPI (1975) from Northern Finland, BLISS and HADLEY (1964) from USA and LANGE (1965) from the Alps, who all found AP values ranging from 0.3 to 0.5 mg CO_2 $g^{-1}h^{-1}$ in these species. The present AP as well as the dark respiration values fit well with earlier findings in the Hardangervidda area (NYGAARD, pers. comm., 1972). In bryophytes the results vary widely. The dominant genera are *Polytrichum* and *Sphagnum* at the willow thicket (KJELVIK and KÄRENLAMPI, 1975), *Philonotis*, *Mnium*, and *Bryum* at the wet meadow and *Pleurozium schreberi* in the birch forest (LYE, 1972). The maximum AP values vary from about 1 mg CO_2 $g^{-1}h^{-1}$ in the birch forest to above 4 mg CO_2 $g^{-1}h^{-1}$ at the wet sites (Table 1). These values are all within the range of those found in the literature (e.g. KALLIO and KÄRENLAMPI, 1975; MAYO et al., 1972). The dark respiration in *Pleurozium schreberi* seems to be rather high

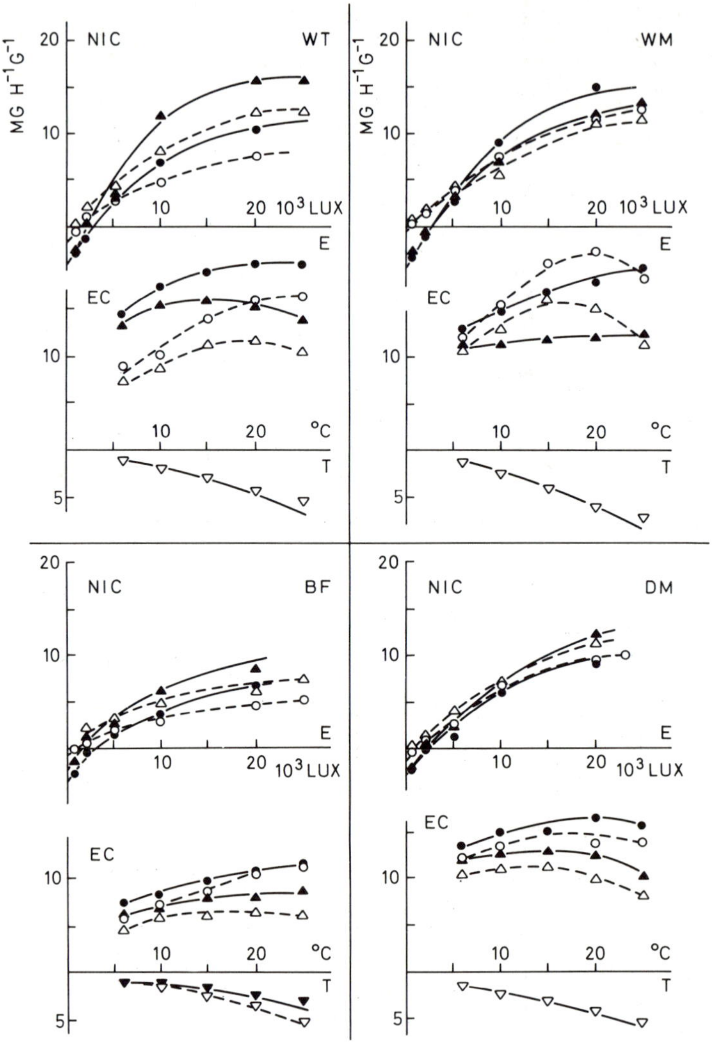

Fig. 3a

Fig. 3a and b. Net input of CO_2 (*NIC*) and CO_2 exchange (*EC*) as functions of light intensity (*E*) and temperature (*T*) for different sites at the first sampling in July (N_1). *WT* willow thicket, *WM* wet meadow, *BF* birch forest, *DM* dry meadow, *LH* lichen heath (*Empetrum hermaphroditum*), *SAL Salix herbacea* from the snowbed (*SB*), *L* lichens from the lichen heath, *BW* bryophytes from the willow thicket. *NIC* as a function of light: At 6° C after low temperature (*LT*) pretreatment (open circles) and high temperature (*HT*) pretreatment (open triangles) and at 20° C after *LT* (filled circles) and *HT* (filled triangles). *EC* as a function of temperature: "Gross" photosynthesis (*GP*) at 20000 lux after *LT* (open circles) and *HT* (filled circles). Apparent photosynthesis (*AP*) at 20000 lux after *LT* (open triangles) and *HT* (filled triangles). Dark respiration (*DR*) after *LT* (open) and *HT* (filled triangles with the points down)

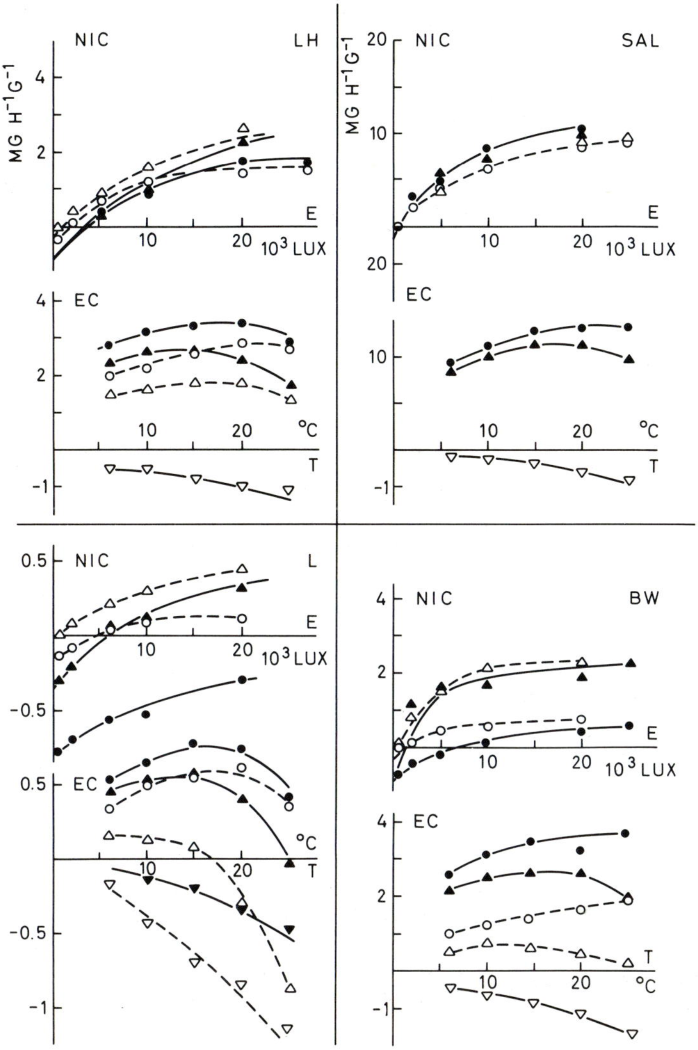

Fig. 3b. Legend see opposite page

in the present study, about three times the values reported by NYGAARD (pers. comm., 1972) which were 0.3–0.4 mg CO_2 $g^{-1}h^{-1}$.

Temperature and Light Reactions. The Q_{10} of dark respiration (calculated as the ratios between DR at 20° and 10° C) is about 2.0 in most cases (Tables 1, 2, and 3), indicating a normal oxidation process. The corresponding Q_{10} values of "gross" photosynthesis (GP = AP+DR) calculated by the ratios between the CO_2 exchange at 15° and 6° C are much lower, usually about 1.3. Exceptions are the lichens from the lichen heath and *Alchemilla glabra*, possessing values of 0.8–1.3. The low Q_{10} of GP and the high Ccp values (Tables 1 and 3) may indicate the presence of photorespiration in the observed plants. The Ccp values are only listed for complete live and green tissue. These ranged from 35 ppm in *Festuca rubra* to 66 ppm in *Carex nigra* (Tables 1 and 3) in good agreement with the Ccp values of detached *Vaccinium myrtillus* and *V. vitis-idaea* shoots (NYGAARD, 1975).

Light strongly influences the temperature reactions of photosynthesis. The observations on optimum temperatures for AP in Tables 1, 2, and 3 are calculated for a light intensity of 20000 lux, but as light decreases this optimum is displaced toward lower temperatures. The optimums are 5–10° lower for AP than for GP as may be seen in Fig. 3. This is a consequence of the higher Q_{10} of DR relative to GP. The optimum temperature for AP in vascular plants varied from about 13° C at the dry meadow to about 23° C at the birch forest site (Table 1) due to higher GP/DR ratios at the latter site. Especially high optimum temperatures were found in *Vaccinium myrtillus* leaves (27° C) and in the wet meadow species *Festuca rubra* and *Carex nigra* (22° C). The high optimum temperature in *Carex nigra* is related to high dark respiration (4.5 mg CO_2 $g^{-1}h^{-1}$), probably as a result of large amounts of reproducing organs in the sample.

In lichens from the lichen heath no positive AP was found at temperatures above 25° C after HT pretreatment and not above 17° C in LT pretreated plants (Fig. 3). Low compensation temperatures (20–30° C) are also reported by LANGE (1965), BÜTTNER (1971) and STÅLFELT (1938). The optimum temperatures in lichens are in the present study found to be between 7° and 13° C (depending on the pretreatment) at light intensities of 20000 lux. For bryophytes the optimum temperature seems to be 12–14° C (Table 1).

The light saturation intensities E_X (in HT pretreated plants), i.e. the light intensities where no further increase in AP is found, vary widely. The figures refer to the amounts of light falling on the upper surface of the vegetation. Bryophytes have relatively low values of E_X ranging from 10000 lux in the birch forest to 22000 lux at the wet meadow site. The low values in the birch forest may be seen as an adjustment to the low light intensity below the trees (STÅLFELT, 1937). The highest (extrapolated) values are found in lichens from the lichen heath (35000 lux). Most E_X values for vascular plants are between 20000 and 25000 lux (Tables 1, 2, and 3) and a light intensity of 20000 lux as used in many of the experiments is thus very close to saturation in most species and sites. These values correspond well with those reported by HADLEY and BLISS (1964) in alpine plants from USA at 1650–1850 m level, but are far lower than those reported by SCOTT and BILLINGS (1964) from the 3500 m level. This indicates a possible adaptation to light *in situ* (BJÖRKMAN and HOLMGREN, 1963).

The light compensation point E_n where AP is not longer positive, is highly dependent on temperature reactions. In the present experiment E_n at 20° C in the flowering stage of HT pretreated vascular plants (Table 1) varied from about 1500 lux to nearly 4000 lux in transections of vegetation and for single species between 1000 lux (*Vaccinium myrtillus* from the birch forest) to 4500 lux (*Salix lapponum* from the wet meadow). At 6° C the E_n values are considerably lower and range from about 500 lux to 800 lux for HT pretreated plants from the first sampling (N_1). In bryophytes E_n at 20° C is mostly within the range of 1000 to 3000 lux and at 6° C down to 200 lux in *Pleurozium schreberi* from the birch forest. In lichens from the lichen heath E_n is extremely high at 20° C, but relatively low at

6° C after HT pretreatment (Table 1). Most of the light compensation values in the present study are lower than those reported elsewhere (HADLEY and BLISS, 1964; SCOTT and BILLINGS, 1964; KALLIO and KÄRENLAMPI, 1975), probably due to seasonal changes. The light reactions may be studied in more detail in Fig. 3.

In the present experiment with vascular plants in the flowering stage (N_1) a significant temperature acclimation effect was found in the samples from the willow thicket, the birch forest and the lichen heath *(Empetrum hermaphroditum)*. The AP was 20–50% higher in plants kept at 48 hrs at high temperature (HT) than in the corresponding LT plants. Strong acclimation affects were also found in the lichens from the lichen heath and the bryophytes from the willow thicket, which was the only bryophyte sample kept at LT. The corresponding reduction of dark respiration (DR) in high temperature as reported e.g. by NYGAARD (1975) and MOONEY et al. (1964) was much weaker and hardly significant, with the exception of lichens (Fig. 3). The small effect of temperature pretreatment on DR is the reason why no significant differences in the optimum temperature of AP could be found. However, a strong increase in E_n is found for LT relative to HT plants, as a result of reduced "gross" photosynthesis and AP.

Of the examined single species, *Salix herbacea* and *Carex nigra* are the only ones which do not show any increase in AP after HT treatment. In *Carex nigra* a clear shift in optimum temperature is found, but at 15° C the apparent photosynthesis in the LT plants is significantly higher than in the HT plants (Table 1). The effect of temperature acclimation seems to be of the same order throughout the season. However, when regarding plants which have been grown under "lowland" conditions like high temperature and omission of UV-light (L) an extremely strong effect is found from the pretreatment. In plants like *Carex nigra* and *Salix herbacea* which do not seem to be affected by LT treatment when grown in natural conditions, the AP is reduced by 50% by LT pretreatment (Table 3). The effect of acclimation on DR is weak in these plants although the respiratory level is high compared to field conditions.

A constant feature is that the Q_{10} of "gross" photosynthesis (GP) was usually 0.1–0.4 units higher at 15° C in plants pretreated at LT than in the corresponding HT plants while the Ccp was 10–30 ppm higher at the same time. A similar effect of temperature pretreatment on Ccp was found by NYGAARD (1975).

Seasonal Variations. Maximum AP rates are generally found at the flowering stage (N_1) when vegetative growth is nearly terminated and the leaves fully expanded (HADLEY and BLISS, 1964). A comparison between the samplings N_1, N_2, and N_3 (Table 1) confirms this theory. A strong decrease in the photosynthetic rates of green tissue is observed during the season, as also reported for net primary production by WIELGOLASKI (1975a). The only exceptions are the bryophytes from the willow thicket, *Salix herbacea* from the snowbed and *Vaccinium myrtillus* from the birch forest, where maximum AP is found in the N_2 plants. In the willow thicket a strong decrease is found in the AP of vascular plants from the first to the second sampling. This decrease may be partly caused by drought effects and damage during transportation, particularly in the taller herbaceous plants like *Geranium silvaticum*. The dark respiration rates decrease strongly in the period, the decrease varying from 10% at the dry meadow to 60% at the willow thicket and 80% at the wet meadow. The relatively slight decrease in DR at the dry meadow site may be the result of an increase in reproductive organs at the time of N_2 (e.g. *Dryas*, *Polygonum viviparum*, *Poa alpina*). At all sites except the birch forest a slight increase (10–20%) in DR takes place in the period between N_2 and N_3 indicating leaf ageing and senescens.

The simultaneous decrease of GP and DR in the period between N_1 and N_2 keeps the AP/DR ratios almost constant as well as optimum temperatures (t_{opt}) for AP and compensation light (E_n). However, a slight decrease in optimum temperatures of AP is found at the birch forest and the wet meadow. On the other hand the slight increase in DR during the last period from N_2 to N_3 at the same time as AP continues to decrease is expected to bring about lower optimum temperatures for AP and increasing E_n. However, no such effects were found and even a slight increase in t_{opt} was observed at most sites (Table 1). Greater availability of light to all leaves in N_3 plants because of smaller amounts of green tissue in the fall may be one reason for the unexpectedly high optimum temperature for AP. Wheather light is a limiting factor under natural conditions depends upon climate, but generally incoming light intensities exceed 20000 lux on sunny days (KJELVIK et al., 1975). At the dry and wet meadows and the willow thicket the amounts of forbs relatively to monocotyledons decreases strongly during the period from N_1 to N_2 and from N_2 to N_3, in good agreement with the general distribution pattern (WIELGOLASKI, 1975b). This may be another reason for the observed changes in photosynthesis and dark respiration during the season.

The relatively high amounts of leaves in *Vaccinium myrtillus* from the birch forest at the time of N_1 and the corresponding high amounts of green stems in the corresponding N_3 plants made it possible to calculate AP in leaves and stems separately. The results indicate that the AP of *Vaccinium myrtillus* stems is only 10–15% of the corresponding leaf photosynthesis. The increase in relative amounts of stems toward the end of the season accounts partly for the observed decrease in AP at the birch forest site. This stem photosynthesis plays an important role in maintaining photosynthetic capacity during the winter season.

A decrease in the AP of lichens is also found during the season, at the same time as the optimum temperature decreases (Table 1), and the compensation temperature after HT pretreatment decreases from 25° to 15° C. This may be a consequence of a lowering of the chlorophyll content (STÅLFELT, 1938; BERG, 1975).

Conclusions and Summary

Maximum values of apparent photosynthesis AP and dark respiration DR of 15–20 mg and 2–3 mg CO_2 $g^{-1}h^{-1}$, respectively, were found in laboratory CO_2 exchange experiments of Norwegian alpine tundra field plants. Among single vascular species *Salix lapponum* leaves reached an AP of 17.8 mg CO_2 $g^{-1}h^{-1}$, while the corresponding value of *Empetrum hermaphroditum* from the lichen heath was 2.5 mg CO_2 $g^{-1}h^{-1}$. In cryptogams AP values were found ranging from 0.3 mg CO_2 $g^{-1}h^{-1}$ in lichens from the lichen heath (*Cladonia* spp. and *Cetraria nivalis*) to 4.0 g CO_2 $g^{-1}h^{-1}$ in bryophytes (mostly *Campylium* and *Bryum*) from the wet meadow. Dark respiration, as a whole, ranged from 0.8 mg at the lichen heath to 3.5 mg CO_2 $g^{-1}h^{-1}$ in the dry meadow.

High temperature pretreatment was found to increase photosynthesis and decrease dark respiration in most cases.

Ecotypic adjustment to light and temperature was observed. Plants from the subalpine birch forest at Maurset (780 m) had lower light compensation points and higher optimum temperatures for photosynthesis than similar vegetation from Stigstuv (1250 m). The light saturation intensities (E_x) varied from 10000 lux in bryophytes from the birch forest to about 35000 lux in lichens from the lichen heath. The light compensation point (E_n) varied from 1400 lux at the birch forest to 3800 lux in the wet meadow in July at a temperature of 20° C. Optimum temperatures ranged from 27° C in *Vaccinium myrtillus* leaves from the birch forest to about 8° C in lichens from the lichen heath at light saturation, but varied with the temperature pretreatment.

As a rule, lower rates of DR were found in monocotyledons than in dicotyledons. The highest values of AP, as well as DR, were found at the time of the first sample collection in July, when most plants were in the flowering stage, while a decrease of 40–50% in AP and 10–50% in DR was found during the period from July to September. However, a slight increase in respiration rates was found from August to September, indicating senescence.

The respiration of roots was measured in separate experiments on single species and was found to be slightly lower than the corresponding respiration of stems and leaves.

Acknowledgements. I wish to thank the technical assistents H. TØNNESEN and M. MORELL-JENSEN for valuable help during the experiment. I am also grateful to Mr. H. S. HØYDAHL and his staff at the phytotrone, Botanical Laboratory, University of Oslo, who have helped me with the technical arrangement and performance of the experiments. Research assistent A. BERG has been responsible for the transport of plants from the field sites to the laboratory and the project leader F. E. WIELGOLASKI has given advise on the experimental design and preparation of the manuscript. I wish to thank them all for valuable help and advice.

References

ARNON, D. I.: Microelements in culture solution experiments with higher plants. Am. J. Botany **25**, 322–325 (1938).

BERG, A.: Pigment structure of vascular plants, mosses and lichens at Hardangervidda, Norway. In: WIELGOLASKI, F. E. (Ed.): Fennoscandian tundra ecosystems, Part 1: Plants and microorganisms, pp. 216–224. Berlin-Heidelberg-New York: Springer 1975.

BERNER, E.: Grønne planters respirasjon i lys. (English summary). Blyttia **30**, 169–186 (1972).

BILLINGS, W. D., GODFREY, P. J.: Acclimation effects on metabolic rates of arctic and alpine *Oxyria* populations subjected to temperature stress. Bull. Ecol. Soc. Am. **49**, 68–69 (1968).

BILLINGS, W. D., SHAVER, G. R., TRENT, A. W.: Temperature effects on growth and respiration of roots and rhizomes in tundra graminoids. In: BLISS, L. C., WIELGOLASKI, F. E. (Eds.): Primary production and production processes, Tundra Biome, pp. 57–64. Edmonton-Oslo: IBP Tundra Biome Steering Committee 1973.

BJÖRKMAN, G., HOLMGREN, P.: Adaptability of the photosynthetic apparatus to light intensity in ecotypes from exposed and shaded habitats. Physiol. Plant. **16**, 889–914 (1963).

BLISS, L. C.: Plant productivity in alpine microenvironments on Mt. Washington, New Hampshire. Ecol. Monogr. **36**, 125–155 (1966).

BLISS, L. C., HADLEY, E. B.: Phytosynthesis and respiration of alpine lichens. Am. J. Botany **51**, 870–874 (1964).

BÜTTNER, R.: Untersuchungen zur Ökologie und Physiologie des Gasstoffwechsels bei einigen Strauchflechten. Flora **160**, 72–99 (1971).

GERASIMENKO, T. V., ZALENSKY, O. V.: Diurnal and seasonal dynamics of photosynthesis in plants of Wrangel Island. Botan. Zhurn. **58**, 1655–1666 (1973) (in Russian).

HADLEY, E. B., BLISS, L. C.: Energy relationships of alpine plants on Mt. Washington, New Hampshire. Ecol. Monogr. **34**, 331–358 (1964).

HEWITT, E. J.: Sand and water culture methods used in the study of plant nutrition. 2ed. Commonwealth Bureau of horticulture and plantation crops **22**. London: Commonwealth Agrigultural Bureau 1966.

JACKSON, W. A., VOLK, R. J.: Photorespiration. Ann. Rev. Plant Physiol. **21**, 385–432 (1970).

JOHANSSON, L. G., LINDER, S.: The seasonal pattern of photosynthesis of some vascular plants on a subarctic mire. In: WIELGOLASKI, F. E. (Ed.): Fennoscandian tundra ecosystems, Part 1: Plants and microorganisms, pp. 194–200. Berlin-Heidelberg-New York: Springer 1975.

KALLIO, P., KÄRENLAMPI, L.: Photosynthetic activity in mosses and lichens. In: COOPER, J. P. (Ed.): Photosynthesis and productivity in different environments. Cambridge: University Press 1975 (in press).

KJELVIK, S., KÄRENLAMPI, L.: Plant biomass and primary production of Fennoscandian subarctic and subalpine forests and of willow and heath ecosystems. In: WIELGOLASKI, F. E. (Ed.): Fennoscandian tundra ecosystems, Part 1: Plants and microorganisms, pp. 111–120. Berlin-Heidelberg-New York: Springer 1975.

KJELVIK, S., WIELGOLASKI, F. E., JAHREN, A.: Photosynthesis and respiration studied by field technique at Hardangervidda, Norway. In: WIELGOLASKI, F. E. (Ed.): Fennoscandian tundra ecosystems, Part 1: Plants and microorganisms, pp. 184–193. Berlin-Heidelberg-New York: Springer 1975.

LANGE, O. L.: Der CO_2-Gaswechsel von Flechten bei tiefen Temperaturen. Planta **64**, 1–19 (1965).

LINDER, S.: Some aspects on pigmentation, photosynthesis and transpiration in nursery-grown seedlings of Scots pine and Norway spruce. Ph.D. Thesis III (1972).

LYE, K. A.: Vegetation of selected localities for IBP investigation in Hardangervidda, southern Norway. IBP i Norden **8**, 101–111 (1972).

MAYO, J. M., DESPAIN, D. G., VAN ZINDEREN BAKKER, E. M., JR.: CO_2 assimilation studies. In: BLISS, L. C. (Ed.): Devon Island IBP project high arctic ecosystem. Project report 1970 and 1971, pp. 217–251. Edmonton: University of Alberta, Dept. of Botany 1972.

MAYO, J. M., DESPAIN, D. G., VAN ZINDEREN BAKKER, E. M., JR.: CO_2-assimilation by *Dryas integrifolia* on Devon Island, Northwest Territories. Can. J. Botany **51**, 581–588 (1973).

MOONEY, H. A., SHROPSHIRE, F.: Population variability in temperature related photosynthetic acclimation. Oecologia **2**, 1–13 (1967).

MOONEY, H. A., WRIGHT, R. D., STRAIN, B. R.: The gas exchange capacity of plants in relation to vegetation zonation in the White Mountain of California. Am. Midland Naturalist **72**, 281–297 (1964).

MOORE, R. T., EHLERINGER, J., MILLER, P. C., CALDWELL, M. M., TIESZEN, L. L.: Gas exchange studies of four alpine tundra species at Niwot Ridge, Colorado. In: BLISS, L. C., WIELGOLASKI, F. E. (Eds.): Primary production and production processes, Tundra Biome, pp. 211–218. Edmonton-Oslo: IBP Tundra Biome Steering Committee 1973.

MOSER, W.: Licht, Temperatur und Photosynthese an der Station »Hoher Nebelkogel« (3184 m). In: ELLENBERG, H. (Ed.): Ökosystemforschung, pp. 203–224. Berlin-Heidelberg-New York: Springer 1973.

NYGAARD, R. T.: Acclimatization effects in photosynthesis and respiration. In: WIELGOLASKI, F. E. (Ed.): Fennoscandian tundra ecosystems, Part 1: Plants and microorganisms, pp. 163–167. Berlin-Heidelberg-New York: Springer 1975.

SCOTT, D., BILLINGS, W. D.: Effects of environmental factors on standing crop and productivity of an alpine tundra. Ecol. Monogr. **34**, 243–270 (1964).

SESTAK, Z., ČATSKY, J., JARVIS, P. G.: Plant photosynthetic production. The Hague: Junk 1971.

SHVETSOVA, V. M., VOZNESENSKII, V. L.: Diurnal and seasonal variations in the rate of photosynthesis in some plants of Western Taimyr. IBP Tundra Biome Translation 2 (1971) from Botan. Zhurn. **55**, 66–76 (1970).

SKARTVEIT, A., RYDÉN, B. E., KÄRENLAMPI, L.: Climate and hydrology of some Fennoscandian tundra ecosystems. In: WIELGOLASKI, F. E. (Ed.): Fennoscandian tundra ecosystems Part 1: Plants and microorganisms, pp. 41–53. Berlin-Heidelberg-New York: Springer 1975.

SKRE, O.: Methods and results of CO_2 exchange studies on Norwegian alpine plants. In: VIK, R. (Ed.): IBP in Norway. Methods and results. Sections PT-UM. Grazing project Hardangervidda. Botanical investigations, pp. 259–295. Oslo: Norwegian National IBP-Committee 1975.

SKRE, O., KJELVIK, S., BERG, A., WIELGOLASKI, F. E.: Nitrogen content in Norwegian alpine plants. In: VIK, R. (Ed.): IBP in Norway. Methods and results. Sections PT-UM. Grazing project Hardangervidda. Botanical investigations, pp. 159–194. Oslo: Norwegian National IBP-Committee 1975.

SONESSON, M., WIELGOLASKI, F. E., KALLIO, L.: Description of Fennoscandian tundra ecosystems. In: WIELGOLASKI, F. E. (Ed.): Fennoscandian tundra ecosytems, Part 1: Plants and microorganisms, pp. 3–28. Berlin-Heidelberg-New York: Springer 1975.

STÅLFELT, M. G.: Der Gasaustausch der Moose. Planta **27**, 30–60 (1937).

STÅLFELT, M. G.: Der Gasaustausch der Flechten. Planta **29**, 11–31 (1938).

SVENSSON, B. H., VEUM, A. K., KJELVIK, S.: Carbon losses from tundra soils. In: WIELGOLASKI, F.E. (Ed.): Fennoscandian tundra ecosystems, Part 1: Plants and microorganisms, pp. 279–286. Berlin-Heidelberg-New York: Springer 1975.

TIESZEN, L. L.: CO_2 exchange in the Alaskan arctic tundra. Measured course of Photosynthesis. In: BOWEN, S. (Ed.): Proceedings 1972 US Tundra Biome Symposium, pp. 29–35. Hannover-New Hampshire: USA CREEL 1972.

WIELGOLASKI, F. E.: Vegetation types and plant biomass in tundra. Arctic and Alpine Res. **4**, 291–305 (1972).

WIELGOLASKI, F. E.: Primary production of tundra. In: COOPER, J. P. (Ed.): Photosynthesis and productivity in different environments, pp. 75–106. Cambridge: University Press 1975a.

WIELGOLASKI, F. E.: Primary productivity of alpine meadow communities. In: WIELGOLASKI, F.E. (Ed.): Fennoscandian tundra ecosystems, Vol. 1, Plants and microorganisms, pp. 121–128. Berlin-Heidelberg-New York: Springer 1975b.

ZELITCH, J.: Increased rate of net photosynthetic carbon dioxide uptake caused by the inhibition of glycolate oxidase. Plant Physiol. **41**, 1623–1631 (1966).

Photosynthesis and Respiration of Plants Studied by Field Technique at Hardangervidda, Norway

S. KJELVIK, F. E. WIELGOLASKI, and A. JAHREN

Gas exchange in plants is measured by several methods, see review in SESTAK et al. (1971). In the present study, photosynthesis and respiration are measured for plant communities by simple cuvette techniques in the field. Based on the CO_2 gas exchange values, rough estimates of the primary productivity of an area over a certain time are found as functions of abiotic parameters.

Results for four of the Norwegian IBP sites at Hardangervidda (SONESSON et al., 1975) are given in the present paper, i.e. dry meadow, lichen heath, wet meadow and subalpine birch forest (the understory only).

Methods

The CO_2 gas exchange is measured in the studies in a closed system with vegetation and soil. A plastic cuvette (diameter 20 cm, height 8 cm) is placed on a cylindrical steel frame which is pressed 15 cm down into the soil (see Fig. 1). A battery-driven fan is placed in the cuvette for air circulation. During the experiment air from the cuvette is sucked through a tube containing a hydrazine compound, giving a blue color according to the CO_2 concentration, which is found by approximately 10% accuracy. Five minutes' experimental time is expected to be reasonable. To compensate for changes in the chamber pressure when air is expelled from the cuvette, a plastic bag is used with free access to the outside atmosphere.

Mean air temperatures from observations at the beginning and end of each experiment are used in the calculations. In bright sunshine, the temperature in the cuvette may rise more than 5–6° C during the experiment. In sunshine, dew inside the cuvette was a problem, reduced however, by use of an anti-dew product. The air humidity was measured within the cuvette at the end of some of the experiments by another test tube, and outside the cuvette at twenty cm height. The soil moisture is measured gravimetrically on a daily basis at each site at 5–10 cm depth. The light intensity is measured in lux by a selenium cell outside the cuvette during the experiments. The cuvette itself has shown some shading effect for the plants studied. On an average, the incoming radiation for these plants amounts to about $^2/_3$ of the light measured, and this is corrected for in the calculations.

Fig. 1. The cuvette, the so-called Dräger apparatus for sucking air out of the cuvette through a tube containing a hydrazine compound, and the steel frame used in the experiments

Measurements are carried out at various times of the day, in various periods of the growing season and under different climatic conditions. Apparent photosynthesis and dark respiration of aboveground parts including all vegetation (both vascular plants and cryptogams) are found from the CO_2 values in the cuvette by experiments with plants respectively in light and dark, and when aboveground plants are removed, although probably some wounding effect may occur. "Total" respiration, including both respiration from plants and soil, is calculated as the difference between CO_2 concentration in the air surrounding the cuvette and the air within the cuvette after dark experiments. Roughly, the amount of CO_2 in mg m^{-2} h^{-1} can be transferred to the amount of dry weight in mg m^{-2} h^{-1} by multiplying with the factor 0.648 (SESTAK et al., 1971).

The correlations between climatic factors and both assimilation and respiration measurements are found to be lower when based on dry weight of vegetation than on ground area.

The material shows that a total of 35–40 experiments with different environmental factors are necessary to give less than 10% variation in the analyses as a function of climatic factors. Four to five measurements are necessary under as similar environmental conditions as possible to find average estimates of assimilation and respiration with 10% standard error. This is caused by the variations in plant cover and soil type within each site, as well as sampling errors in the experiments.

Results and Discussion

The values given in Table 1 for rates of "total" respiration of soil and plants (TR), dark respiration of above ground parts of the plants (DR) and apparent photosynthesis (AP) are means of several observations within the ranges given. The rates are lower than reported by SKRE (1975), partly because the dry weight of green includes cryptogams as well. In addition, many of the experiments were probably carried out under water conditions unfavorable to the plants. The average values for AP in the wet meadow, however, seem to be of the same order as found for forbs with a similar technique in a north Swedish bog (JOHANSSON and LINDER, 1975). The dark respiration generally increases, as expected, in the present studies more with temperature than "total" respiration and apparent photosynthesis.

Apparent photosynthesis was highest in the wet meadow (Table 1), lower in the dry meadow and still lower in the understory in the birch forest, the snow bed

Table 1. "Total" respiration of soil and plants (TR), dark respiration of aboveground plants (DR) and apparent photosynthesis (AP) at the sites (1973) at different temperatures (average light intensity) and at different light intensities (average temperatures) within the cuvettes. TR in mg CO_2 m^{-2} h^{-1} and DR and AP in mg CO_2 m^{-2} h^{-1} and mg CO_2 g^{-1} d.w. h^{-1} (weight of total mixed green material of cryptogams and vascular plants). Dry meadow = average *Dryas* and mixed vegetation. Mean temperature in °C

	TR	DR		AP		Mean light intensity
	per m^2	per m^2	per g d.w.	per m^2	per g d.w.	lux
Wet meadow						
10–15° C	236	45	0.6	213	2.8	10600
20–25° C	587	141	1.9	440	5.8	46500
Dry meadow						
10–15° C	297	70	0.7	243	2.6	12600
20–25° C	777	310	3.3	292	3.1	39200
Lichen heath						
10–15° C	86	14	0.03	30	0.08	9500
20–25° C	227	122	0.3	52	0.13	41100
Birch forest (understory)						
10–15° C	535	174	1.1	73	0.5	6200
20–25° C	795	261	1.6	140	0.9	16700
Snow bed						
10–15° C	170	36	0.6	13	0.2	4600
20–25° C	418	152	2.5	77	1.3	30100

	AP with light 10–20000 lux		Mean temp.	AP with light 20–30000 lux		Mean temp.
	per m^2	per g d.w.		per m^2	per g d.w.	
Wet meadow	245	3.2	14.6	370	4.9	16.9
Dry meadow	292	3.1	16.3	360	3.8	17.0
Lichen heath	23	0.06	15.6	45	0.11	17.4
Birch forest	145	0.9	20.4	218	1.3	21.5

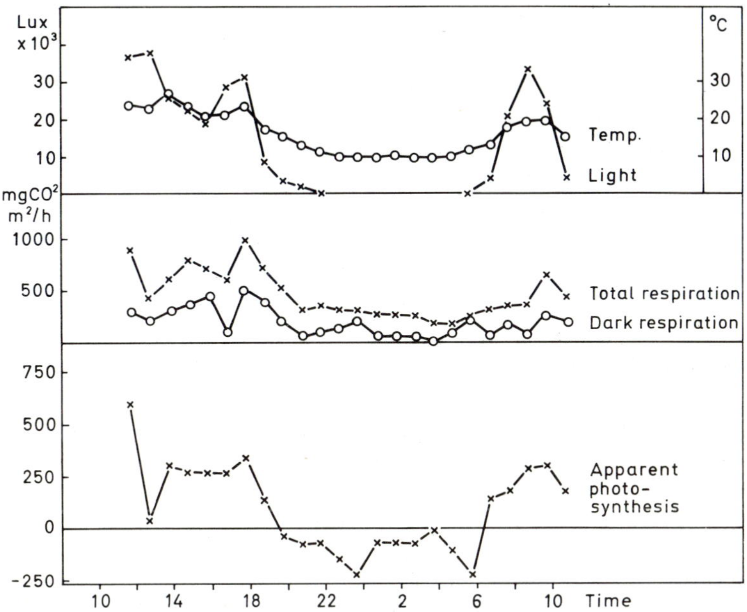

Fig. 2. Photosynthesis and respiration at the dry meadow (total vegetation) 6–7 July 1973 in mg CO_2 m^{-2} h^{-1}. Temperature and light intensity within the cuvette

and the lichen heath. The daily production at the sites calculated from harvesting results follows the same pattern (WIELGOLASKI and KJELVIK, 1973). The mean AP values given in this paper are much lower than the examples of maximum AP values reported by WIELGOLASKI (1975a) for different tundra areas, but seem to be reasonable in calculating daily and seasonal CO_2 exchanges.

The "total" respiration is mostly influenced by temperature in the dry meadow, both in plots dominated by *Dryas octopetala* ($r = 0.79$***, $n = 59$) (*** = sign. 0.1%) and by mixed vegetation of grasses and forbs ($r = 0.90$***, $n = 60$). Temperature influence is improved by including soil water in multiple correlations with "total" respiration, and there is also a weak effect of soil temperature and air humidity in multiple analyses. There are reduced "total" and dark respiration activities around 12.30 and 16.30 on a clear day (Fig. 2), which may be expressed as drought effects. The apparent photosynthesis also decreases at 12.30, but possibly the respiration drop at 16.30 prevents a decrease in apparent photosynthesis at that time. The low activity at 12.30 may be caused by reduced stomata openings (UNGERSON and SCHERDIN, 1962).

The multiple correlation between "gross" photosynthesis (AP + DR) as a dependent variable, and temperature and soil water as independent variables was lower in the dry meadow in 1972 ($R = 0.61$***, $n = 60$) than1973 ($r = 0.84$***, $n = 52$). The lower correlation in 1972 may have been caused by a greater decrease in the daily photosynthesis curve because of water stress. There was a long dry period in July 1972; in 1973, the July precipitation was higher and more evenly distributed throughout the month.

Apparent photosynthesis at this site showed a highly significant correlation with light intensity, particularly in the last year; $r=0.54^{***}$, $n=51$ for *Dryas*-dominated plots an $r=0.71^{***}$, $n=51$ for mixed vegetation. The correlation for AP in the wetter season was not much improved by inroducing variables related to the water situation of the plants, in contrast to the drier year. This demonstrates that particularly in dry years the production is also influenced by lack of water at this site. The regression equations are $y=0.0025x+126.9$ and $y=0.0081x+146.6$ for the two types of plots where $y=\text{mg } CO_2 \text{ m}^{-2} \text{ h}^{-1}$ and x = light in lux. Mean values of AP in the experiments carried out are 186 mg CO_2 m^{-2} h^{-1} (1.9 mg CO_2 g^{-1} d.w. h^{-1}) and 361 mg CO_2 m^{-2} h^{-1} (3.8 mg CO_2 g^{-1} d.w. h^{-1}) in the *Dryas*-dominated plots and mixed vegetation, (mean climatic conditions 24100 lux and 16.9° C and 26400 lux and 18.0° C, respectively within the cuvettes). The values for the *Dryas*-dominated plots seem to be somewhat lower than averages calculated by SHVETSOVA and VOZNESENSKII (1970), but compare well with MAYO et al. (1973) for closely related *Dryas* species in the arctic.

Apparent photosynthesis in the wet meadow in the drier year (Fig. 5) seems to be most dependent on light intensity and relative air humidity ($R=0.54^{*}$, $n=20$) (* = sign. 5%), while in the wetter year air humidity had only a minor influence on AP. The simple correlation with light was then found to be ($r=0.60^{***}$, $n=31$). The regression equation for light intensity is $y=0.00537\,x+175.6$, where $y=\text{mg } CO_2 \text{ m}^{-2} \text{ h}^{-1}$ and x = light in lux. Mean values for AP (mixed vegetation of vascular plants and cryptogams) in the experiments in the wetter season were 318 mg CO_2 m^{-2} h^{-1} or 4.2 mg CO^2 g^{-1} d.w. h^{-1} (mean light intensity 26400 lux, and mean temperature 17.0° C within the cuvettes). In a smaller number of experiments vascular plants and bryophytes were treated separately in the wetter season, giving mean AP values in the experiments of 104 mg CO_2 m^{-2} h^{-1} and 252 mg CO_2 m^{-2} h^{-1} for bryophytes and vascular plants, respectively (mean light intensity 20000 lux and mean temperature 12.0° C). On a dry weight basis, this means AP about 4.6 mg CO_2 g^{-1} h^{-1} for vascular plants, and a similar value for bryophytes when based on only visually green material. Probably, however, some of the visually non-green bryophytes also assimilate.

The diurnal pattern of the parameters measured is given for a bright and warm day in the wet meadow in Fig. 3. There was always access to soil water, and temperature and light intensity explain most of the variation observed, except for a mid-day depression. Even at this site where the water content in the soil is always high, this may be seen as an effect of reduction in stomata opening caused by water-stress in the plants. The "total" respiration in correlation analyses was highly dependent on temperature in the wetter year ($r=0.87^{***}$, $n=39$), but no significant correlation with temperatures is found in the drier year. The multiple analysis on "gross" photosynthesis shows (as for AP) a significant correlation with light intensity and relative air humidity ($R=0.50^{***}$, $n=49$) in the drier year, while in the wetter year (as for respiration), temperature gives a highly significant simple correlation ($r=0.81^{***}$, $n=31$).

At the lichen heath dark respiration of lichens is low, particularly at high temperatures (34 mg CO_2 m^{-2} h^{-1} or 0.09 mg CO_2 g^{-1} d.w. h^{-1} at 20–25° C),

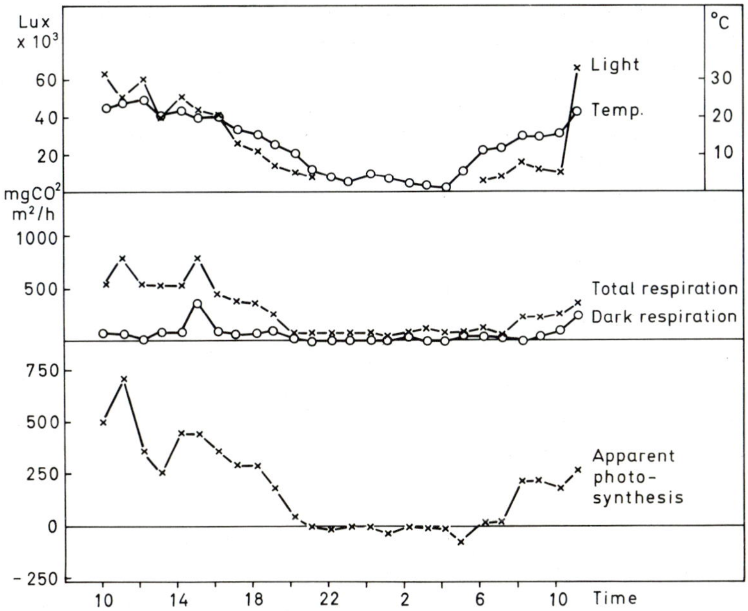

Fig. 3. Photosynthesis and respiration at the wet meadow (total vegetation) 12–13 July 1973 in mg CO_2 m^{-2} h^{-1}. Temperature and light intensity within the cuvette

probably a drought effect. It was found that dark respiration of lichens depends to a great extent on the water percentage of the plants. Apparent photosynthesis in lichens is very low and sometimes even negative in dry periods and relatively high temperatures, which explains the low productivity at the site (KJELVIK and KÄRENLAMPI, 1975). As an average of all measurements, the apparent photosynthesis of lichens is 4 mg CO_2 m^{-2} h^{-1} or 0.01 mg CO_2 g^{-1} h^{-1} (at 27900 lux and 19.8° C within the cuvettes). For the few vascular plants at the lichen heath (mostly *Empetrum hermaphroditum*), the mean AP in the experiments is calculated to be about 30 mg CO_2 m^{-2} h^{-1} or 1.1 mg CO_2 g^{-1} d.w. h^{-1} (at 29100 lux and 19.6° C as an average within the cuvettes). This is of the same order as calculated by JOHANSSON and LINDER (1975) for *Empetrum hermaphroditum* in northern Sweden.

Diurnal variation in climate, photosynthesis and respiration for lichens at the lichen heath is shown in Fig. 4. Although the water effect on lichen plant activity was obviously most important at the dry lichen heath, the "total" respiration from soil and plants was mostly temperature-dependent, both in pure lichen plots ($r = 0.63$***, $n = 43$) and in plots with mixed vegetation in both years studied (respectively, $r = 0.74$***, $n = 28$, and $r = 0.78$***, $n = 42$ in the drier and wetter year). In the wetter year the mean value of "total" respiration at plots with mixed vegetation was 140 mg CO_2 m^{-2} h^{-1} at 15.6° C, and at plots with only lichens 100 mg CO_2 m^{-2} h^{-1} at 15.6° C. In the drier year, stepwise multiple analysis showed significant correlation for "gross" photosynthesis with light intensity and relative air humidity ($R = 0.55$**, $n = 33$) (** = sign. 1%), and in the wetter year (where air humidity was not observed) with light and temperature ($R = 0.53$*,

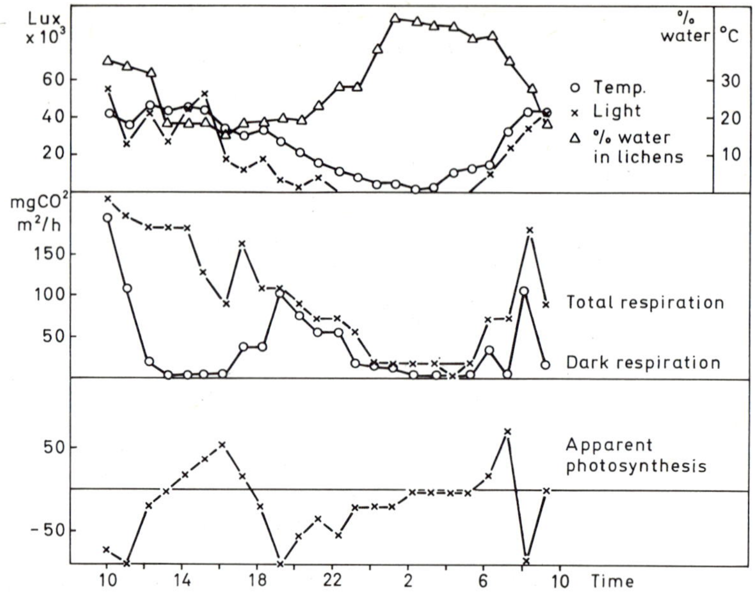

Fig. 4. Photosynthesis and respiration in lichens at the lichen heath 4–5 July 1973 in mg CO_2 m^{-2} h^{-1}. Temperature and light intensity within the cuvette and percent water in the lichens

$n = 23$). The partial effect of temperature is negative because of negative intercorrelation with the water conditions. Apparent photosynthesis in the mixed vegetation showed (as in the wet meadow) a tendency to the same pattern as "gross" photosynthesis, but no significant correlations were found.

For the understory at the birch forest, the increase in dark respiration and apparent photosynthesis with increasing temperature seemed to be of the same order within the temperature range given in Table 1. This may be an adaptation to the generally higher temperatures in the growing season at this lower elevated site. Even in the relatively dry year, the "total" respiration was mainly temperature-dependent ($r = 0.61$***, $n = 51$) with a small improvement achieved by adding soil water in a multiple analysis, and still more temperature-dependent in a wetter year ($r = 0.79$***, $n = 28$). In the birch forest, some areas are dominated by *Deschampsia flexuosa* and others by *Vaccinium* spp., but they are very similar in respect to "total" respiration. "Gross" photosynthesis is, in the results from the relatively dry year, dependent on light, temperature and soil water; in the wetter year soil water had no effect.

Apparent photosynthesis in the dry year was dependent on light and soil water ($R=0.70$**, $n=16$). The regression found was $y=0.0019x_1+24.213x_2-422.17$ where $y=$ mg CO_2 m^{-2} h^{-1} and x_1 is light in lux and x_2 soil water in percent. The mean AP in the experiments (mixed understory vegetation) is 151 mg CO_2 m^{-2} h^{-1}, or about 1 mg CO_2 g^{-1} d.wt.h^{-1} (at 25400 lux within the cuvettes and 22% water in the soil calculated on fresh weight). The importance of water on photosynthetic activity at the birch forest site is in accordance with the results from

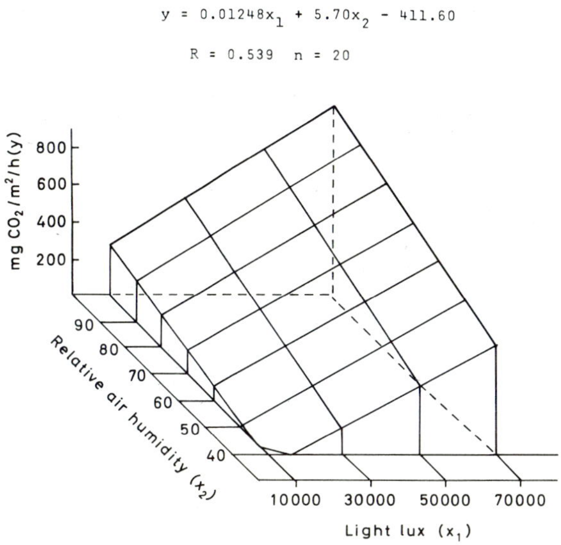

Fig. 5. Apparent photosynthesis per unit ground area in a relatively dry year at the wet meadow as a function of light intensity and relative air humidity

productivity studies at harvestings (WIELGOLASKI and KJELVIK, 1973). The light intensity at the birch forest site is reduced in the understory vegetation because of shading from the trees. This may reduce the production and cause similar photosynthesis in the birch forest understory and in the dry meadow, although the temperature is higher in the lower elevated birch forest.

The regression for apparent photosynthesis on relative air humidity and light intensity in the wet meadow in the drier year 1972 (Fig. 5) was used for calculation of AP in July and August 1970. The calculations are based on mean hourly values of relative humidity of the air and light intensity in hours with over 4000 lux outside the cuvettes (on an average 15 hrs per day), as this intensity was found to be approximately the compensation point (Fig. 3). The averages for the two months were 83.8% relative air humidity, 20000 lux and 8° C. The calculations show an average AP of 444 mg CO_2 m^{-2} h^{-1} or 6.7 g CO_2 m^{-2} day^{-1}. These are somewhat lower values than found in an arctic sedge meadow in Alaska (TIESZEN, 1972), probably partly caused by the shorter days in southern Norway. By multiplying the AP per day with the factor 0.648 (SESTAK et al., 1971), a production of 4.3 g dry matter m^{-2} day^{-1} is estimated for the wet meadow in July and August 1970. The estimate by harvesting is 8.3 g dry matter m^{-2} day^{-1} as a sum for cryptogams and vascular plants above and below ground when compensated for decomposition, according to the method described by WIELGOLASKI (1975b).

The estimated primary production from the measured apparent photosynthesis is thus lower than found from harvesting data. Reasons for this may e.g. be heterogeneity in the vegetation at the plots used for harvesting and photosynthesis studies, respectively. At the peaty wet meadow, especially, there were also problems in measuring the root production by harvesting. The "wounding" effect from removal of aboveground vegetation may result in increased respiration and thus underestimation of the apparent photosynthesis. If the regressions found in the wetter year are used, even greater underestimates of the production in 1970 are found, probably because of lack of experiments in 1973 on days with heavy cloud cover.

The "total' respiration calculated for the same periods in 1970 on basis of the regressions was found to be somewhat higher in the birch forest and the wet meadow (about 11 g CO_2 m^{-2} day^{-1}) than in the dry meadow (about 8.5 g CO_2 m^{-2} day^{-1}) and the lichen heath (about 5 g CO_2 m^{-2} day^{-1}). One reason for this may be especially high respirational activity taking place under the usually favorable moisture conditions in the wet meadow in warm and dry periods, and because of higher temperatures at the birch forest site. "Total" respiration is also high at the wet meadow because of the very high root biomass at this site (WIELGOLASKI, 1975b). In 1970, drought may have caused both reduced apparent photosynthesis and respiration at the drier sites. The low "total" respiration at the lichen heath is also caused by low root biomass.

At all the sites studied there is, as expected, usually a clear general temperature dependency for "total" respiration and dark respiration of aboveground plant parts. Even at the dry lichen heath, temperature is the main factor in predicting the "total" respiration. "Gross" and apparent photosynthesis are more dependent on light intensity, but temperature may also in some cases give significant correlation, probably partly because of the intercorrelation between light and temperature. It is obvious that the low temperature in the growing season at the sites studied is astrong limiting factor for apperent photosynthesis and production. However, the water conditions may also limit the plant activities in dry years. In wetter years the water supply is more sufficient, except for shorter periods during the daily course of photosynthesis.

Acknowledgements. The field measurements were mostly carried out by the technical assistants in Norwegian IBP sections PT/UM, A. POLDER and E. M. LUNDE, and the biological student R. LIEN. The stepwise multiple regression analysis was carried out at the Computer Centre at the Norwegian Agricultural College. We wish to thank all who have made this study possible.

References

JOHANSSON, L. G., LINDER, S.: The seasonal pattern of photosynthesis of some vascular plants on a subarctic mire. In: WIELGOLASKI, F. E. (Ed.): Fennoscandian tundra ecosystems, Part 1: Plants and microorganisms, pp. 194–200. Berlin-Heidelberg-New York: Springer 1975.

KJELVIK, S., KÄRENLAMPI, L.: Plant biomass and primary production of Fennoscandian subarctic and subalpine forests and of alpine willow and heath ecosystems. In: WIELGOLASKI, F. E. (Ed.): Fennoscandian tundra ecosystems, Part 1: Plants and microorganisms, pp. 111–120. Berlin-Heidelberg-New York: Springer 1975.

MAYO, J. M., DESPAIN, D. G., VAN ZINDEREN BAKKER, E. M., JR.: CO_2 assimilation by *Dryas integrifolia* on Devon Island, Northwest Territories. Canad. J. Botan. **51**, 581–588 (1973).

SESTAK, Z., ČATSKY, J., JARVIS, P. G.: Plant photosynthetic production. The Hague: Junk 1971.

SHVETSOVA, V. N., VOZNESENSKII, V. L.: Diurnal and seasonal variations in the rate of photosynthesis in some plants of western Taimyr. Botan. Zhur. **55**, 66–76 (1970). (Int. Tundra Biome Transl. **2** (Ed. BOWEN, S.) 1971.

SKRE, O.: CO_2 exchange in Norwegian tundra plants studied by infrared gas analyzer technique. In: WIELGOLASKI, F. E. (Ed.): Fennoscandian tundra ecosystems, Part. 1: Plants and microorganisms, pp. 168–183. Berlin-Heidelberg-New York: Springer 1975.

SONESSON, M., WIELGOLASKI, F. E., KALLIO, P.: Description of Fennoscandian tundra ecosystems. In: WIELGOLASKI, F. E. (Ed.): Fennoscandian tundra ecosystems, Part. 1: Plants and microorganisms, pp. 3–28. Berlin-Heidelberg-New York: Springer 1975.

TIESZEN, L. L.: CO_2 exchange in Alaskan arctic tundra: Measured course of photosynthesis. In: BOWEN, S. (Ed.): Proceedings 1972 U.S. Tundra Biome Symposium, pp. 29–35. Hanover, New Hampshire: USA CRREL 1972.

UNGERSON, J., SCHERDIN, G.: Untersuchungen über den Tagesverlauf der Photosyntese und der Atmung unter natürlichen Bedingungen in der Subarktis (Finnish-Lappland). Ann. Botan. Soc. "Vanamo" **32**, 1–22 (1962).

WIELGOLASKI, F. E.: Primary production of tundra. In: COOPER, J. P. (Ed.): Photosynthesis and productivity in different environments, pp. 75–106. Cambridge: University Press 1975a.

WIELGOLASKI, F. E.: Primary productivity of alpine meadow communities. In: WIELGOLASKI, F.E. (Ed.): Fennoscandian tundra ecosystems Part. 1: Plants and microorganisms, pp. 121–128. Berlin-Heidelberg-New York: Springer 1975b.

WIELGOLASKI, F. E., KJELVIK, S.: Production of plants (vascular and cryptogams) in alpine tundra, Hardangervidda. In: BLISS, L. C., WIELGOLASKI, F. E. (Eds.): Primary production and production processes, Tundra Biome, pp. 75–86. Edmonton-Oslo: IBT Tundra Biome Steering Committee 1973.

The Seasonal Pattern of Photosynthesis of Some Vascular Plants on a Subarctic Mire

L.-G. JOHANSSON and S. LINDER

During the summer of 1972, eco-physiological investigations were added to the research programme of the Swedish IBP Tundra Biome mire project at Stordalen (JOHANSSON et al., 1973). The mire site is situated north of the Arctic Circle; the plant cover is characterised by the bryophytes (300 g d.w.m^{-2} live phytomass) and the dwarf shrubs (140 g d.w.m^{-2} live aboveground phytomass). Graminids, herbs and lichens contribute with about 18, 18, and 12 g m^{-2}, respectively (SONESSON et al., 1973). Since the production rates for most of the species concerned are low, especially at the beginning and end of the growing season, it is difficult to measure production as phytomass changes during short periods by means of the harvesting technique.

The main purpose of the eco-physiological studies was to obtain detailed information about the primary production of the dominant vascular plant species on the mire, in relation to short-term variations in the climatic conditions.

Methods and Material

The species investigated were; *Rubus chamaemorus* L., *Betula nana* L., *Andromeda polifolia* L. and *Empetrum hermaphroditum* Hagerup. On the last two species, which are evergreen, both current (c) and one-year-old (c+1) leaves were studied. The species studied represent 80% of the standing crop of vascular plants on the mire (FLOWER-ELLIS, 1973).

The rate of photosynthesis was measured as the rate of CO_2-fixation, using labelled CO_2 according to the method described by SHIMSHI (1969), somewhat modified by JOHANSSON et al. (1973). The labelled carbon dioxide was supplied to the leaf for 20 seconds, by means of a leaf clamp forming a small transparent assimilation chamber when closed. For practical reasons the leaves (except for *Rubus*) were detached from the shoot immediately before exposure to $^{14}CO_2$. The four species were treated somewhat differently owing to differences in leaf size and morphology.

The leaves or leaf disks, after exposure to the radioactive gas mixture (0.5 Ci/mol CO_2), were immediately immersed in liquid nitrogen, then stored in scintillation vials. Labelled samples were combusted by wet-combustion, using a concentrated sulphuric acid-dichromate solution. The CO_2 evolved during combustion was absorbed in 0.25 M NaOH, and the amount of ^{14}C absorbed in the solution was then determined by means of a liquid scintillation counter.

The photosynthetic rate of the species studied was determined every third to tenth day throughout the growing season. The measurements started as soon as the snow cover had melted away from the investigation area. Some measurements were performed on plants which were still covered with snow. These plants (*Empetrum*) were dug out from the snow cover immediately before the measurements were made. On every occasion, ten replicates were taken at random of each species and leaf age-class. The measurements were always carried out between 10.00–12.00 h.

Simultaneously with the determination of photosynthetic rates, the water potential of the plants was measured, using the pressure bomb technique (SCHOLANDER et al., 1964, 1965). The chlorophyll content in the leaves of the species studied was also determined throughout the season (JOHANSSON, 1974a).

An attempt was made to circumvent the difficulty that the values obtained from different species may not be fully comparable by the technique described. Instead of exposing single leaves to $^{14}CO_2$, an area of about 5 dm^2 of the vegetation was enclosed in a cube-shaped transparent perspex assimilation chamber. This had no bottom and could therefore be pressed down into the peat. The chamber was flushed with air containing $^{14}CO_2$ supplied from a compressed air cylinder. The plants within the chamber were exposed to the labelled CO_2 for five minutes, after which the chamber was removed. Only relative values of photosynthetic rates were obtained from these gas exchange experiments.

The photosynthetic rate of one or two species present in the chamber was simultaneously determined using the "leaf clamp technique" for conversion from relative to "absolute" values. These measurements were made on plants outside the assimilation chamber, but forming part of the same populations as the enclosed plants. When the "absolute" values for one or two of the species and the relative values for all species enclosed in the chamber were known, the "absolute value of the photosynthetic rate" could be calculated for all species. Using this method, it was possible to obtain some values for the CO_2-fixation rate of the dominant bryophytes. For more detailed information concerning material and methods, see reports by JOHANSSON et al. (1973) and JOHANSSON (1974b). The meteorological data used in the report were obtained from the meteorological station at Stordalen (RYDÉN, 1973; 1974).

Results and Discussion

Seasonal Course of Photosynthesis. The photosynthetically active period for *Rubus* and *Betula* was about three months, from the second week in June to the middle of September (Fig. 1). The start of the period was well defined, with most of the leaves developing within a few days. The end of the period was far more diffuse. The senescence of *Rubus* leaves started at the beginning of August, and most of them were dead by 10th September (JOHANSSON, 1974a). However, in the

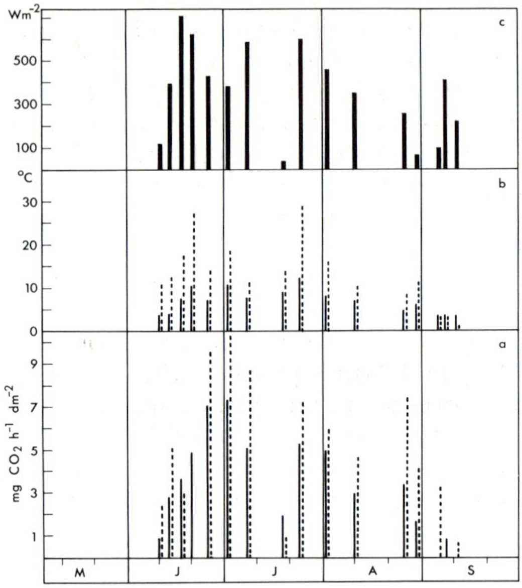

Fig. 1. (a) Seasonal course of photosynthesis for *Rubus chamaemorus* (solid column) and *Betula nana* (broken column). The values are means of ten samples per occasion. Maximum standard error 10%. (b) Air temperature at ground level (broken column) and peat temperature 5 cm below ground level (solid column). (c) Solar radiation. The climatic data are hourly means for the hour of measurement

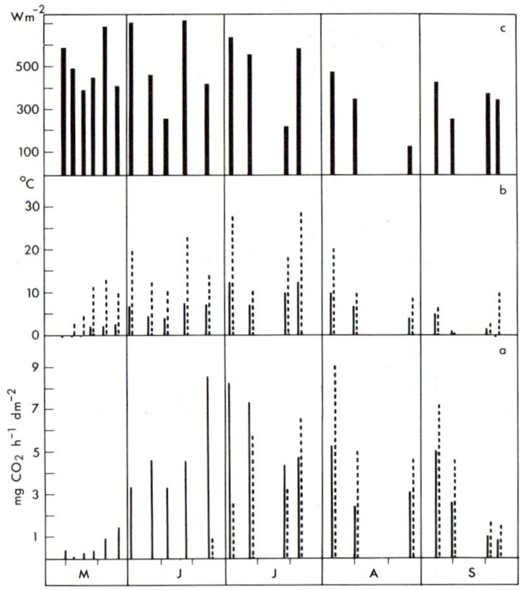

Fig. 2. (a) Seasonal course of photosynthesis for *Andromeda polifolia*, one-year-old leaves (solid column) and current year leaves (broken column). The values are means of ten samples per occasion. Maximum standard error 10%. (b) Air temperature at ground level (broken column) and peat temperature 5 cm below ground level (solid column). (c) Solar radiation. The climatic data are hourly means for the hour of measurement

middle of September, green leaves of *Rubus*, with a high photosynthetic capacity, could still be found (JOHANSSON, 1974b).

The senescence of *Betula* leaves started one to two weeks later, but was almost complete by the second week of September. For both species the photosynthetic rate increased rapidly during June, both when expressed on the basis of leaf area (Fig. 1) and dry weight (Fig. 3). After maximum rates had been attained at the beginning of July, the photosynthetic rates gradually decreased for both species until the end of the active period in September. The low rates recorded in the middle of July can be explained by the low light intensity for the day on which the measurements were carried out.

Andromeda and *Empetrum* showed positive photosynthesis from immediately after snowmelt in May, until 24th September, when the series of measurements ended because of snow cover on the mire (Figs. 2 and 3). The current year's leaves of both species developed during late June, and from the middle of July their photosynthetic rates exceeded those of the one-year-old leaves, both when expressed on a leaf area and on a dry weight basis (Figs. 2 and 3).

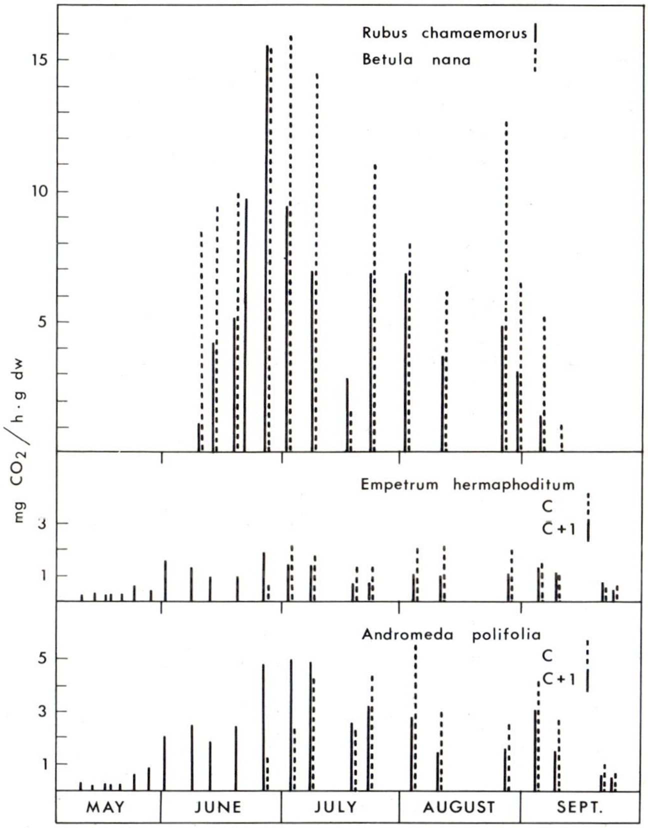

Fig. 3. Seasonal course of photosynthesis for four vascular plant species expressed as mg CO_2 g d.w.$^{-1}$h^{-1}. The values are means of ten samples per occasion. Climatic data for *Rubus* and *Betula* are given in Fig. 1, and for *Andromeda* and *Empetrum* in Fig. 2

Table 1. The photosynthetic rate of different plant species growing on an elevated area of the mire at Stordalen, measured using the "chamber technique". Values for the reference species outside the chamber are given in italics (see text)

Date Time of day	21. June 1200–1300	5. July 1400–1500	22. July 1400–1500	9. Sept. 1000–1100
Light intensity, and air temperature measured outside the chamber	100000 lux 25° C	50000 lux 15° C	8000 lux 24° C	40000 lux 8° C
Species:	mg CO_2 $h^{-1} \cdot g^{-1}$ d.w.	mg CO_2 $h^{-1} \cdot g^{-1}$ d.w.	mg CO_2 $h^{-1} \cdot g^{-1}$ d.w.	mg CO_2 $h^{-1} \cdot g^{-1}$ d.w.
Andromeda polifolia one-year-old leaves	3.4	2.6	0.6	0.8
Andromeda polifolia current year leaves	0.7	2.9	1.0	*3.6*
Empetrum hermaphroditum one-year-old leaves	1.8	1.6	0.5	—
Empetrum hermaphroditum current year leaves	—	2.1	0.5	—
Rubus chamaemorus	*6.7*	*8.6*	*1.8*	1.3
Betula nana	11.6	12.3	1.2	3.3
Vaccinium uliginosum	—	—	2.9	—
Vaccinium microcarpum	—	—	1.2	—
Bryophytes				
Sphagnum fuscum	0.4	0.2	0.4	0.8
Sphagnum lindbergii	—	0.6	—	—

It is evident that the photosynthetic capacity changes during the growing season, when the photosynthetic rates recorded in spring and autumn are compared with the climatic data for the same periods.

When the photosynthetic capacity of plants still covered with snow was measured, it was found that plants covered with 0.5 m of snow had a positive CO_2-fixation when exposed to light, while plants with a snow cover of 1 m had not. These findings indicate that the plants in question had the capacity to begin photosynthesising as soon as the snow cover disappeared. Some CO_2-fixation may even take place under a snow cover thin enough to let the required amount of light through.

The basis for comparison of the photosynthetic rate of different plant species is important. The result may differ markedly if the photosynthetic rates are expressed on a dry weight or a leaf area basis. When comparing the evergreen species (*Andromeda*, *Empetrum*) with *Rubus* (herb) and *Betula* (deciduous dwarf shrub), the two last-mentioned species had a far higher rate of photosynthesis than had the evergreen species on a dry weight basis (Fig. 3). If expressed on a leaf area basis instead, the evergreen species had rates of the same magnitude—sometimes even higher—as those of *Rubus* and *Betula* (Figs. 1 and 2).

The results from the "assimilation chamber experiments" (Table 1) gave the same picture as the results from the leaf clamp method. The "chamber experiments" also gave some values for the CO_2-fixation rate of the dominant bryophyte *Sphagnum fuscum*. For this species fixation rates of 0.2–0.8 mg CO_2

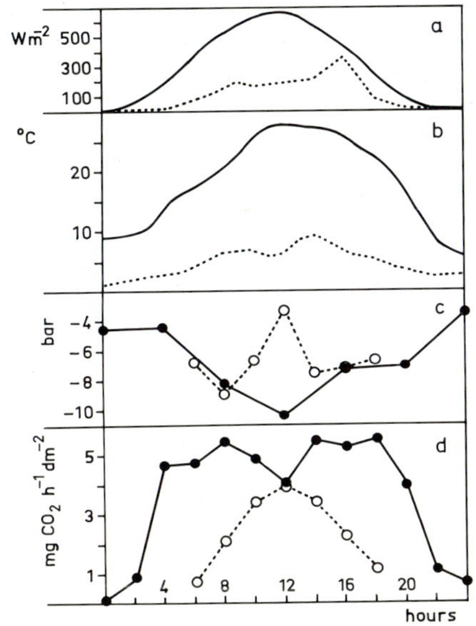

Fig. 4a–d. The diurnal course of photosynthesis of *Rubus chamaemorus* on 20th June (solid lines) and 24th August (broken lines). (a) Solar radiation, (b) temperature at ground level, (c) plant water potentials, (d) photosynthetic rate

g d.w.$^{-1}$ h^{-1} were found, the higher rate being recorded in September, at a time of the season when the rate of the vascular plants had already declined considerably. In spite of the low photosynthetic rate of the bryophytes, they are the dominating primary producers on the mire, owing to their high biomass (SONESSON et al., 1973).

Diurnal Course of Photosynthesis. In an earlier investigation it was found that the diurnal pattern of photosynthesis was very similar for all species investigated (JOHANSSON et al., 1973); the diurnal course of photosynthesis will therefore only be exemplified by two 24-hrs periods recorded for *Rubus* (Fig. 4).

On 20th June the weather was unusually warm, and the sky completely clear. The course of photosynthesis was very symmetrical, with a steep rise at "sunrise" and a steep decline at "sunset". Except for a temporary midday depression, the photosynthetic rate was constant for almost 15 hrs, and the total photosynthetically active period was close to 24 hrs. This supports earlier findings that the maximum rates of photosynthesis are found in the morning, followed by a midday depression. During the morning hours, light is the ratelimiting factor, but at noon other environmental or endogenous factors may be limiting, for example supraoptimal temperatures, water deficits, accumulation of assimilates or endogenous rhythms (SCOTT and BILLINGS, 1964; NEALES and INCOLL, 1968; HODGES, 1967). Since water potentials down to −10 bar were recorded at noon, the midday depression recorded could probably be explained by water stress. The results indicate that even on a mire, which supposedly has good access to water, plants may be exposed to water stress on a hot day. This stress may be due to poorly developed root systems or to low rates of conduction in prostrate stems and rhizomes, which cannot supply water at a sufficient rate to compensate for the loss of water through transpiration. Other measurements during the summer showed that the plant water potentials were very sensitive to weather conditons. A few hours of sunshine gave low water potentials, and a rain shower immediately resulted in higher values.

On 24th August the sky was heavily overcast during most of the day, with occasional drizzle. The curve of photosynthesis was symmetrical with a maximum at noon (Fig. 4). The photosynthetic rate at noon was the same as that on 20th June, in spite of quite different climatic conditions. The photosynthetic production on that day was only one-third of the production recorded on 20th June; one of the main reasons for this was the much shorter day in August. The results presented show clearly that neither the seasonal (Figs. 1–3) nor the diurnal course of photosynthesis (Fig. 4) can be explained by using only the climatic parameters solar radiation, soil and air temperature. To make it possible to predict the photosynthetic production of the plants on the mire, more must be known about the relationship between photosynthesis and factors such as stomatal movements, water relations, accumulation of assimilates and translocation rates. To convert the photosynthetic production into primary production figures, information about the seasonal pattern of respiration and translocation is needed.

References

FLOWER-ELLIS, J. G. K.: Growth and morphology in the evergreen dwarf shrubs *Empetrum hermaphroditum* and *Andromeda polifolia* at Stordalen. Progress Report 1972. Swedish IBP Tundra Biome Project Tech. Rep. **14**, 47–61 (1973).

HODGES, J. D.: Patterns of photosynthesis under natural environmental conditions. Ecology **48**, 234–242 (1967).

JOHANSSON, L.-G.: Chlorophyll content of four species of vascular plant at Stordalen. Progress Report 1973. Swedish IBP Tundra Biome Tech. Rep. **16**, 157–163 (1974a).

JOHANSSON, L.-G.: Photosynthetic rates of some vascular plants on a subarctic mire at Stordalen. Progress Report 1973. Swedish IBP Tundra Biome Project Techn. Rep. **16**, 145–155 (1974b).

JOHANSSON, L.-G., ERIXON, P., LINDER, S.: The use of labelled CO_2 in field studies of photosynthesis and translocation of assimilates. Progress Report 1972. Swedish IBP Tundra Biome Project Tech. Rep. **14**, 78–89 (1973).

NEALES, T. F., INCOLL, L. D.: The control of leaf photosynthesis rate by the level of assimilate concentration in the leaf: A review of the hypothesis. Botan. Rev. **34**, 107–125 (1968).

RYDÉN, B. E.: Abiotic studies, Stordalen 1972. Progress Report 1972. Swedish IBP Tundra Biome Project Tech. Rep. **14**, 14–25 (1973).

RYDÉN, B. E.: Abiotic studies, Stordalen 1973. Progress Report 1973. Swedish IBP Tundra Biome Project Tech. Rep. **16**, 185–202 (1974).

SCHOLANDER, P. F., HAMMEL, H. T., BRADSTREET, E. D., HEMMINGSEN, E. A.: Sap pressure in vascular plants. Science **148**, 339–346 (1965).

SCHOLANDER, P. F., HAMMEL, H. T., HEMMINGSEN, E. A., BRADSTREET, E. D.: Hydrostatic pressure and osmotic potential in leaves of mangroves and some other plants. Proc. Nat. Acad. Sci. U.S. **52**, 119–125 (1964).

SCOTT, D., BILLINGS, W. D.: Effects on environmental factors on standing crop and productivity of an alpine tundra. Ecol. Monogr. **34**, 243–270 (1964).

SHIMSHI, D.: A rapid method for measuring photosynthesis with labelled carbon dioxide. J. Exp. Botany **20**, 381–401 (1969).

SONESSON, M., FLOWER-ELLIS, J. G. K., JONSSON, S., ROSSWALL, T., RYDÉN, B. E.: Word model for Abisko (Stordalen), Sweden. Progress Report 1972. Swedish IBP Tundra Biome Project Tech. Rep. **14**, 4–13 (1973).

Energy Content and Use of Solar Radiation of Fennoscandian Tundra Plants

F. E. Wielgolaski and S. Kjelvik

Introduction and Methods

The calorific content of plant material from Finnish and Norwegian IBP tundra sites described elsewhere (Sonesson et al., 1975), and photosynthetic efficiency of plants at five of the Norwegian sites are presented. Results from other plant calorific investigations in alpine and arctic tundra are given by Bliss (1962a); Hadley and Bliss (1964); Brzoska (1971); Svoboda (1972); Muc (1972, 1973), and Pakarinen and Vitt (1974), while efficiency estimates are given by Bliss (1962b), Webber (1972), and Brzoska (1973).

Harvesting and sorting of plant material are described by Wielgolaski and Kjelvik (1972) and Kjelvik and Kärenlampi (1975). Analyses of various fractions of the most important plant species were carried out in an "adiabatic bomb-calorimeter" on plant material dried at 80° C using the method described by Lieth (1968). The results presented in this paper are based on from three to seven analyses of material harvested at different times throughout mostly two seasons, *Empetrum* from northern Finland from four season.

Incoming radiation approximately within the wavelengths 400–700 nm may be used in photosynthesis. According to e.g. Anderson (1967) 45% of the global radiation is said to be within these wavelengths. Calculations of the average photosynthetic efficiency on a ground area basis per day of the growing season for the total plant production carried out in Norway is therefore based on this percentage of the average radiation measured (Skartveit et al., 1975). The production in kcal m^{-2} day^{-1} for each of the compartments studied are based on the production values estimated by the authors elsewhere (Wielgolaski, 1975; Kjelvik and Kärenlampi, 1975), which are multiplied by calculated average calorific content for the various living compartments at each site. The radiation utilization efficiency by the plants is estimated as the ratio between the sum of calorific contents of living compartments per ground area and the calculated incoming active radiation.

Results and Discussion

In most cases the calorific values were somewhat lower in roots (especially when based on total dry weight and not ashfree weight) than in above-ground

Table 1. Calorific contents as an average of all Norwegian tundra sites in the various compartments of the main plant groups. Kcal g^{-1} dry weight (the values given for pteridophytes are for plants in the wet meadow only and the values for forbs only for the dry meadow). 1 kcal g^{-1} = 4186.8 kJ kg^{-1}

	Shrubs	Forbs	Monocot.	Pteridoph.	Bryoph.	Lichens
Green	4.85	4.53	4.42	4.15	4.08	3.87
Non-green ab. l. root	4.74	4.42	4.39	4.14	3.92	—
Roots	4.61	4.29	4.27	4.10	—	—
Dead	4.47	4.47	4.34	3.39	—	—

Table 2. Calorific contents per g dry weight of some woody plants and lichens from northern Finland. Sampling early June for lichens, mid August for pine and an average of several harvestings in June and July for birch and *Empetrum*

Pinus sylvestris	5357 ± 33
Betula tortuosa (lowest elevation)	5082 ± 35
Betula tortuosa (mountain site)	5178 ± 56
Empetrum hermaphroditum	5660 ± 30
Non-nitrogen fixing lichens	4333 ± 82
Nitrogen fixing lichens	4800 ± 82

non-green organs and highest in green material (Tables 1 and 3). BRZOSKA (1971) and MUC (1972) also generally found higher calorific content in green leaves than in living roots of vascular plants, but SVOBODA (1972) does not report clear trends between vascular plant compartments in his studies on dry Canadian high arctic communities.

The shrubs showed in the present studies clearly higher calorific values for all sites and compartments than the herbaceous vascular plants (both dicotyledons and monocotyledons). The energy in bryophytes was significantly lower than in vascular plants and in lichens still lower (see Table 1). The differences between woody and non-woody plants are also known from other tundra studies (BLISS, 1962a; HADLEY and BLISS, 1964; SVOBODA, 1972; MUC, 1973). The values seem generally to be somewhat lower in the present Fennoscandian tundra studies, especially data from alpine southern Norway (both for the average groups of woody and herbaceous vascular plants) than for the same groups in the Canadian high arctic studies. Similar differences were also found for tundra bryophytes (PAKARINEN and VITT, 1974) and for lichens (SVOBODA, 1972). The calorific values found in high arctic Canada for some lichen species were also higher than those reported from the Austrian alps for species of the same genera (BRZOSKA, 1971). The Norwegian lichens showed, however, the same energy content as found in the Austrian studies.

Both the woody plants and lichens studied in northern Finland showed higher calorific values (Table 2) than in southern Norway (Table 3). The values were lower in non-nitrogen fixing lichens and particularly *Cetraria nivalis* (4020 ± 64 cal g^{-1} in the Finnish studies) than in nitrogen fixing species (Table 2).

There seems to be a general tendency to higher calorific content with latitude. The everlasting days in summer in the polar regions will cause a longer period of apparent photosynthesis per day and may influence the energy transformation in the plants. The more energy-rich compounds in plants may increase compared to other substances by decreasing temperature. EAGLES (1967), SVOBODA (1972), and BERG et al. (1973) all report that the relative and often also the actual content of soluble carbohydrates increases with stress situations as e.g. lower temperatures (while the total carbohydrate content decreases). A possible tendency to higher lipid content with decreasing temperature (HADLEY and BLISS, 1964; MCCOWN and TIESZEN, 1972) may be still more important to the increase of the calorific content with latitude.

The Norwegian tundra data indicate a similar trend to increasing calorific content with elevation and thus decreasing temperature and still more with increasing radiation. The average energy values during the seasons for the species *Deschampsia flexuosa* were e.g. for all compartments lower at the Norwegian subalpine birch forest site than at the low alpine dry meadow (Table 3). They are also lower for birch in southern Norway than in the Finnish birch forest and lower at this low elevation than at the Finnish mountain site (Table 2). Similarly the calorific content of all compartments of e.g. *Empetrum hermaphroditum* increased from the birch forest sites to the alpine heaths. The highest energy content was found in green material of the species at the Finnish sites (Tables 2 and 3) because of the longer summer days in this sub-arctic area. The somewhat shaded understory vegetation in the Norwegian willow thicket showed lower calorific content than similar species at the alpine meadows e.g. for *Carex* spp. and *Festuca rubra* (Table 3). Although the species composition of bryophytes is different at the various sites and the results therefore hardly directly comparable, a tendency could possibly also be traced to an increase in the average calorific content of green bryophyte parts with increasing light intensity in the Fennoscandian tundra studies, at least on an ashfree weight basis. The highest energies in green bryophytes were found in plants at the Norwegian wet meadow (average 4.52 kcal g^{-1} ashfree), at the dry meadow (average 4.49 kcal g^{-1} ashfree) and at the lichen heath (average 4.45 kcal g^{-1} ashfree), lowest in the plants at the birch forest (average 4.35 kcal g^{-1} ashfree), and at the willow thicket (average 4.19 kcal g^{-1} ashfree). The calories in lichens were considerably higher in northern Finland than in the mountain areas studied in southern Norway (Tables 2 and 3).

The differences were very small between the calorific content found for vascular plants at the Norwegian less shaded alpine sites. As for the bryophytes a weak tendency may be traced to higher values at the dry meadow than at the lichen heath by comparison of the same compartments of *Carex bigelowii* and *Vaccinium vitis-idaea*, especially when based on ashfree dry weight. Analyses of green material of *Carex bigelowii* at the dry meadow showed the value 4.76 kcal g^{-1} ashfree and at the lichen heath 4.55 kcal g^{-1} ashfree, of *Vaccinium vitis-idaea* respectively 5.07 kcal g^{-1} ashfree and 5.05 kcal g^{-1} ashfree at the two sites. These values are of the same order as found for the species by BLISS (1962a) in eastern American mountains. There was, however, a tendency to higher calorific values for *Salix herbacea* at the dry meadow than at the wet meadow and also for the forbs analyzed at the two sites.

Table 3. Calories per g dry weight in plants at the Norwegian tundra sites. ± standard error

	Green material	Non green above l. root	Living roots	Dead material
Dry meadow				
Salix herbacea	4679 ± 38	4784 ± 29	4731 ± 27	4334 ± 59
Salix reticulata	4712 ± 187	4650 ± 24	4693 ± 21	4265 ± 56
Vaccinium vitis-idaea	4895 ± 25	4734 ± 32	4731 ± 52	4634 ± 83
Anthoxanthum alpinum	4388 ± 34	4304 ± 66	4300 ± 44	4280 ± 59
Carex bigelowii	4462 ± 22	4392 ± 69	4421 ± 25	4369 ± 62
Deschampsia flexuosa	4514 ± 72	4432 ± 47	4523 ± 38	4476 ± 60
Dryas octopetala	4986 ± 71	4773 ± 74	4805 ± 36	4482 ± 72
Astragalus alpinus	4402 ± 82	4550 ± 84	4511 ± 107	4598 ± 174
Polygonum viviparum	4641 ± 32	4289 ± 63	4072 ± 16	4333 ± 104
Alchemilla vulgaris coll.	4140 ± 87	4250 ± 45	4186 ± 74	4171 ± 39
Ranunculus acris	4174 ± 32	4213 ± 10	4102 ± 37	4122 ± 64
Bryophytes	4195 ± 97	4024 ± 60		
Lichens	3735 ± 43			
Wet meadow				
Salix herbacea	4604 ± 36	4613 ± 39	4388 ± 58	4437 ± 80
Salix lapponum	4692 ± 30	4573 ± 32	4308 ± 34	4364 ± 66
Carex nigra	4437 ± 33	4364 ± 26	4199 ± 29	4404 ± 94
Festuca rubra	4554 ± 62	4176 ± 70	4131 ± 37	4264 ± 44
Ranunculus acris	4149 ± 192	3961 ± 59	4288 ± 90	—
Polygonum viviparum	4435 ± 56	3929 ± 15	4077 ± 63	3912 ± 62
Equisetum palustre	4150 ± 21	4144 ± 33	4102 ± 120	4389 ± 93
Bryophytes	4136 ± 76	3968 ± 62		
Lichen heath				
Salix herbacea	4817 ± 49	4879 ± 42	4602 ± 86	4290 ± 66
Arctostaphylos alpina	4609 ± 43	4567 ± 62	4641 ± 38	4134 ± 77
Empetrum hermaphroditum	5290 ± 58	5123 ± 82	4801 ± 156	4732 ± 94
Vaccinium vitis-idaea	4906 ± 61	4702 ± 39	4452 ± 36	4167 ± 79
Carex bigelowii	4447 ± 33	4423 ± 29	4084 ± 100	4326 ± 45
Bryophytes	3813 ± 120	3556 ± 131		
Lichens	3968 ± 17			
Birch forest				
Betula tortuosa	4844	5049 ± 121	4860	5795
Empetrum hermaphroditum	5108 ± 62	4907 ± 42	4769 ± 76	4874 ± 62
Vaccinium myrtillus	4757 ± 28	4722 ± 46	4656 ± 35	4665 ± 59
Deschampsia flexuosa	4282 ± 33	4425 ± 49	4397 ± 26	4363 ± 44
Bryophytes	4086 ± 47	4083 ± 36		
Lichens	3900 ± 43			
Willow thicket				
Salix lapponum	4585 ± 56	4659 ± 48	4390 ± 36	4418 ± 72
Festuca rubra	4175 ± 30	4148 ± 34	4170 ± 18	4046 ± 54
Carex spp.	4281 ± 10	4204 ± 31	4134 ± 19	4133 ± 26
Deschampsia caespitosa	4017 ± 14	3951 ± 50	4097 ± 40	4061 ± 45
Alchemilla vulgaris coll.	3956 ± 25	3966 ± 42	4005 ± 54	3888 ± 34
Ranunculus acris	3991 ± 43	3986 ± 66	3909 ± 16	3921 ± 13
Geum rivale	3851 ± 37	4131	3832 ± 34	3800 ± 26
Geranium silvaticum	4060 ± 60	3925 ± 47	3874 ± 13	3956 ± 55
Sphagnum spp.	3967	3741 ± 46		
Polytrichum spp.	4042	3964 ± 27		
Pleurozium schreberi	4087 ± 39	3895 ± 51		

Table 4. Daily means of the growing seasons 1969–1971 for Norwegian tundra plant production and estimated global radiation within the wavelengths 400–700 nm, both in kcal m^{-2}. Photosynthetic efficiency in % of the available radiation

	45% of global radiation kcal m^{-2} day	Mean daily increase of energy in plants kcal m^{-2}					Effic. %
		Vascular plants			Cryptogams	Sum	
		Green	Non-green above l. root	Roots			
Wet meadow	1539	8.5	3.1	17.9	6.8	36.3	2.3
Dry meadow	1565	8.2	2.3	10.3	1.9	22.7	1.4
Lichen heath	1588	3.0	1.0	3.6	3.2	10.8	0.7
Birch forest:	1581					28.2	1.7
Understory		10.4	2.8	3.7	1.6	18.5	1.1
Trees		3.8	4.6	1.3		9.7	0.6
Willow thicket:	1539					38.1	2.4
Understory		6.9	1.2	9.1	8.3	25.5	1.6
Willows		3.3	5.0	4.3		12.6	0.8

The highest calorific content of all species both in the Norwegian and Finnish studies was (as in the studies of BLISS, 1962a) found in green material of *Empetrum hermaphroditum* (Tables 2 and 3). *Dryas* had as high calorific content as real shrubs. This was also found by SVOBODA (1972) in the Canadian high arctic.

BRZOSKA (1971) has shown that the energy content of various organs of alpine plants varies with climatic conditions and that there is a seasonal variation dependent on the phenophases of the plants. He found usually high calorific values in the leaves just after bud break in spring. Similar was found also in the present studies as well as generally higher values in a summer with relatively high temperature than in a cooler year. The variation in the content through the season was, however, less than 250 calories for 36% of the analyses carried out and less than 500 calories for 81%. The seasonal variation seemed to be more obvious at the Norwegian dry meadow than at the other Norwegian sites studied with a summer decrease in energy especially in the leaves in July and usually a higher calorific content again in the fall. This was also found in many plants in northern Finland. Max. calorific value of leaves of *Empetrum* were e.g. as high as 6.5 ± 0.04 kcal g^{-1}. Generally, however, the dwarf shrubs showed a weaker seasonal variation tendency than the herbaceous vascular plants in the Norwegian studies, and the variation in cryptgams is very unclear.

The photosynthetic efficiencies calculated for the various Norwegian sites are shown in Table 4 and were within the range 0.6–1.9% found by other authors in connection with tundra areas (BLISS, 1962b; WEBBER, 1972; BRZOSKA, 1973) except for the Norwegian wet meadow and the willow thicket sites which showed somewhat higher values. At those eutrophic sites the soil water is never limiting the production wherefore a high efficiency percentage is found, particularly at the oxygen rich willow thicket. The lichen heath with its low production and also low calorific values of lichens, showed of course, the lowest efficiency. For only the understory in the birch forest, the efficiency was found to be lower than in the dry meadow. The true efficiency in the use of radiation in the understory may, however, be somewhat higher, because of lower radiation below the trees and shrubs than measured above the canopy and used in the calculations.

Acknowledgements. The technical assistants A. POLDER and E. M. LUNDE have carried out the Norwegian calorimeter studies. We wish to thank them and others who have taken part in the work for valuable help and the Norwegian IBP for the funding of the project. We are specially grateful to Professor L. KÄRENLAMPI, Kuopio, Finland, for the kindness of submitting unpublished Finnish material for our comparisons.

References

ANDERSON, M. C.: Photon flux, chlorophyll content and photosynthesis under natural conditions. Ecology **48**, 1050–1053 (1967).

BERG, A., SKRE, O., WIELGOLASKI, F. E., KJELVIK, S.: Leaf areas and angles, chlorophyll and reserve carbon in alpine and subalpine plant communities, Hardangervidda, Norway. In: BLISS, L. C., WIELGOLASKI, F. E. (Eds.): Primary production and production processes, Tundra Biome, pp. 239–254. Edmonton-Oslo: IBP Tundra Biome Steering Committee 1973.

BLISS, L. C.: Caloric and lipid content in alpine tundra plants. Ecology **43**, 753–757 (1962a).

BLISS, L. C.: Net primary production of tundra ecosystems. In: LIETH, H. (Ed.): Die Stoffproduktion der Pflanzendecke, pp. 35–46. Stuttgart: Fischer 1962b.

BRZOSKA, W.: Energiegehalte verschiedener Organe von nivalen Sprosspflanzen im Laufe einer Vegetationsperiode. Photosynthetica **5**, 183–189 (1971).

BRZOSKA, W.: Stoffproduktion und Energiehaushalt von Nivalpflanzen. In: ELLENBERG, H. (Ed.): Ökosystemforschung, pp. 225–233. Berlin-Heidelberg-New York: Springer 1973.

EAGLES, C. F.: Variation in the soluble carbohydrate content of climatic races of *Dactylis glomerata* (cocksfoot) at different temperatures. Ann. Botany **31**, 645–651 (1967).

HADLEY, B., BLISS, L. C.: Energy relations of alpine plants on Mt. Washington, N. H. Ecol. Monogr. **34**, 331–357 (1964).

KJELVIK, S., KÄRENLAMPI, L.: Plant biomass and primary production of Fennoscandian subarctic and subalpine forests and of alpine willow and heath ecosystems. In: WIELGOLASKI, F. E. (Ed.): Fennoscandian tundra ecosystems. Part 1: Plants and microorganisms, pp. 111–120. Berlin-Heidelberg-New York: Springer 1975.

LIETH, H.: The measurement of calorific values of biological material and the determination of ecological efficiency. In: ECKHARDT, F. E. (Ed.): Functioning of terrestrial ecosystems at the primary production level, pp. 233–242. Paris: UNESCO 1968.

MCCOWN, B. H., TIESZEN, L. L.: A comparative investigation of periodic trends in carbohydrate and lipid levels in arctic and alpine plants. In: BOWEN, S. (Ed.): Proceedings 1972 US Tundra Biome Symposium, pp. 40–45. Hanover-New Hampshire: USA CRREL 1972.

MUC, M.: Vascular plant production in the sedge meadows of the Truelove lowland. In: BLISS, L. C. (Ed.): Devon Island IBP project high arctic ecosystem. Project Report 1970 and 1971, pp. 113–145. Edmonton: University of Alberta, Dept. of Botany 1972.

MUC, M.: Primary production of plant communities of the Truelove lowland, Devon Island, Canada. Sedge meadows. In: BLISS, L. C., WIELGOLASKI, F. E. (Eds.): Primary production and production processes, Tundra Biome, pp. 3–14. Edmonton-Oslo: IBP Tundra Biome Steering Committee 1973.

PAKARINEN, P., VITT, D. H.: The major organic components and caloric contents of high arctic bryophytes. Can. J. Botan. **52**, 1151–1161 (1974).

SKARTVEIT, A., RYDÉN, B. E., KÄRENLAMPI, L.: Climate and hydrology of some Fennoscandian tundra ecosystems. In: WIELGOLASKI, F. E. (Ed.): Fennoscandian tundra ecosystems. Part 1: Plants and microorganisms, pp. 41–53. Berlin-Heidelberg-New York: Springer 1975.

SVOBODA, J.: Vascular plant productivity studies of raised beach ridges (semi polar desert) in the Truelove lowland. In: BLISS, L.C. (Ed.): Devon Island IBP project high arctic ecosystem. Project Report 1970 and 1971, pp. 146–185. Edmonton: University of Alberta, Dept. of Botany 1972.

WEBBER, P.J.: Comparative ordination and productivity of tundra vegetation. In: BOWEN, S. (Ed.): Proceedings 1972 US Tundra Biome Symposium, pp. 55–60. Hanover-New Hampshire: USA CRREL 1972.

WIELGOLASKI, F.E.: Primary productivity of alpine meadow communities. In: WIELGOLASKI, F.E. (Ed.): Fennoscandian tundra ecosystems. Part 1: Plants and microorganisms pp. 121–128. Berlin-Heidelberg-New York: Springer 1975.

WIELGOLASKI, F.E., KJELVIK, S.: The methodology of net primary production investigations in Norwegian IBP tundra studies. Rep. Kevo Subarctic Res. Stat. **9**, 24–27 (1972).

WIELGOLASKI, F.E., KJELVIK, S.: Production of plants (vascular plants and cryptogams) in alpine tundra, Hardangervidda. In: BLISS, L.C., WIELGOLASKI, F.E. (Eds.): Primary production and production processes, Tundra Biome, pp. 75–86. Edmonton-Oslo: IBP Tundra Biome Steering Committee 1973.

Distribution of ^{14}C Photosynthates in Norwegian Alpine Plants

A. BERG, S. KJELVIK, and F. E. WIELGOLASKI[1]

Introduction and Methods

Only limited knowledge exists on problems of quantitative and even qualitative transport of photosynthates within plants in tundra areas, while somewhat more is known of agricultural plants (LUPTON, 1966) and forest trees (URSINO and PAUL, 1973). In an arctic sedge meadow community at Barrow, Alaska, ALLESSIO and TIESZEN (1973) have studied the translocation pattern in the gras *Dupontia fischeri*; at Abisko, northern Sweden, some ^{14}C translocation studies are reported on the dwarf shrub *Andromeda polifolia* and the deciduous species *Rubus chamaemorus* by JOHANSSON (1974). The present studies were carried out in the field both on woody plants, forbs and monocotyledons at the Norwegian IBP alpine sites at Hardangervidda, southern Norway (the sites described by SONESSON et al., 1975).

Plants within a transparent cuvette described by KJELVIK et al. (1975) were labelled with ^{14}C from a $Na^{14}CO_3$ solution (0.1 ml to 1.6 l cuvette, activity 0.1 m Ci ml^{-1} solution) upon acidification by HCl (KJELVIK and WIELGOLASKI, 1972) for 15 min, after which the cuvette was removed. This procedure was carried out at different plots three to five times throughout the growing season, generally at two to three weekly intervals. A number of individuals (mostly 4–8) of each of the main species were harvested (total plants above- and belowground) at various time intervals after labelling (from about 2 hrs to 2 weeks), the live material separated into green, non-green above first root, and below first root (in *Polygonum viviparum* the whole bulbous rhizome is taken to be above first root), and fresh weight measured for the fractions. The activity in all plant compartments from the various harvestings was measured by scintillation counting after wet combustion of plant subsamples according to modifications of methods described by SHIMSHI (1969), JOHANSSON et al. (1973) and JOHANSSON (1974). The amount of ^{14}C assimilated by the plants is not measured immediately after labelling. By adding the counts from the various plant compartments of each species in the different harvestings, however, an estimate is found for the total ^{14}C in the plants, and also for losses of ^{14}C between harvestings.

To study the translocation in early spring, plots from vegetation mats placed in isopor boxes in the field (SKRE, 1975) assimilated in ^{14}C atmosphere through most

---

[1] Responsible for the manuscript.

of September, and were left in the field over winter. In early spring the boxes were placed in a cooled daylight phytotron chamber for growth. Whole plants of various species were harvested immediately after snow melt (before any new green tissue was formed) to determine the activity in overwintering material. As soon as new green tissue developed in other plants, this material was harvested at three day intervals to estimate the amount of new material originating from the over-wintering (storage) organs. About two weeks after budbreak, whole plants were again harvested to determine the total radiocarbon activity in the plant, and to calculate the ^{14}C losses in the period from before budbreak.

It is difficult to determine exactly the ^{14}C losses e.g. by respiration during the experiments, and it is therefore also difficult to find absolute values for the incorporated ^{14}C translocated to the various compartments of the plants. The relative value of ^{14}C in the various plant compartments as a percentage of the total radiocarbon recovered in the plant at a given time, will be an estimate of the amount of ^{14}C translocated to the various fractions, and thus is used in this paper. The method has, however, several limitations, as mentioned by e.g. JOHANSSON (1974).

Results and Discussion

It was found by URSINO and PAUL (1973), JOHANSSON et al. (1973) and JOHANSSON (1974) that the highest respiratory losses occur during the first days after labelling. The same is found in the present study. As an average for all plant species studied and for the whole season, about 75% (± 4.9, $n=89$) of the radiocarbon activity (sum of all compartments) measured at the first harvesting (on an average 5 hrs from labelling) was recovered in the second harvesting (on an average about 24 hrs from labelling). After about 48 hrs the activity was reduced to approximately 65% ($\pm 6.3, n=89$) of the value at the first harvesting, and after on an average 8 days, 40% ($\pm 4.5, n=48$) of this value was recovered. These losses are probably mostly caused by respiration, but other factors such as e.g. transport to fine root-hairs not harvested may also have influenced the observed total values. The respiration from the plants will vary strongly with climatic conditions as well as species; and also from green to non-green, and from above to belowground organs of the same species.

The amount of ^{14}C recovered in different parts of the plants in percentage of the total ^{14}C in the plant at harvesting is often relatively constant after some days. This is, in the present study, found for e.g. *Anthoxanthum alpinum* and *Ranunculus acris* in early to mid August (Fig. 1), and is also reported by URSINO and PAUL (1973) and JOHANSSEN (1974). The flattening of the relative translocation curve after some days is partly caused by transformation to more stable compounds.

The evergreen species *Vaccinium vitis-idaea* (and also *Empetrum hermaphroditum*) showed a weak translocation out of the leaves for most of the growing season at the lichen heath (Fig. 1). Usually, more than 80% of the total labelled photosynthates found in the plants remained within the leaves for about two weeks after labelling. The species has no need of large reserves in the non-green organs as most of the photosynthesis apparatus is intact throughout the winter. Only studies in

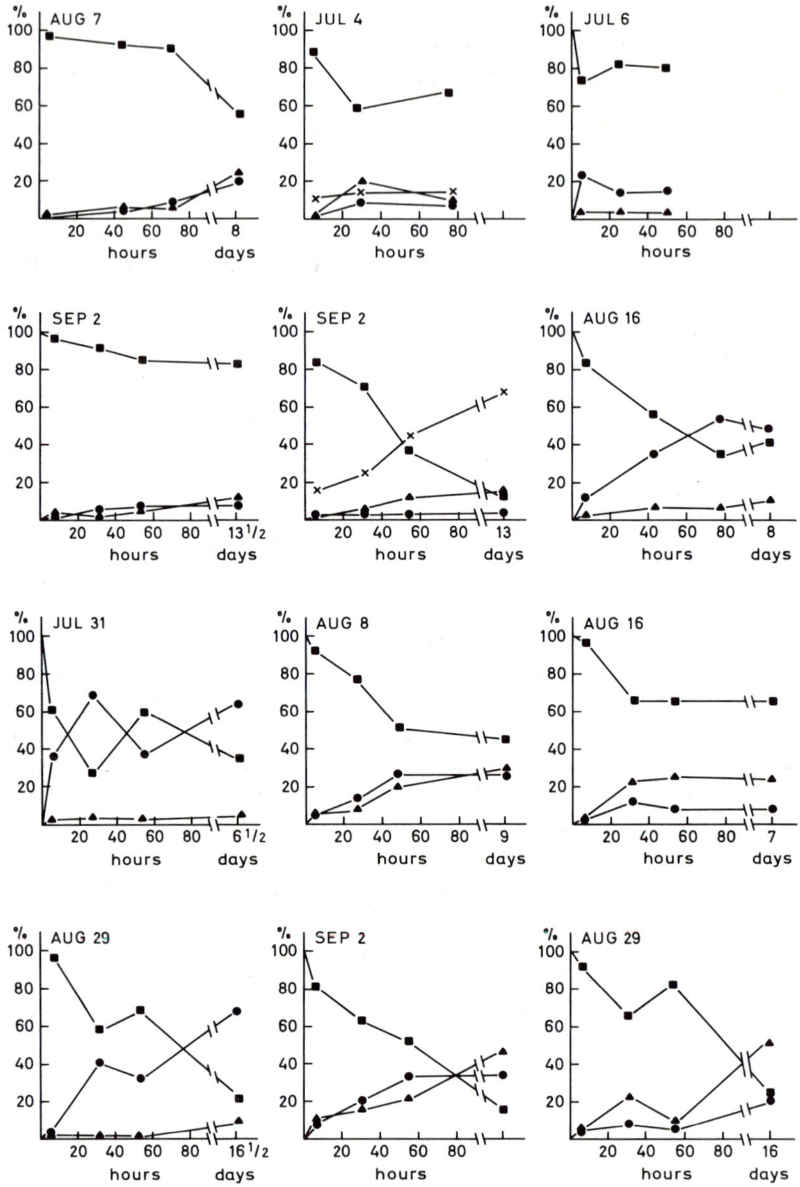

Fig. 1. Relative amounts of radiocarbon (%) recovered in various plant compartments at different time intervals from labelling. Squares = green leaves; Crosses = green stems (*Vaccinium myrtillus*); Dots = non-green above first root; Triangles = below first root. Upper left half of the figure: *Vaccinium vitis-idaea* from the alpine lichen heath; upper mid half: *Vaccinium myrtillus* from the subalpine birch forest; upper half to the right: *Poa alpina* from the alpine dry meadow; lower half to the right: *Anthoxanthum alpinum* from the low alpine willow thicket; lower mid half: *Ranunculus acris* from the wet meadow; lower half to the left: *Polygonum viviparum* from the wet meadow

the first part of August showed stronger translocation out of leaves, indicating a stronger growth of woody parts above and below ground at that time. A similar percentage of the total ^{14}C in the plant was found in non-green, both above and below the first root. Generally, less of the labelled photosynthates were recovered in underground parts in *Vaccinium vitis-idaea* than found by JOHANSSON (1974) for the evergreen *Andromeda polifolia* in northern Sweden. In this species, a rapid and considerable translocation is found; more than 50% of the activity found is recovered below ground after three weeks in early spring and about 50% in fall, but only about 20% in July, a great part of this after a few hours (JOHANSSON et al., 1973).

The translocation in deciduous shrubs in the present study was considerably higher than in evergreens. In *Vaccinium myrtillus*, 50–60% of the total radiocarbon in the plants was translocated out of the leaves in about two days in mid to late July. Much of the labelled CO_2 was recovered in green stems after this time lapse (30–35%), although a relatively low percentage of the total was found in this fraction shortly after labelling. This indicates that the green stems are important storage organs in the plants. This hypothesis is supported by the results of the labelling experiments in fall, where only very little of the total labelled carbon in the plant was recovered in green leaves, and a much higher percentage recovered in green stems than in non-green tissue (Fig. 1). In the dwarf shrub *Salix reticulata* in the Hardangervidda dry meadow, more than 60% seemed to be translocated out of the green material in about two days in mid July. More than 50% was recovered in aboveground woody material, less than 10% in belowground fractions. Most of the shoot length growth is observed to end by mid July in this species in the dry meadow. The climate was favorable for photosynthesis in the period, and therefore "excess" of assimilates were translocated to storage organs. At the phenologically later wet meadow, translocation in *Salix herbacea* was also studied in mid July. Only about 20% of the total radiocarbon photosynthates in plants was then recovered in the non-green material of the species after two days, and mostly in belowground parts. The shoot growth did not end at the time of experiment, and most of the new assimilates were probably directly used for new growth of green material. The non-green fraction above the first root is small in *Salix herbacea* and is the main reason for the higher percentage recovered in fractions below the first root.

In herbaceous plants, a rapid translocation from green material is also usually found for most species (Fig. 1). In *Carex nigra* and *Poa alpina*, however, during the strong growth phase in July and partly also in the bulbous forb *Polygonum viviparum* (Fig. 1), a somewhat lower percentage of the total radiocarbon in the plants is recovered in non-green material after one to two days than shortly after labelling. This may be caused by a relatively higher respiration in belowground than in green material; in *Carex nigra*, particularly, there is also a possibility that the radiocarbon is translocated to rhizomes and fine roots far from the top, and therefore not harvested. The total translocation out of the green material is usually lower in the first harvesting in spring (e.g. in *Poa alpina* as shown in Fig. 1) than later in the season, probably because of the strong new shoot growth at that time using most of the photosynthates produced. This was also found by JOHANSSON (1974) for *Rubus chamaemorus*.

In late July and in August, often more than 50% of labelled photosynthates is recovered in non-green fractions after two days, e.g. *Carex bigelowii* (lichen heath), *Carex nigra* (wet meadow), *Poa alpina* (dry meadow), *Anthoxanthum alpinum* (willow thicket), *Polygonum viviparum* (dry and wet meadow) and *Ranunculus acris* (wet meadow and willow thicket). In most of the herbaceous plants studied, less than 60% of the total activity is found in green leaves after 24 hrs in several harvestings. In *Poa alpina* only 50% and in *Polygonum viviparum* 60% (Fig. 1) of the total amount of radioactive arbon was recovered in the leaves after 5 hrs in, respectively mid and late July. This leaves a lower level of radiocarbon activity remaining in the portion treated with $^{14}CO_2$ than observed by e.g. ALLESSIO and TIESZEN (1973) and NELSON and GORHAM (1957). The most rapid reduction in recoverable amounts in the green fraction is found in the warmest period of summer with high respiratory losses.

As mentioned for *Vaccinium vitis-idaea*, a prolonged translocation out of the leaves may be seen to be caused by new production in non-green compartments. The same tendency was found e.g. in forbs in the present study in the first part of August, and in *Carex bigelowii* and particularly *Carex nigra* somewhat later in the month. In *Polygonum*, the production of non-green compartments in early August seemed to be concentrated in the bulbous rhizomes, while in *Ranunculus* and the two *Carex* species much of the radiocarbon is recovered in the roots at these harvestings (more than 60% in *Ranunculus* roots on August 6th, 6 days after labelling, and more than 20% in *Carex bigelowii* roots on August 15 th, and in *Carex nigra* roots on August 23 rd, 7–8 Days after labelling, while 10–15% was found in *Carex* roots in other harvestings).

By senescence of herbaceous plants in the first two weeks of September, a strong translocation out of the leaves is found to take place both in monocotyledons and forbs, in agreement with the increase in storage polysaccharides found in fall, particularly in some forb roots in the present study (SKRE et al., 1975). About two weeks after labelling only about 20% of the radiocarbon activity was recovered in the leaves of most species. In *Ranunculus* and *Anthoxanthum* most of the ^{14}C-material is found in the compartments below first root (Fig. 1), while very little of the total ^{14}C in the plant seemed to be stored in roots of either *Polygonum*, *Poa alpina* or the two *Carex* species. In *Carex bigelowii* more than 70% was recovered in the rhizomes at harvesting in mid September, two weeks after labelling, and in *Poa alpina* even 80%. These two species have a swollen base of each of the tillers, which obviously serve as storage organs during winter. The storage organ of *Polygonum* is, as expected, the bulbous rhizome; 60–70% of the total of labelled photosynthates in the plants was recovered in this organ in mid September (two weeks after labelling) both in the dry and wet meadow.

The relative amount of radiocarbon in the various compartments found in spring is shown in Table 1 for some species. It is obvious that the translocation to leaves in spring is smaller in *Vaccinium* than in most other species studied. This is caused by the evergreen leaves of *V. vitis-idaea* and the green young stems of *V. myrtillus* which are active for photosynthesis as soon as snow melts in spring. On a relative scale a higher amount of radiocarbon seems to have been translocated to the roots in fall in the smaller plants of *V. vitis-idaea* from the lichen heath site than in the larger plants of the

Table 1. Relative amounts of radiocarbon in various plant compartments in spring, at budbreak and about two weeks later

Species (site)	Percent of total radiocarbon recovered in plants					
	At budbreak			About two weeks after budbreak		
	Green	Non-green above 1. root	Below ground	Green	Non-green above 1. root	Below ground
Vaccinium vitis-idaea (lichen heath)	22	23	55	37	24	39
Vaccinium vitis-idaea (birch forest)	39	25	35	49	30	21
Vaccinium myrtillus (birch forest)	—	—	—	26 (leaves) 25 (green stems)	26	23
Poa alpina (dry meadow)	53	35	11	—	—	—
Carex nigra (wet meadow)	65	31	2	71	15	14
Carex bigelowii (lichen heath)	36	45	19	41	33	26
Anthoxanthum alpinum (willow thicket	—	—	—	31	23	46
Polygonum viviparum (wet meadow)	75	13	10	69	31	0.1
Ranunculus acris (willow thicket)	14	51	35	48	51	0.4

Table 2. Radiocarbon activity in 1000 counts per gram dry weight in spring after incorporation in fall; average of various species of alpine woody plants, monocotyledons and forbs

	At budbreak			4 days after budbreak	7 days after budbreak	10 days after budbreak	Two weeks after budbreak			Activity two weeks after budbreak in % of counts at budbreak		
	Green	Non-green above 1. root	Below ground	Green	Green	Green	Green	Non-green above 1. root	Below ground	Green	Non-green above 1. root	Below ground
Woody plants	167	83	139	396	289	132	161	87	90	96	104	65
Monocotyledons	308	473	135	373	334	288	259	280	125	84	59	92
Forbs	38	48	34	69	84	78	82	30	15	218	62	44
Average all observ.	178	264	98				152	125	95	86	48	98

same species from the birch forest. In accordance with the results of the fall experiments reported earlier small relative amounts of radiocarbon were recovered also in spring in the roots of e.g. *Poa alpina*, *Carex nigra*, *Carex bigelowii* and *Polygonum viviparum* (Table 1). The relative increase of the activity in the roots of the two *Carex* species from budbreak to two weeks later may be seen as a transport from the storage rhizomes to active growing roots in spring. The translocation seems to be strong from roots to green tops of *Ranunculus acris*, while the activity in roots of *Anthoxanthum* is high even two weeks after budbreak.

The radiocarbon activity in spring is generally lower in forbs than in woody plants and highest in monocotyledons (Table 2). More plants of forbs probably ended their growth before the time of ^{14}C incorporation in fall than plants of other groups, and therefore the activity recovered is smallest in forbs. On an average the activity in forb roots is at least as much reduced as the activity in the non-green aboveground parts during the two weeks from budbreak. In monocotyledons, however, most of the ^{14}C recovered in green parts seems to originate from the aboveground storage rhizomes (Table 2). It is obvious that the highest activity in green plant parts is found less than a week after budbreak indicating a short, highly active upward translocation period in spring. In woody plants and monocotyledons the radiocarbon activity in green material is considerably lower again already one to two weeks after budbreak, when the first one to two leaves are fully developed, while the activity in green parts of forbs is not significantly reduced at the last harvesting (Table 2), probably due to late budbreak of some of the species, e.g. *Saussurea alpina* and *Maianthemum bifolium*. The fresh weight per plant, average for all species, was 0.4 g two weeks after budbreak.

The overall reduction in radiocarbon activity in the plants from budbreak to two weeks later indicates, as expected, relatively strong respiration losses in the early growth period in spring. If these losses are taken into account, about 40% of the radiocarbon activity present in the storage organs at budbreak may, on an average, be estimated to have been translocated to growing plant tissue during the two weeks. When high activity is found in the new green material of forbs, monocotyledons and deciduous woody plants (e.g. *Salix herbacea*) also at budbreak, and particularly a few days later, the total upward translocation after ^{14}C incorporation in fall must be even higher. Based on the present observatons, this total upward translocation of radiocarbon may, on an average for all species studied, be estimated to be in the order of 60% of the ^{14}C incorporated when respiration losses are accounted for, but is obviously higher for many species. In the present studies 60% upward translocation is used in the primary production calculations (WIELGOLASKI, 1975), while somewhat higher values are used in the simulation model for the area (SANDHAUG et al., 1975).

Acknowledgements. Cand. mag. G.T. BERG and R. LIEN have assisted with the field and laboratory work and we wish to thank them for valuable help.

References

ALLESSIO, M. G., TIESZEN, L. L.: Patterns of translocation and allocation of ^{14}C photoassimilate *in situ* studies with *Dupontia fischeri*. R. Br. Barrow, Alaska. In: BLISS, L. C., WIELGOLASKI, F. E. (Eds.): Primary production and production processes, Tundra Biome, pp. 219–230. Edmonton-Oslo: IBP Tundra Biome Steering Committee 1973.

JOHANSSON, L. G.: The distribution and fate of ^{14}C photoassimilated by plants on a subarctic mire at Stordalen. Progress report 1973. Swedish IBP Tundra Biome project Tech. Rep. **16**, 165–172 (1974).

JOHANSSON, L. G., ERIXON, P., LINDER, S.: The use of labelled CO_2 in field studies of photosynthesis and translocation of assimilates. Progress report 1972. Swedish IBP Tundra Biome Project Tech. Rep. **14**, 78–89 (1973).

KJELVIK, S., WIELGOLASKI, F. E.: Photosynthesis at Hardangervidda measured by ^{14}C-method (in Norwegian, English summary). IBP Norden **8**, 91–100 (1972).

KJELVIK, S., WIELGOLASKI, F. E., JAHREN, A.: Photosynthesis and respiration of plants studied by field technique at Hardangervidda, Norway. In: WIELGOLASKI, F. E. (Ed.): Fennoscandian tundra ecosystems, Part 1: Plants and microorganisms, pp. 184–193. Berlin-Heidelberg-New York: Springer 1975.

LUPTON, F. G. H.: Translocation of photosynthetic assimilates in wheat. Ann. Appl. Biol. **57**, 355–364 (1966).

NELSON, C. D., GORHAM, P. R.: Uptake and translocation of C-14 labelled sugars applied to primary leaves of soybean seedlings. Can. J. Botan. **35**, 339 (1957).

SANDHAUG, A., KJELVIK, S., WIELGOLASKI, F. E.: A mathematical simulation model for terrestrial tundra ecosystems. In: WIELGOLASKI, F. E. (Ed.): Fennoscandian tundra ecosystems, Part 2: Animals and systems analysis, in press. Berlin-Heidelberg-New York: Springer 1975.

SHIMSHI, D.: A rapid method for measuring photosynthesis with labelled carbon dioxide. J. Exp. Botan. **20**, 381–401 (1969).

SKRE, O.: CO_2-exchange in Norwegian tundra plants studied by infrared gas analuzer technique. In: WIELGOLASKI, F. E. (Ed.): Fennoscandian tundra ecosystems, Part 1: Plants and microorganisms, pp. 168–183. Berlin-Heidelberg-New York: Springer 1975.

SKRE, O., BERG, A., WIELGOLASKI, F. E.: Organic compounds in alpine plants. In: WIELGOLASKI, F. E. (Ed.): Fennoscandian tundra ecosystems, Part 1: Plants and microorganisms, pp. 339–350. Berlin-Heidelberg-New York: Springer 1975.

SONESSON, M., WIELGOLASKI, F. E., KALLIO, P.: Description of Fennoscandian tundra ecosystems. In: WIELGOLASKI, F. E. (Ed.): Fennoscandian tundra ecosystems, Part 1. Plants and microorganisms, pp. 3–28. Berlin-Heidelberg-New York: Springer 1975.

URSINO, D. J., PAUL, J.: The long term fate and distribution of ^{14}C photoassimilated by young white pines in late summer. Can. J. Botan. **51**, 683–687 (1973).

WIELGOLASKI, F. E.: Primary productivity of alpine meadow communities. In: WIELGOLASKI, F. E. (Ed.): Fennoscandian tundra ecosystems, Part 1: Plants and microorganisms, pp. 121–128. Berlin-Heidelberg-New York: Springer 1975.

Pigment Structure of Vascular Plants, Mosses and Lichens at Hardangervidda, Norway

A. BERG

Introduction and Methods

The pigment structure of plants at the Norwegian IBP sites described by SONESSON et al. (1975) is determined on a ground area unit harvested at two weekly intervals and consisting of 10 circular areas, each 0.02 m^2 (random choice). Sampling and harvesting of green material were carried out as described by WIELGOLASKI and KJELVIK (1972). The chlorophyll content of the shrub and tree layer is determined on a dry weight (85° C) unit from twig samples harvested at random throughout the summer. Conversion to chlorophyll content per unit ground area is carried out by using the destructive woody plant analysis (KJELVIK, 1973) and the leaf area index measurements (BERG et al., 1975).

The vascular plant material was subdivided into three subsamples (2–3 g) directly after harvesting, packed in aluminium foil and frozen down to about −20° C. At the same time subsamples were dried to determine dry weight percentage. After a period of from one to four days the subsamples were quantitatively extracted in ice-cold 80% acetone (MACKINNEY, 1941; ARNON, 1949; BRUINSMA, 1963), and the extracted pigment was filtered through a sintered glass filter (Porosity 3).

Bryophytes and lichens were dried in the dark at room temperature, and three subsamples (1–3 g) were thoroughly crushed in a ball mill for pigment analyses. For lichens the subsamples were soaked in 80% pyridine, and then the pigments extracted in methanol (HILL and WOOLHOUSE, 1966). Bryophytes were extracted directly in methanol without any preliminary soaking in pyridine. Before filtering, 80% acetone was added to extracts of both lichens and bryophytes.

All pigment extraction was carried out under minimal light conditions and at about 4° C; $CaCO_3$ was added to buffer the solution (BERG, 1974). The extractions were immediately placed in a dark water bath at 20° C in volumetric flasks, and exact volume measurements were made at this temperature before the absorbance was measured at 475 nm, 645 nm, 652 nm, and 663 nm in a Beckmann DB-G spectrophotometer (chamber cooled to 20° C).

Results and Discussion

The highest chlorophyll content per m^2 for the total vegetation is found around August 1st. In the birch forest, trees and shrubs account for about 870 and the understory 630 mg chlorophyll per m^2, while in the willow thicket the shrubs yield 600 mg m^{-2} and the understory an equal amount. For dry meadow, wet meadow and lichen heath the results are, respectively, 630, 480, and 160 mg m^{-2} (Berg et al., 1973).

A well developed lichen vegetation is found at the lichen heath site, and 60% of the chlorophyll is concentrated here. At the other sites lichens are scarce and represent less than 1% of the chlorophyll total.

The bryophyte cover is relatively well developed at the wet meadow, where it accounts for about 20% of the total chlorophyll. At the willow thicket, dry meadow and birch forest sites the results are, respectively, 15, 10, and 7% chlorophyll from the bryophytes.

The amount of chlorophyll per unit dry weight varies greatly throughout the growing season. Maximum values in vascular plants were found around July 15th (Fig. 1). The dry meadow, wet meadow and willow thicket sites, high in forbs and sedge and grass species have the highest chlorophyll values per unit dry weight.

The birch forest, and particularly the lichen heath sites with a relatively high percentage of heath species, show lower chlorophyll content per g dry weight

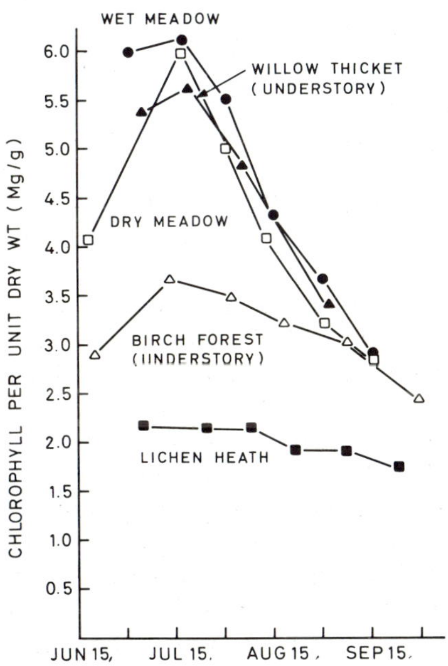

Fig. 1. Seasonal variation in amount of chlorophyll per unit dry tissue in representative transactions of all vascular plants from one sub-alpine (birch forest) and four low mid-alpine sites

Table 1. The pigment structure in some species at different sites at Hardangervidda (each value is an average of measurements from harvestings around July 5th, July 20th and August 15th). C_a = chlorophyll a, C_b = chlorophyll b, OD_{475}/OD_{652} = carotenoid ratio. *Empetrum* shoots = last two year's shoots

	Compartments	$C_a + C_b$ mg per g dry weight	mg chlorophyll per g nitrogen	mg chlorophyll per 100 cm^2 leaf area	Ra/b	OD_{475}/OD_{652}	% nitrogen[a]
Birch forest							
Betula tortuosa	leaves	6.71	336	5.37	4.07	1.98	2.00
Betula nana	leaves	6.45	273	2.80	3.99	1.93	2.36
Empetrum hermaphroditum	shoots	3.11	259	—	3.38	2.00	1.20
Maianthemum bifolium	green	5.24	—	1.62	3.48	1.83	—
Deschampsia flexuosa	green	3.68	361	—	3.30	1.97	1.02
Nardus stricta	green	4.70	336	—	3.60	1.82	1.40
Lichen heath							
Empetrum hermaphroditum	shoots	1.76	154	—	3.54	2.21	1.14
Carex bigelowii	green	4.80	244	5.39	4.04	2.58	1.97
Festuca ovina	green	4.00	288	—	3.58	2.33	1.39
Dry meadow							
Salix reticulata	leaves	6.26	331	5.44	3.77	1.95	1.89
Dryas octopetala	green	4.56	290	4.75	3.75	2.01	1.57
Carex bigelowii	green	4.63	210	3.26	3.82	2.68	2.21
Wet meadow							
Salix lapponum	leaves	5.85	297	4.12	4.27	2.23	1.97
Salix herbacea	leaves	7.37	286	3.86	3.80	1.91	2.58
Ranunculus acris	green	9.41	466	7.97	3.55	1.98	2.02
Carex nigra	green	4.66	197	4.09	4.12	2.60	2.36
Willow thicket							
Salix lapponum	leaves	5.60	235	5.83	4.33	2.11	2.38
Alchemilla vulgaris	green	6.34	409	3.43	3.77	2.04	1.55
Geranium silvaticum	green	9.00	479	7.14	3.71	1.90	1.88

[a] Data submitted by WIELGOLASKI and KJELVIK (pers. comm.).

(Fig. 1). This tendency to lower chlorophyll content on a weight basis for ericaceous material than for non-ericaceous plants is also shown by TIESZEN and JOHNSON (1968) and BLISS (1970). Some of the most important species have been investigated separately for the chlorophyll content, and a wide variation is found (Tables 1 and 2). The amount of chlorophyll per g dry weight is, for instance, three times greater in the forbs *Ranunculus acris* and *Geranium silvaticum* than in the evergreen heath species *Vaccinium vitisidaea* and *Empetrum hermaphroditum.*

Table 2. The pigment structure in the three *Vaccinium* species measured at a sub-alpine site (the birch forest) and two low-mid alpine sites at Hardangervidda (each value is an average of measurements from harvestings around July 5th, July 20th, and August 15th). C_a = chlorophyll a, C_b = chlorophyll b, OD_{475}/OD_{652} = carotenoid ratio

Species	Compartments	Site	C_a+C_b in mg per g dry weight	mg chlorophyll per g nitrogen	mg chlorophyll per 100 cm² leaf area	Ra/b	OD_{475}/OD_{652}	% nitrogen
Vaccinium myrtillus	leaves	Birch forest	6.02	267	2.31	3.64	1.75	2.34
Vaccinium myrtillus	leaves	Lichen heath	4.73	204	2.38	3.85	1.99	2.90
Vaccinium myrtillus	leaves	Dry meadow	5.21	171	2.00	3.99	1.99	3.07
Vaccinium uliginosum	leaves	Birch forest	6.26	234	3.41	3.91	1.93	2.67
Vaccinium uliginosum	leaves	Lichen heath	5.60	195	4.55	4.10	2.08	2.82
Vaccinium uliginosum	leaves	Dry meadow	5.72	186	—	4.31	2.08	3.09
Vaccinium vitis-idaea	last year leaves	Birch forest	3.37	176	2.06	3.74	1.82	2.31
Vaccinium vitis-idaea	last year leaves	Lichen heath	2.20	123	2.37	3.93	2.09	1.87
Vaccinium vitis-idaea	last year leaves	Dry meadow	2.29	106	—	4.10	2.10	2.67
Vaccinium vitis-idaea	earlier yrs leaves	Birch forest	3.24	227	4.94	3.54	1.87	1.56
Vaccinium vitis-idaea	earlier yrs leaves	Lichen heath	2.18	157	3.93	3.68	1.97	1.03
Vaccinium vitis-idaea	earlier yrs leaves	Dry meadow	2.74	147	4.17	3.77	1.96	1.81

Marked differences also exist from site to site. Plants which grow in the understory of the sub-alpine birch forest have a much higher chlorophyll content both per unit dry weight and per unit nitrogen, compared with the same species growing on low-mid alpine sites (Tables 1 and 2). It looks as though the quantitative chlorophyll pattern, besides being strongly species-influenced, is also to a certain degree modified by environment. Berner (1949) showed that the amount of chlorophyll per unit fresh weight increases in young coniferous plants with decreasing light intensity. Billings and Mooney (1968) also demonstrated a decrease in chlorophyll per g dry weight with increase in height above sea level. The higher chlorophyll amount in the understory at the birch forest site, which lies 780 m.a.s.l. compared with the lichen heath (1220 m) and the dry meadow (1275 m), is taken to represent mainly a combination of these two effects.

The chlorophyll content per unit nitrogen for species at the lichen heath (generally low N-content is higher than in the dry meadow (Table 2). The chlorophyll

Table 3. Amount of chlorophyll per unit dry weight, chlorophyll a:b ratio (Ra/b) and carotenoid ratio ($OD_{475}:OD_{652}$) in representative transactions of all vascular plants, bryophytes and lichens from different sites at Hardangervidda (each value is an average of measurements taken around July 5th, July 20th, and August 15th)

	Vascular plants					Bryophytes				Lichens	
	Birch forest (under-story)	Lichen heath	Dry meadow	Wet meadow	Willow thicket (under-story)	Birch forest	Dry meadow	Wet meadow	Willow thicket	Birch forest	Lichen heath
mg chlorophyll per g dry weight	3.5	2.1	5.0	5.5	5.3	2.7	2.9	2.0	1.5	0.7	0.3
Ra/b	3.17	3.31	3.44	3.60	3.59	2.64	2.52	2.53	2.77	3.13	2.25
OD_{475}/OD_{652}	1.86	2.26	2.02	2.15	1.92	1.90	1.93	1.61	2.14	—	—

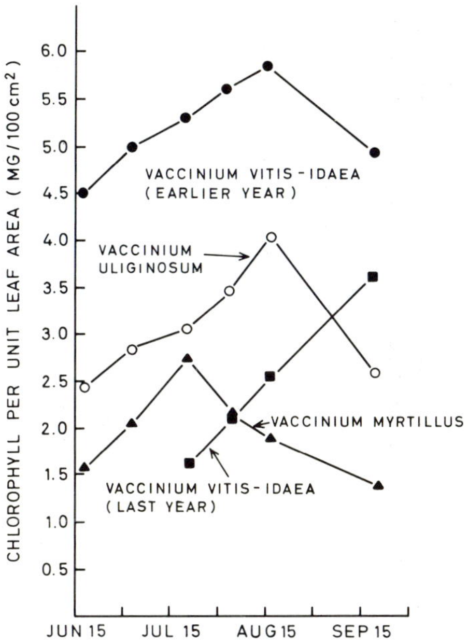

Fig. 2. Amount of chlorophyll per unit leaf area in the *Vaccinium* species at the birch forest site

content per unit dry weight is, on the other hand, lower than could be expected from light conditions, compared to the dry meadow (Table 2). The explanation may be that plants at the lichen heath are more xeromorphous than the same species in the dry meadow, because of water and nutrition stress (BERG et al., 1975).

The chlorophyll content in the bryophytes was generally somewhat lower than in vascular plants (Table 3). The amount per dry weight unit seems to decrease with decreasing soil moisture and/or nutrients. In lichens, the chlorophyll content per unit dry weight is much lower than in both vascular plants and bryophytes. Values of the order of 0.2 mg to 1.0 mg chlorophyll per g dry weight are reported in *Cladonia alpestris* (KÄRENLAMPI, 1970) and in other species (WILHELMSEN, 1959), and similar chlorophyll values are also found in lichens from the different sites at Hardangervidda (Table 3).

The chlorophyll content per unit leaf area in different species varies greatly (Tables 1 and 2 and Fig. 2), from 1.6 mg per 100 cm^2 leaf area in *Maianthemum bifolium* to 8.0 mg in leaves of *Ranunculus acris*. *Vaccinium myrtillus* shows a maximal amount of chlorophyll per unit leaf area around August 1st at all the sites included in the study, while *V. uliginosum* and earlier years' leaves of *V. vitis-idaea* show a maximum somewhat later (Fig. 2). Current years' leaves of *V. vitis-idaea* have an almost linear increase in chlorophyll throughout the course of the summer (Fig. 2). The same general picture as is shown in Fig. 2 for *Vaccinium*

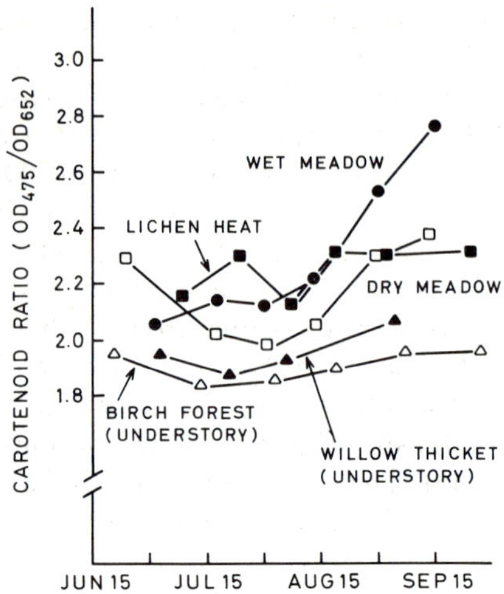

Fig. 3. Seasonal variation in the carotenoid ratio (OD_{475}:OD_{652}) in representative transactions of all vascular plants

species at the birch forest site, is true also for these species at the lichen heath and dry meadow.

The ratio between chlorophyll a and chlorophyll b (*Ra/b*) is given by BLISS (1966) to be between 3.0 and 3.3 for some common alpine plant communities. Several research workers cite corresponding values. In representative transections of vascular plants from Hardangervidda the *Ra/b* lies between about 3.2 and 3.6, values which represent an average from harvestings around July 5th, July 20th, and August 15th (Table 3). The ratio is lower in the subalpine birch forest than in the alpine communities for the same species (Table 2). The species variation is fairly high (Tables 1 and 2), but the seasonal variation is relatively low. For vascular plants at the birch forest site, for example, *Ra/b* varies from 3.0 to 3.3 in six measurements from June 21st to September 28th.

The chlorophyll a to chlorophyll b ratio seems to be dependent on the light conditions at the sites. In addition to the altitudinal increases, the influence of light is indicated by a gradual decrease in *Ra/b* values from layer to layer downwards in the vegetative profile (Table 1). For example, *Ra/b* is about 4.1 in leaves of *Betula tortuosa*, 4.0 in leaves of *Betula nana* and 3.3 in *Deschampsia flexuosa* at the birch forest site. The same pattern is shown even in the low vegetation in the wet meadow. For example, in *Salix lapponum* leaves, which are mostly found above the other vegetation, the *Ra/b* is 4.3, in *Carex nigra* 4.1, and in *Salix herbacea* leaves, which are part of the lowest layer of the vegetational cover, the *Ra/b* is only 3.8.

For the bryophyte cover *Ra/b* lies between about 2.5 and 2.8, and the ratio is relatively constant from site to site (Tab. 3), while seasonal variations are also small. For lichens, different research workers report highly varying *Ra/b* from about 2.0 in *Cladonia alpestris* (KÄRENLAMPI, 1970) to 3.0 and up to 6.5 in *Parme-*

lia species and *Xanthoria* species (WILHELMSEN, 1959). For all the sites investigated at Hardangervidda, the *Ra/b* in lichen species is between 2.0 and 3.5. The variation between species is greater for lichens than for vascular plants.

A careful analysis of the carotenoids is very time-consuming, but the ratio of light absorbtion at 475 nm to absorbtion at 652 nm ($OD_{475}:OD_{652}$) is an estimate of the amount of carotenoids relative to chlorophylls. TIESZEN and JOHNSON (1968) report a carotenoid ratio of between 1.68 and 2.43 in vascular plants from four arctic tundra communities. At the different sites at Hardangervidda, the carotenoid ratio is found to be of the same order (Tabs. 1 to 3). The ratio is (as found for *Ra/b*) lower in vascular plants of the subalpine birch forest than in low-midalpine vegetation at the other sites. This implies that plants which are adapted to higher light intensities have a lower concentration of oxidized pigments (RABINOWITCH, 1945). The carotenoid ratio is for all plant types studied usually somewhat higher in spring and autumn, compared to midsummer (Fig. 3). The reason is that the seasonal variations in the carotenoid content are smaller than the corresponding chlorophyll variations.

Acknowledgements. My wife, Cand. mag. G.T. BERG, assisted with most of the field and laboratory work. Mrs. M. ESPELAND translated the manuscript. The author wishes to thank them both for their valuable help.

References

ARNON, D. J.: Copper enzymes in isolated chloroplasts. Polyphenoloxidase in *Beta vulgaris*. Plant Physiol. **24**, 1–15 (1949).

BERG, A.: A description and critical evaluation of methodology used in pigment analyses on alpine plants at Hardangervidda, Norway. Rapp. Høyfjellsøkol. Forsk. Stn, Finse, Norge **1**, 20–35 (1974).

BERG, A., KJELVIK, S., WIELGOLASKI, F. E.: Measurements of leaf areas and leaf angles of plants at Hardangervidda, Norway. In: WIELGOLASKI, F. E. (Ed.): Fennoscandian tundra ecosystems, Part 1: Plants and microorganisms, pp. 103–110. Berlin-Heidelberg-New York: Springer 1975.

BERG, A., SKRE, O., WIELGOLASKI, F. E., KJELVIK, S.: Leaf areas and angles, chlorophyll and reserve carbon in alpine and subalpine plant communities, Hardangervidda, Norway. In: BLISS, L. C., WIELGOLASKI, F. E. (Eds.): Primary production and production processes, Tundra Biome, pp. 239–254. Edmonton-Oslo: IBP Tundra Biome Steering Committee 1973.

BERNER, E., JR.: Chlorophyll production by young coniferous plants at different light intensities. Univ. of Bergen Nat. Sci. series **6**, 1–32 (1949).

BILLINGS, W. D., MOONEY, H. A.: The ecology of arctic and alpine plants. Biol. Rev. **43**, 481–529 (1968).

BLISS, L. C.: Plant productivity in alpine microenvironments on Mt. Washington, New Hampshire, Ecol. Monogr. **36**, 125–155 (1966).

BLISS, L. C.: Primary production within arctic tundra ecosystems. In: FULLER, W. A., KEVAN, P. G. (Eds.): Proceedings Conferance productivity and conservation in northern circumpolar lands, Edmonton 1969, pp. 77–85. Morges: IUCN Publ. new ser. **16**, 1970.

BRUINSMA, J.: The qualitative analysis of chlorophylls a and b in plant extracts. Photochem. Photobiol. **2**, 241–249 (1963).

HILL, D. J., WOOLHOUSE, H. W.: Aspects of the autecology of *Xanthoria parietina* agg. Lichenologist **3**, 207–214 (1966).

KÄRENLAMPI, L.: Distribution of chlorophyll in the lichen *Cladonia alpestris*. Rep. Kevo Subarctic Res. Stat. **7**, 1–8 (1970).

KJELVIK, S.: Biomass and production in a willow thicket and subalpine birch forest, Hardangervidda, Norway. In: BLISS, L.C., WIELGOLASKI, F.E. (Eds.): Primary production and production processes, Tundra Biome, pp. 115–122. Edmonton-Oslo: IBP Tundra Biome Steering Committee 1973.

MACKINNEY, G.: Absorption of light by chlorophyll solutions. J. Biol. Chem. **140**, 315–322 (1941).

RABINOWITCH, E.I.: Photosynthesis and related processes. Vol. 1. New York: Interscience Publ. 1945.

SONESSON, M., WIELGOLASKI, F.E., KALLIO, P.: Description of Fennoscandian tundra ecosystems. In: WIELGOLASKI, F.E. (Ed.): Fennoscandian tundra ecosystems, Part 1: Plants and microorganisms, pp. 3–28. Berlin-Heidelberg-New York: Springer 1975.

TIESZEN, L.L., JOHNSON, P.L.: Pigment structure of some arctic tundra communities. Ecology **49**, 370–373 (1968).

WIELGOLASKI, F.E., KJELVIK, S.: The methology of net primary production investigations in Norwegian IBP tundra studies. Rep. Kevo Subarctic Res. Stat. **9**, 24–27 (1972).

WILHELMSEN, J.B.: Chlorophylls in the lichens *Peltigera*, *Parmelia* and *Xanthoria*. Botan. Tidskr. **55**, 30–36 (1959).

4. Decomposers

Introduction

T. ROSSWALL

The biomass of microorganisms is a small component of most ecosystems, but is of fundamental importance in regulating ecosystem function. In terrestrial ecosystems, the greater proportion of primary production passes directly through the decomposer chain (GOKSØYR, this section), while in some aquatic systems 90% passes through the herbivores (PETRUSEWICZ and MACFADYEN, 1970; SATCHELL, 1974).

The high relative activity of the decomposer organisms is shown by the ratio of energy flow (kcal m^{-2} day^{-1}) to biomass (g m^{-2}), which has been calculated to 1000 for soil bacteria, but only 0.5 for deer (ODUM, 1968). Not only do microorganisms have a high general activity, but they have also a rapid turnover rate. At the Swedish site, the ratio of yearly production to biomass was found to be 6.3 for bacteria and only 0.2 for the vascular plants (but varies strongly between vegetation types). This stresses the difference in life strategies of the various biotic components of the ecosystem, where the use of biomass values overemphasizes the importance of large organisms, the use of numbers overemphasizes the importance of small organisms (ODUM, 1971).

The study of energy flow was a major goal of the IBP tundra studies (DAHL and GORE, 1968). This was therefore studied indirectly at the Fennoscandian tundra sites by several methods. The energy flow through the soil microbial population was estimated by considering the productivity of the fungi (HANSSEN and GOKSØYR, this section) and bacteria (CLARHOLM, LID-TORSVIK and BAKER, this section) by measuring the decomposition of plant litter (ROSSWALL et al., this section) or model substances (BERG et al., this section) or by determining the amounts of carbon released from the soil system by respiration (SVENSSON et al., this section).

There are inherent difficulties in the estimation of production of microbial biomass in soil. The lack of precise information is the more serious, as the importance of bacteria and fungi lies in their rapid turnover rather than a large biomass. The productivity of soil bacteria was estimated at the Swedish tundra site (CLARHOLM et al., this section) to 22.3 g m^{-2} yr^{-1}; calculations of fungal productivity for the Norwegian sites have indicated values ranging from 31–36 g m^{-2} yr^{-1} (dry lichen heath) to 265–305 g m^{-2} yr^{-1} (birch forest) (HANSSEN and GOKSØYR, this section). These investigations cover topics hitherto largely neglected, and stress the point made by ODUM (1971) that as the size of the organisms under study decreases, the study of rates of function rather than measurements of standing crop becomes increasingly necessary.

Although productivity of microorganisms is a good general measure of activity, most ecological studies on the activity of these organisms have been concerned with measurements of process rates, e.g. litter decomposition, nitrification and nitrogen fixation. Plant litter decomposition studies were an important part at the Fennoscandian IBP tundra sites (ROSSWALL et al., this section). They were concerned with all major plant species, but unfortunately largely restricted to aboveground litter; root decomposition studies were only performed at the Norwegian sites. The yearly decomposition rates varied between 0.3% for *Sphagnum fuscum* at Stordalen to 42% for *Carex nigra* at the Hardangervidda wet meadow. Moisture, being either too high or too low, seemed to be the main factor limiting decomposition rates at the Fennoscandian tundra sites. A high soil moisture content sometimes results in accumulation of organic matter as peat, thus making important nutrients unavailable to plants, whereas litter in the litter layer often is dry, resulting in low decomposition rates (KÄRENLAMPI, 1971; LOHAMMAR and ROSSWALL, 1974).

The litter bag studies were supplemented with soil respiration measurements (SVENSSON et al., this section) giving direct measurements of the carbon flow from the soil to the atmosphere. The soil respiration rates were significantly correlated with barley straw decomposition data at the five Hardangervidda sites ($r = +0.96$, $p < 0.02$) and indicated that the same factors limited decomposition both above and below ground.

Most earlier studies concerning the carbon balance of ecosystems have only considered the evolution of carbon dioxide. Although the production of methane amounts to an average of only a few per cent of the annual carbon loss to the atmosphere (EHHALT, 1973) this could make up a major part in wet ecosystems like mires. At the Swedish mire site methane accounted for nearly half the yearly carbon loss in the wet minerotrophic areas (SVENSSON et al., this section). This methane production was equivalent to 54 g C m^{-2} yr^{-1} which can be compared with the data given by EHHALT (1973) of 158 g C m^{-2} yr^{-1} for paddy fields and 79–158 g for swamps. The latter data are probably too high as a result of the experimental procedure with laboratory measurements, whereas the Swedish data were from field studies. Soil respiration measurements are valuable for assessing the total biological activity of soils, but the production of other gaseous products than carbon dioxide should always be considered.

Although energy is driving ecological processes, the rates at which they occur are controlled in natural systems by nutrient availability (REICHLE et al., 1974). Nutrient status in the Fennoscandian tundra regions is covered in another section of this volume. Papers of decomposition and soil processes in the tundra will also be found in HOLDING et al. (1974).

References

DAHL, E., GORE, A.J.P. (Eds.): Proceedings Working Meeting on Analysis of Ecosystems, Tundra Zone. Oslo: Norwegian National IBP Committee 1968.

EHHALT, D.H.: Methane in the atmosphere. In: Carbon in the Atmosphere. U.S. Atomic Energy Commission. Symposium Series **30**, 144–158 (1973).

HOLDING, A. J., HEAL, O. W., MACLEAN, S. F., FLANAGAN, P. W. (Eds.): Soil Organisms and Decomposition in Tundra. Stockholm: IBP Tundra Biome Steering Committee 1974.

KÄRENLAMPI, L.: Weight loss of leaf litter on forest soil surface in relation to weather at Kevo Station Finnish Lapland. Rep. Kevo Subarctic Res. Stat. **8**, 101–103 (1971).

LOHAMMAR, T., ROSSWALL, T.: The use of the "Abisko model" for simulating *Rubus chamaemorus* litter decomposition. In: Progress Report 1973. Swedish IBP Tundra Biome Tech. Rep. **16**, 97–107 (1974).

ODUM, E. P.: Energy flow in ecosystems: A historical review. Am. Zool. **8**, 11–18 (1968).

ODUM, E. P.: Fundamentals of Ecology. 3rd edition. Philadelphia-London-Toronto: W. B. Saunders Co. 1971.

PETRUSEWICZ, K., MACFADYEN, A.: Productivity of terrestrial animals: Principles and Methods. IBP Handbook No. 13. Oxford-Edinburgh: Blackwell Scientific Publications 1970.

REICHLE, D. E., O'NEILL, R. V., SHUGART, H. H., HARRIS, W. F.: Principles of energy and material exchange in ecosystems (Abstract). In: CAVE, A. J. (Ed.): Proceedings of the First International Congress of Ecology, p. 11. Wageningen: Centre for Agricultural Publishing and Documentation 1974.

SATCHELL, J. E.: Introduction. Litter-Interface of animate/inanimate matter. In: DICKINSON, C. H., PUGH, G. J. F. (Eds.): Biology of Plant Litter Decomposition, Vol. 1, pp. xiii–xliv (1974).

Decomposition, Microbiology, and Ecosystem Analysis

J. GOKSØYR

In natural, unharvested terrestrial ecosystems, more than 90% of the carbon assimilated by plants will eventually be released as carbon dioxide by decomposition in litter and soil. The need for a detailed analysis and understanding of this process as part of the functioning of the whole ecosystem is obvious. This discussion of some of the major problems in this special field of ecosystem analysis is especially related to the situation in areas where invertebrates play a minor role in the energy transfer process from litter to carbon dioxide (mainly due to absence of lumbricides). These transfers are then predominantly carried out by microorganisms. To such areas belong tundra regions (WHITFIELD, in prep.; HEAL et al., in prep., cf. also Table 1).

Definitions

A number of different methods have been used to assay decomposition rates within IBP. When comparing these methods, they seem to rely upon different implicit assumptions as to what decomposition is. Thus, a number of different operational definitions of decomposition have existed, such as: Using the litter bag weight loss technique; the rate at which litter disintegrates to the extent that it disappears from the bag. Using chemical analysis of litter components; the rate of disappearance of major chemical compounds (e.g. cellulose) of plant origin. Using measurements of carbon dioxide evolution rate; the total mineralization rate of organic carbon compounds. Using measurements of heat evolution (microcalorimetry); total energy loss rate.

Of these alternative definitions, the last two have the advantage that they define decomposition as a process with a well-characterized end product. It is also likely that ecologists today use decomposition and (carbon) mineralization as synonymous concepts. None of the above-mentioned definitions contain biological implications. According to them, decomposition might just as well be a purely non-biological process.

Decomposition can be defined explicitly by stating that it is the sum of biological, chemical and physical processes going on in litter and soil, and by which organic carbon compounds are transformed to carbon dioxide, water und nutrient ions, with concomitant liberation of heat energy and increase in entropy. The biological part of these processes is predominantly carried out by microorganisms, which are aided by, or predated upon, by invertebrate animals. The chemical processes are for the greatest part oxidations and polymerizations of compounds either released by microbial activity on the litter, or formed as metabolic products. Among the physical processes, leaching and comminution are probably the most important.

Decomposition is a very essential part of the soil formation processes, so that a soil profile may be considered as an expression of the decomposition history of the site.

The Biological Processes

Normally, living plants are quite resistant to attack from saprophytic microorganisms, although plant surfaces are the habitat for several types of epiphytic bacteria and fungi. Plants may, however, become attacked by parasitic forms, especially fungi. Characteristic of most of these parasites are their relatively slow growth, and the ability to utilize pectin and partly cellulose in the cell walls without disturbing the plant cells to a high degree. These parasitic forms thus rarely consume more than a very small part of the biomass of the plant. As the plant organs become senescent and dying, either because of parasitic attack or as part of their natural development, the resistance towards saprophytic or weakly parasitic forms is reduced. Standing dead material is quite heavily invaded by fungi, and partly by bacteria; so are freshly fallen litter and dead plant roots.

Among both the fungi and the bacteria, numerous types are found which possess the enzymes required to attack cellular or structural components in plants. In so far as they are present and the environmental conditions permit growth, these organisms will compete for the nutrients in the litter. It can be argued that fungi must be the most efficient type in this competition, as long as the plant material is structurally fairly intact. A mycelial organism can spread its hyphae, covering a large volume of substrate with a small expenditure of cellular material, and do this in a stationary surface film of water, where spreading of bacteria (with the possible exception of actinomycetes) will be considerably hampered. A large number of fungi, including cellulolytic forms, possess efficient proteolytic enzymes. It is thus a fallacy to assume that the nutrients in the litter are distributed in a sort of predetermined manner between bacteria and fungi. Both groups of organisms compete for all there is available to them. The success of each species is dependent on uptake efficiency, growth potential, spreading efficiency and potential to inhibit competitors under the environmental conditions that exist when the nutrient supply is made available.

The situation will become somewhat different when the litter has passed through the gut of a saprovore invertebrate. This passage will result in a mechanical disintegration of the plant material, probably removal of some of the nutrients, but addition of mucous material from the gut and also some nitrogenous compounds. By this passage, the material will become thoroughly inoculated with soil bacteria and fungal spores. This process creates conditions where bacteria can compete more successfully, especially as the spreading problem is reduced. In finely ground, lignified plant material, microbial cellulose decomposition proceeds more rapidly than in intact material (VIRTANEN and NIKKILÄ, 1946).

An early biological step in the decomposition process is the utilization of plant chemical compounds to build up microbial tissue, with simultaneous liberation of carbon dioxide and heat energy. From the preceding, it can be assumed that fungi play a dominating role in this step. Excluding the parasitic and semiparasitic forms, soil-inhabiting fungi are traditionally divided into three main groups, according to the types of carbon compounds that they can utilize (BURGES, 1958). These are the sugar fungi, being only able to utilize simple sugars and in addition proteins and organic acids; the cellulolytic fungi, which in addition can also utilize cellulose in fairly pure forms as in parenchymatic cell walls; and the xylolytic fungi, which in addition to what the former can utilize, can also use lignin and lignified cellulose. The growth rates of these different forms decrease considerably in the order mentioned, and this determines their relative competitive abilities for simple substrates and protein.

The mycorrhizal fungi represent an ecological group of their own, or perhaps several groups. The Hymenomycetes forming ectotrophic mycorrhiza on forest trees are, as a rule, very weakly cellulolytic or ligninolytic, and are considered to receive most of their organic nutrients from the tree, which they in turn supply with inorganic nutrient salts (MELIN, 1959, 1963; BJÖRKMAN, 1949). There are, however, a number of other mycorrhizal types, of which far less is known than the ectotrophic mycorrhiza with Hymenomycetes. In the tundra context, especially the ericacean mycorrhiza (NIEUWDORP, 1969) and the vesicular-arbuscular (Endogone) types on a number of plant species (e.g. ALI, 1969 for *Nardus stricta*) should be mentioned. Even though the fungi of some of these mycorrhizas may be expected to be litter-decomposing, the ecological importance of mycorrhizal fungi in general probably lies on the primary production side, in relation to mineral nutrition. These fungi should then, in an ecological sense, be considered as extensions of the plant root system, and their biomass and growth (including the fructifications) as an energy expenditure by the plant, necessary to obtain a sufficient supply of inorganic nutrients. The microbiological problem in this context is how to identify hyphae of mycorrhizal fungi in the soil.

The role of bacteria in the decomposition process is more difficult to assess than that of the fungi. It is important to bear in mind the large difference in dimension between bacterial cells and fungal hyphae. Scanning electron micrography and fluorescence antibody technique indicate that fungal hyphae may be quite densely covered with bacteria. Hyphae invaded by bacteria are also quite frequently seen in the light microscope. This indicates that some of the bacteria species receive their nutrients from hyphal exudates or from hyphal material. In the microstructure of soil, it is tempting to talk about the "hyphosphere" in a similar sense as the rhizosphere. But it must be stressed that basically, there is a free-for-all competition for all material which each individual species register as a nutrient. And most probably, the largest part of bacteria live in aggregates dispersed on or between the soil particles (ZVYAGINSEV, 1973).

One of the major obstacles in attempting to assess the role of bacteria is that we at present know extremely little about both their biomass and their overall activity in intact soil. Characteristic for natural soil systems, and especially for soils in cold climates, is the very great difference between values obtained by plate counts and by microscopic counts (CLARHOLM et al., 1975). The plate count values are certainly too low to account for all the potentially metabolic active bacteria. On the other hand, total count values evidently take in a substantial amount of bacterial "ghosts". There is, however, electron-microscopic evidence, also from our laboratory, that bacterial forms in the soil may be too small in size to be easily identified in the light microscope. Thus, we can merely state that we do not know, and do not at present have appropriate techniques to study, the major part of the bacterial populations in the soil.

WINOGRADSKY (1925) divided the soil inhabiting bacteria into two major groups, the zymogenous bacteria and the autochthonous. The first ones have a rapid growth response when supplied with external nutrients. The amounts of these bacteria in soil will therefore show major fluctuations in the microhabitats they occupy. The autochtonous types have a lower growth potential, and do not respond to the supply of external nutrients with population fluctuations. Although the classical example of autochthonous bacteria are *Arthrobacter* forms which may be isolated on plates, and although it means stretching WINOGRADSKY's concept to a certain extent, it is tempting to assume that the plate count values mainly represent the zymogenous population, while the total (direct) counts (minus the plate counts) represent the autochtonous populations. If so, the number of autochthonous bacteria in alpine and tundra soils must be about 10^3–10^4 times that of zymogenous bacteria. It is only possible by inference to guess at the main carbon and energy sources for the autochthonous forms. They are found associated with soil organic matter, and in numbers fairly proportional to the mass of organic matter in the soil. This makes it tempting to assume that they utilize humic acids and related complex aromatic compounds, and thus represent the final step in the decomposition process. If these assumptions are true, it also follows that they must be very slow-growing, with a correspondingly low metabolic rate. Probably most of them are aerobic, since breakdown of soil organic matter is very much slowed down under anaerobic conditions.

Kinetics of the Decomposition Process

The assimilatory efficiency of a microorganism lies between 35 and 50% under aerobic conditions. This would mean that in the first step of conversion of plant compounds to microbial tissue, at least 50% is simultaneously converted to carbon dioxide. The upper level is more difficult to assess, because microorganisms may have a loose coupling between anabolic and catabolic metabolism, and can respire without growing (resting metabolism, maintenance metabolism) (STEPHENSON, 1930). This is, however, only possible when growth is inhibited, e.g. due to lack of an essential nutrient. Assuming that nitrogen is the growth-limiting nutrient in litter, we may conclude that the fungal population will grow to a level that is determined by the amount of nitrogen available, and it will convert 12 to 15 times as much organic carbon to carbon dioxide. If all nitrogen in the litter is available (no leaching loss), the C/N ratio should thus fall near 12 to 15, meaning again that a reutilization of nitrogen at least twice is necessary to reduce the C/N ratio from that normally found in fresh litter (40 to 50) to the values around 20, which are normal in the A_0 horizons in the soil. If the leaching loss is substantial, the degree of recycling must be higher, whereas if there is a high degree of resting metabolism and thus a lower assimilatory efficiency, the degree of recycling is lower. This recycling can occur either as part of a nutrient chain, or it can with fungi occur as an intracellular process. These reasonings are complicated by the fact that there are very great differences in the utilization velocities of the different types of plant compounds. Thus the primary decomposition step is spread over a span of time, which may last several years for strongly lignified litter. This could mean that the late primary decomposers might risk a severe nitrogen shortage, bearing in mind the high C/N ratio in litter. But during this slow process of lignocellulose decomposition, the secondary processes of decomposition are simultaneously going on, with microbial tissue breaking down and utilized anew as nutrients. This results in a continous low supply of nitrogenous compounds, for which the fungi growing on lignified plant tissue can also compete. It is perhaps significant that these fungi almost exclusively require organic growth factors which they hardly can receive directly from the litter, but which may be supplied by bacteria (FRIES, 1938). These fungi, which mostly are basidiomycetes, also seem to have very efficient nitrogen saving systems. As the hyphae grow, the older parts are autolyzed and the nutrients can be transported to the younger parts. Thus, with an efficient turnover of a limited nitrogen pool, these fungi may be able to decompose a substantial amount of lignified cellulose. Related forms living in wood are thus able to decompose material with a C/N ratio of up to 700 to 1000 (HUNGATE, 1940; GOKSØYR, 1965).

The overall decomposition rate follows a curve which is superficially exponential, but which it is more reasonable to assume as being composed of several curves added together, all based on the general expression:

$$-\frac{dC_n}{dt} = k_n \cdot C_n \qquad (1)$$

where k_n is the decomposition rate constant for either a chemical component of the litter, or a major metabolic product from the decomposer organisms, and C_n is the mass of this component. The biological assumption behind this function is that the activity of a decomposer organism is proportional to the mass of substrate it acts upon (JENNY et al., 1949; OLSON, 1963). This may be more evident if taken in two steps: The decomposition velocity is proportional to the microbial biomass, and this again is proportional to the amount of decomposing material. Thus:

$$-\frac{dC}{dt} = p \cdot M, \quad \text{and} \quad M = q \cdot C, \qquad (2)$$

where M is the biomass of the decomposer microorganisms. The coefficient p reflects strongly the "quality" of the material to be decomposed, but will of course also be a function of external conditions, especially temperature and moisture, q is the relative amount of microbial biomass in the soil organic matter. The data available from the IBP tundra sites indicate that the microbial biomass based on soil organic matter (q) varies less than when expressed per g dry soil (CLARHOLM et al., 1975). Factors that may affect q are: Internal space (soil structure), antagonism and predation (predation may also influence p in a positive manner) and nutritional quality.

HEAL et al. (in prep.) presents a comprehensive discussion, on how temperature, moisture and litter "quality" affects decomposition rates in tundra regions.

If C is a metabolic product (e.g. microbial mass; monomers from polymeric compounds in the litter), one will have a consecutive reaction sequence, of the type:

$$\frac{dC_2}{dt} = k_1 C_1 - k_2 C_2 \tag{3}$$

where C_1 is the compound preceding C_2 in the reaction sequence. In litter decomposition, the physical process of leaching may also be considered to follow first order reaction kinetics. Otherwise, there are obviously long series of consecutive reactions and also side reactions that make the total kinetics especially of the first stages of the process extremely complicated.

At the steady state situation, all decomposition velocities will remain constant, and likewise all masses (compartments) that are engaged in the process. For simplicity, this is treated in a chemically closed system, and the annual fluctuations are left out. We will then have:

$$L - \frac{dC}{dt} = p \cdot q \cdot C \tag{4}$$

where L is the input of litter or a litter component, and C is the mass of this component. In steady state, we must have $dC/dt = 0$, which gives

$$p \cdot q = \frac{L}{C}. \tag{5}$$

If L is litter input and C is soil organic matter, $p \cdot q$ will be the decomposition coefficient (JENNY et al., 1949). Simplified like this, one can state that the ratio between primary production and soil organic matter is a measure of overall decomposition efficiency in a system. There are, however, two points to be made. One is that (5) is valid only for a system in steady state. For an accumulating system (4) must be solved:

$$C = \frac{L}{p \cdot q}(1 - e^{-p \cdot q \cdot t}) \quad \text{for} \quad C = 0 \quad \text{at} \quad t = 0. \tag{6}$$

By this equation, the accumulation rate can be estimated if $p \cdot q$ is known or vice versa.

The other point is whether it is correct to use litter input in (4), since the first decomposition processes including leaching have such complicated kinetics. It may be more appropriate to enter that fraction of the litter which has the lowest decomposition rate, i.e. the lignocellulose fraction.

If C (= soil organic matter) is considered to be one homogenous compartment, the ratio L/C is analogous to the dilution rate in a chemostat. The average retention time of organic carbon in this compartment will then be C/L, which for the Stigstuv (Hardangervidda) dry meadow and lichen heath (SONESSON et al., 1975) gives values around 100 years. This is probably far too low. (Carbon dating figures exist only for the wet meadow, which obviously is not in steady state. The age found here by radiocarbon dating is about 5000 years at 50 cm depth and 8000 years in the bottom layers of the peat at about 75 cm depth.) The conclusion must be that the soil organic matter is composed of several compartments, with a successively decreasing decomposition rate.

Conclusions

Much of the understanding or biological interpretation of the decomposition process is qualitative, and cannot be transcribed with profit into a simulation model. On the other hand, a formalized, symbolic expression of the main processes and the populations taking part in them, e.g. in the form of a block diagram, may prove very useful, both because if presented in an appropriate manner, it will help to clarify the complex pattern of reaction sequences in the decomposition process, and because it leads to considerations about which parts of the process are of major importance, and which ones should be neglected at the level of resolution on which the model operates.

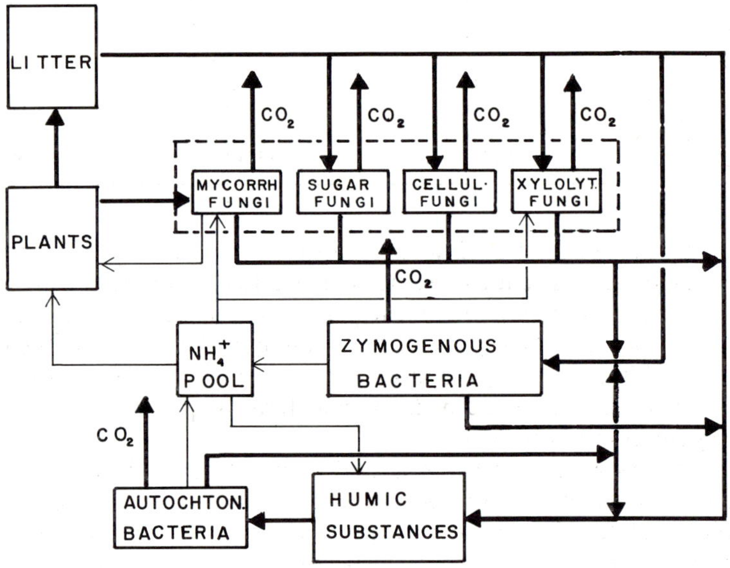

Fig. 1. Transfers of carbon and nitrogen during decomposition of litter. Only microbial processes are included. Thick lines indicate transfer of carbon or of carbon and nitrogen. Thin lines indicate transfer of nitrogen only

Table 1. Energy transfers and biomass compartments for the decomposition process at the Stigstuv (Hardangervidda) sites. Invertebrate data are from SOLHØY (personal communication), litter production data from WIELGOLASKI and KJELVIK (personal communication), fungal and bacterial data from HANSSEN, LID-TORSVIK and TORSVIK (1972). Protozoan values are guesses, based upon Devon Island estimates. The energy transfer model is according to WHITFIELD (in prep.). All values are given in cal. 10^3 m^{-2} $year^{-1}$. Wet meadow has been calculated as being in steady state

	Lichen heath	Dry meadow	Wet meadow
Transfers			
Litter input	860	2390	3480
To saprovores	1.6	5.6	11.0
To fungi	774	2150	3130
To bacteria	621	1702	2500
To protozoa	26	52	52
To fungivores	16	30	23
To carnivores	0.4	0.5	0.5
Saprovore respir.	0.6	2.2	4.4
Fungivore respir.	6.4	12	9.2
Carnivore respir.	0.2	0.2	0.2
Microbial respir.	853	2375	3465
(Fungal resp.	464	1290	1880
Bacterial resp.	371	1049	1549
Protozoan resp.)	18	36	36
Compartments (Biomasses)			
Saprovores	0.8	1.8	5.5
Fungivores	1.6	3.0	2.3
Carnivores	0.4	0.5	0.5
Protozoa	2	4	4
Fungi	260	1290	1880
Bacteria (total c.)	2750	3100	4700
Bacteria (plate c.)	0.03	2.5	0.1
Soil organic matter	43000	151000	>240000

For basic purposes, a model describing carbon and nitrogen transfers seems best suited, especially since it can easily be expanded to cover also the transfer of other nutrients and their recycling in the plant-decomposer system. Further, such a model describes as nearly as possible "real" biological processes in the system. It will also—and this is probably the most important part of using it—clarify where information is most urgently needed, and which techniques have to be developed in order to gain this information.

Figure 1 shows a model for the microbiological parts of the decomposition process. In podsolic, organic and polar soils it will cover something like 80–90% of the carbon transfers. Table 1 gives estimated energy transfer rates and compartment values for the three Stigstuv (Hardangervidda) sites, expressed and calculated according to WHITFIELD's (in prep.) energy transfer scheme for decomposition.

To sum up, there is according to the author's opinion little real difference between decomposition and other nutrient webs in an ecosystem. But whereas ecologists for other nutrient webs tend to put emphasis on the organisms, the tendency when it comes to decomposition is to place the emphasis on the process, and to leave out the organisms. This has aggravated in an appalling lack of precise information of the biology of the decomposition process—and thus in reality of a very prominent part of the whole ecosystem.

References

ALI, B.: Occurrence and characteristics of the vesicular-arbuscular endophyte of *Nardus stricta*. Nova Hedwigia **17**, 409–425 (1969).

BJÖRKMAN, E.: The ecological significance of the ectotrophic mycorrhizal association in forest trees. Svensk Botan. Tidskr. **43**, 223–262 (1949).

BURGES, A.: Microorganisms in the soil. London: Hutchinson Univ. Library 1958.

CLARHOLM, M., LID-TORSVIK, V., BAKER, J. H.: Bacterial populations of some Fennoscandian tundra soils. In: WIELGOLASKI, F. E. (Ed.): Fennoscandian tundra ecosystems, Part 1: Plants and microorganismus, pp. 251–260. Berlin-Heidelberg-New York: Springer 1975.

FRIES, N.: Über die Bedeutung von Wuchsstoffen für das Wachstum verschiedener Pilze. Symb. Botan. Upsalienses **3**, 1–189 (1938).

GOKSØYR, J.: Wood-decomposing fungi and their adaptation to life in wood. Advan. Science **1965**, 147–156.

HANSSEN, J. F., LID-TORSVIK, V., TORSVIK, T.: Microbiology. In: Annual Report 1972, pp. 75–92. Oslo: Norsk IBP 1973. (In Norwegian).

HEAL, O. W., FLANAGAN, P. W., FRENCH, D. D., GOKSØYR, J., HOLDING, A. J., MACLEAN, S. F., WHITTAKER, J. B.: Decomposition and soil processes in Tundra. In: MOORE, J. J. (Ed.): Tundra and related habitals. Cambridge: University Press (In prep.).

HUNGATE, R. E.: Nitrogen content of sound and decayed coniferous woods and its relation to loss in weight during decay. Botan. Gaz. **102**, 382–392 (1940).

JENNY, H., GESSEL, S. P., BINGHAM, F. T.: Comparative studies of decomposition rates of organic material in temperate and tropical regions. Soil Sci. **68**, 419–432 (1949).

MELIN, E.: Mycorrhiza. In: RUHLAND, W. (Ed.): Handbuch der Pflanzenphysiologie, Vol. XI, pp. 605–638. Berlin-Heidelberg-New York: Springer 1959.

MELIN, E.: Some effects of forest tree roots on mycorrhizal Basidiomycetes. In: NUTMAN, P. S., MOSSE, B. (Eds.): Symbiotic Associations XIII Symposium Soc. Gen. Microbiol. pp. 125–145. Cambridge: University Press 1963.

NIEUWDORP, P. J.: Some investigations on the mycorrhiza of *Calluna*, *Erica* and *Vaccinium*. Acta Botan. Neerl. **18**, 180–196 (1969).

OLSON, J. S.: Energy storage and the balance between producers and decomposers in ecological systems. Ecology **44**, 322–321 (1963).

SONESSON, M., WIELGOLASKI, F. E., KALLIO, P.: Description of Fennoscandian tundra ecosystems. In: WIELGOLASKI, F. E. (Ed.): Fennoscandian tundra ecosystems, Part 1: Plants and microorganisms, pp. 3–28. Berlin-Heidelberg-New York: Springer 1975.

STEPHENSON, M.: Bacterial Metabolism. London: Longmans, Green and Co. 1930.

VIRTANEN, A. I., NIKKILÄ, O. E.: Cellulose fermentation in wood-dust. Suomen Kemistilehti **19B**, 3–4 (1946).

WHITFIELD, D.: Energy flow through tundra ecosystems. In: MOORE, J. J. (Ed.): Tundra and related habitats. Cambridge: University Press (in prep.).

WINOGRADSKY, S.: Études sur la microbiologie du sol. I. Sur la methode. Ann. Inst. Pasteur **39**, 299–354 (1925).

ZVYAGINSEV, D. G.: Microbial ecology as studied by luminescence microscopy in incident light. In: ROSSWALL, T. (Ed.). Modern Methods in the Study of Microbial Ecology. Bull. Ecol. Res. Comm. (Stockholm) **17**, 61–65 (1973).

Biomass and Production of Soil and Litter Fungi at Scandinavian Tundra Sites

J. F. HANSSEN and J. GOKSØYR

Most soil fungi are decomposer organisms, using plant litter compounds as their carbon and energy source. In order to understand the biology and the dynamics of the decomposition process, knowledge is necessary of the species composition, properties, biomass, and activities of the fungal populations. This paper deals with the estimates of the biomass of mycoflora in the Norwegian (Hardangervidda) and Swedish (Abisko, Stordalen) IBP tundra sites (SONESSON et al., 1975), and a discussion of fungal production. Details of the Swedish investigations were reported by HAYES (1973) and RHEINBERG (1975).

Methods

Soil samples at the Hardangervidda sites were taken as 2 cm dia. cores, inserted horizontally into the soil at different depths. At the Abisko sites, samples were taken as 10 cm dia. cores, inserted vertically in the soil.

Measurements of hyphal lengths at Abisko were carried out according to JONES and MOLLISON (1948), whereas for the Norwegian sites a variant of this method, using membrane filtration of the homogenized soil suspension was used (HANSSEN et al., 1974). For this technique, a statistical analysis of the microscopic measurements was developed. Normally, the number of hyphal fragments measured was sufficient to give an accuracy of 10–15%.

At Abisko, a dilution plate technique as modified by JENSEN (1962) was used for quantitative and qualitative estimates of the fungal populations (HAYES and RHEINBERG, 1975). The plates were incubated for 8–10 days at 25° C before assay.

Conversion of hyphal lengths to biomass requires a number of assumptions about average properties of fungal hyphae in soil. Only to a very rough extent can these properties be quantified by direct measurements or laboratory studies. The most crucial value is the average hyphal diameter, since this figure is squared when calculating hyphal volumes. Other values required are specific gravity of the mycelium and the percent dry weight. In order to obtain uniformity in the data presented here, all hyphal lengths have been converted to biomass using the same set of values. These are:

Hyphal diameter: 3.5 μm (PARKINSON et al., 1971; SAITO, 1955).

Specific gravity: 1.3 (PARKINSON et al., 1971; GRAY and WILLIAMS, 1971). SAITO (1955) gives the figure 1.12, while the others give values between 1.0 and 1.5.

Percent dry weight: 17% (HANSSEN, unpublished results).

For conversion of biomass to caloric values, it is in addition necessary to know the caloric value of the mycelium. This has been estimated from the assumption that the lipid content of the mycelium is 10%, and the rest mainly composed of carbohydrates and proteins (COCHRANE, 1958). This gives a value of 4500 calories per g dry weight. It should be noted that the lipid content and thus the caloric value, will vary with nutrient and growth conditions for the same species, and is also different for different species.

Results and Discussion

Table 1 shows mean values for total hyphal lengths and fungal biomass at the different sites. Generally, fungi can be stated to be the dominating non-photosynthetic group of organisms at all sites (with the possible exception of bacteria, if

Table 1. Fungal biomass in Scandinavian tundra soils. Mean values calculated per m^2 down to 10 cm soil depth

Sites	Hyphal length in 10^6 meters	Fungal biomass in grams	Fungal biomass in % of organic matter	Fungal biomass in Kcalories
Hardangervidda, Norway				
Wet meadow	87	182	1.08	820
Dry meadow	54	113	0.53	600
Lichen heath	24	50	0.77	225
Birch forest	107	204	1.26	1000
Snow bed	33	69	1.27	310
Abisko, Sweden				
Stordalen				
Hummock	28	58	0.71	260
Depression	28	58	0.71	260
Njulla	11	22		100

Table 2. Hyphal lengths at different depths at the Norwegian Hardangervidda IBP tundra sites. Average values, given as m hyphae g^{-1} dry soil

Site	0–2 cm	2–7 cm	7–10 cm	Soil type
Wet meadow	6850	4500	2400	Deep organic soil
Dry meadow	2900	1400	350	Podzol
Lichen heath	730	150	50	Shallow podzol
Birch forest	3600	1300	250	Podzol
Snow bed	850	325	nil	Shallow podzol

Table 3. Hyphal lengths and number of propagules at different soil depths at four Swedish IBP tundra sites, Stordalen, Abisko (H: hummocks, D: depressions). Hyphal lengths given as 10^2 m g^{-1} dry soil, numbers as 10^4 g^{-1} dry soil (from RHEINBERG, 1975)

Depth (cm)	79 H		79 D		15 H		15 D	
	Length	No	Length	No	Length	No	Length	No
0– 3	22	106	13	13	13	108	18	154
3– 5	14	44	33	34	24	54	11	44
7–10	16	17	24	9	22	2	28	9
12–15	20	8	26	1	39	2	16	2
18–21	26	5	28	0	29	2	15	3
25–28	31	2	Frozen		10	0	0.3	3
Number of propagules per m of hyphae								
0– 3		480		100		830		860
3– 5		320		103		225		400
7–10		106		37		9		32
12–15		40		4		5		12
18–21		19		—		7		20
25–28		6		—		—		90

estimates are based on microscopic counts). The highest fungal biomass per area unit was found in the Norwegian birch forest (Maurset) and the Stigstuv (Hardangervidda) alpine wet meadow.

At the podzolic sites, the highest values were found in the organic (A_{0-1}) horizons, and with a drastic reduction in the lower horizons. Such a reduction with depth was not found in the organic soils (Tables 2 and 3). When evaluating these figures, it should be born in mind that the technique used does not distinguish between live and dead hyphae, and that the ratio between these may differ with depth. At Abisko, Stordalen where dilution counts were made, the number of propagules was found to decrease rapidly below the top 5 cm (Table 3), while the hyphal lengths remained more or less constant (RHEINBERG, 1975). This may, however, also be caused by a species differentiation or a physiological differentiation with depth, with relatively less sporulation in the deeper layers.

The data available do not permit very definite conclusions to be drawn about seasonal variations in fungal biomass. At the Stordalen site, samples were taken only in July and August, while at Hardangervidda the samplings for one year (1972) were extended to start in May under the snow cover and to last until just before the snow covered the sites in late September. However, only monthly samplings were carried out. The results obtained for the top two cm for the Hardangervidda sites are shown in Table 4. At all the sites, a seasonal variation takes place. Generally, there is a spring or summer bloom in June or July, and a less marked autumn bloom in August or September. The data are too sparse to permit valid statistical correlations with climatic factors, but superficially the bloom in the drier sites seems to be correlated with high moisture.

At Stordalen, the data available from hummocks and depressions in 1972 and 1973 (HAYES, 1973; RHEINBERG, 1975) may be interpreted as being related to the moisture conditions of these two years (Table 5).

Table 4. Seasonal variations of the hyphal lengths in the top two cm of soil at Hardangervidda, Norway during 1972. Hyphal lengths given as $m\,g^{-1}$ dry soil, soil moisture in % of dry soil

	May 5		June 10		July 9		August 1		September 12	
	Hyphal length	Moist. (%)	Hyphal length	Moist. (%)	Hyphal length	Moist. (%)	Hyphal length	Moist. (%)	Hyphal length	Moist. (%)
Wet meadow	5040	1000	4900	388	13200	615	4050	488	7100	1000
Dry meadow	2120	264	3680	217	1680	95	3590	211	2940	227
Lichen heath	350	171	600	77	220	16	360	65	905	112
Birch forest	4370	456	5800	181	2720	126	2100	198	3000	198
Snow bed					705	40			1000	18

Table 5. Hyphal lengths in the top 5 cm of soils from Abisko, Sweden (Stordalen and Njulla). Seasonal variations in 1972 and 1973. Hyphal lengths given as $m\,g^{-1}$ dry soil, water content as % of dry soil (from HAYES, 1973; RHEINBERG, 1975)

		1972		1973		
		July	August	July 5	July 25	Aug. 4
Stordalen						
hummocks	lengths	3350	3170	2500	1800	4950
%	water	508	466	183	625	494
depressions	lengths	2310	4200	3250	1950	3520
%	water	1240	1180	879	1044	814
					July 19	
Njulla						
	lengths	1960	1490		830	
%	water	465	320		183	

No direct observations are available for the growth rate of hyphae or production of mycelium. The seasonal shifts show that the annual production must be of the same order of magnitude as the standing crop. From litter production data it is possible to estimate the annual fungal production if some assumptions are made. Thus, if it is assumed that 90% of the litter carbon is utilized by fungi, with an assimilatory efficiency of 40%, the annual production of fungal mycelium (exclusive mycorrhizal fungi) will be 36% of the litter production. Table 6 shows results from the three alpine Stigstuv (Hardangervidda) sites.

The production ratios given in Table 6 mean that for each meter of hyphae present, about 1.4 m are produced per year. Bearing in mind that hyphae have apical growth, it may be illuminating to estimate how many growth-points per m are required for this growth rate. Assuming an average growth rate of $0.1\,\mu m\,min^{-1}$, which is typical for slow-growing fungi under laboratory conditions (BURNETT, 1968) and a growth season of 100 days, it is seen that about 100 growth-points per meter are needed. It is interesting to compare this figure with observations from Stordalen, where the number of propagules per m of hyphae was calculated, and values between 800 and 10 were found (Table 3).

Table 6. Fungal biomass, litter production and estimated fungal production at the Stigstuv (Hardangervidda) alpine-tundra sites. All values in Kcal m^{-2} $year^{-1}$

	Lichen heath	Dry meadow	Wet meadow
Litter production	860	2390	3480
Fungal biomass	225	600	820
Fungal production	310	860	1250
Annual production ratio	1.38	1.44	1.52

These calculations do not permit definite conclusions to be drawn as to fungal productivity, but they seem to indicate that a production ratio (annual production/average standing crop) of about 1 is physiologically feasible. The litter formation data do not permit production values much higher than 1.3–1.5 times the apparent standing crop (including dead, morphologically intact hyphae). On the other hand, the fluctuations in hyphal lengths observed during the growth season seem to indicate values in excess of this, at least if it is accepted that the monthly samplings carried out have only uncovered part of the fluctuations.

Thus, with some assurance it can be stated that at the Stigstuv sites, the annual fungal production is about 1.0–1.5 times the average apparent standing crop of mycelium, and that this production requires between 25 and 35% of the litter carbon, and practically all the litter nitrogen. At the same time, between 50 and 60% of the litter carbon is converted to carbon dioxide by fungal respiration.

References

BURNETT, J. H.: Fundamentals of mycology. London: Edward Arnold Publ. Ltd. 1968.

COCHRANE, V. W.: Physiology of fungi. New York: Wiley 1958.

GRAY, T. R. G., WILLIAMS, S. T.: Soil microorganisms. Edinburgh: Oliver and Boyd 1971.

JENSEN, V.: Studies on the microflora of Danish beech forest soils. 1. The dilution plate count technique for the enumeration of bacteria and fungi in soil. Zentralbl. Bakt. Parasitenkunde, Abt. II, **116**, 13–32 (1962).

JONES, P. C. T., MOLLISON, J. E.: A technique for the quantitative estimation of soil microorganisms. J. Gen. Microbiol. **2**, 54–69 (1948).

HANSSEN, J. F., THINGSTAD, T. F., GOKSØYR, J.: Evaluation of hyphal lengths and fungal biomass in soil by a membrane filter technique. Oikos **25**, 102–107 (1974).

HAYES, A. J.: Studies on the microfungi occurring at Stordalen and Njulla, 1972. Swedish IBP Tundra Biome Project Tech. Rep. **15**, 1–22 (1973).

HAYES, A. J., RHEINBERG, P.: Microfungal populations of the Abisko area, northern Sweden. In: WIELGOLASKI, F. W. (Ed.): Fennoscandian tundra ecosystems, Part 1: Plants and microorganisms, pp. 244–250. Berlin-Heidelberg-New York: Springer 1975.

PARKINSON, D., GRAY, T. R. G., WILLIAMS, S. T.: Methods for studying the ecology of soil microorganisms. IBP Handbook No. 19. Oxford: Blackwell Scientific Publications 1971.

RHEINBERG, P.: Microfungal studies at Stordalen and Njulla, 1973. Swedish IBP Tundra Biome Project Tech. Rep. **17** (1975) in press.

SAITO, T.: The significance of plate counts of soil fungi and the detection of their mycelia. Ecol. Rev. **14**, 69–74 (1955).

SONESSON, M., WIELGOLASKI, F. E., KALLIO, P.: Description of Fennoscandian tundra ecosystems. In: WIELGOLASKI, F. E. (Ed.): Fennoscandian tundra ecosystems, Part 1: Plants and microorganisms, pp. 3–28. Berlin-Heidelberg-New York: Springer 1975.

Microfungal Populations of the Abisko Area, Northern Sweden

A. J. Hayes and P. Rheinberg

While a number of studies have taken place on the microfungi in northern temperate soils (Dale, 1912; Campbell, 1938; Ellis, 1939) and also those forms colonising decaying plant remains at these latitudes (Kendrick and Burges, 1962; Gremmen, 1957; Hayes, 1965a; Hering, 1965, 1967; Frankland, 1966, 1969), the role of microfungi in soils and on decaying plant remains in north European tundra areas has received little or no attention, apart from a small number of taxonomic studies (Hagem, 1907, 1910a, b; Traaen, 1914; Nielsen, 1927).

During 1972, the opportunity arose as part of the Swedish IBP tundra project to initiate an examination of the microfungi of the Abisko area in northern Sweden (Hayes, 1973), which study was later continued in 1973 by Rheinberg (1975). The aims of these studies were to describe the microfungal populations qualitatively, noting wherever possible changes in composition associated with time, soil type and depth; to describe the successions of microfungi occurring on decomposing plant material of species representative of the different vegetation types examined, and finally, to estimate fungal biomass (Hanssen and Goksøyr, 1975).

Methods

Soil samples for assessing fungal populations were taken on different occasions during June, July, and August 1972–1973 in northern Sweden. Two distinct sites were examined, the Stordalen mire (400 m a.s.l.) and the summit slopes of Njulla (1000 m a.s.l.). The former site was the intensive study area for the Swedish IBP Tundra Biome investigations. At Stordalen, a wet (small depression) and a drier (hummock) subtype of the mire were investigated (Bergman et al., 1970), whereas on Njulla, samples were taken on *Dryas octopetala* L. heaths (Widden et al., 1972; Hayes, 1973).

In 1972, both the 0–5 and 5–10 cm soil layers were examined, but in 1973 attention was concentrated upon the 0–5 cm layer. Changes in types and numbers of organisms lower down the soil profile were monitored by collecting from a number of different depths (Hayes, 1973).

Examination of the microfungi on leaves and litter of *Empetrum hermaphroditum* (Lange) Hagerup, *Rubus chamaemorus* L., *Eriophorum vaginatum* L., *Carex rostrata* Stokes and *Dryas octopetala* L. was carried out to determine successional changes.

The types of organisms present were determined, using the dilution plate technique as modified by JENSEN (1962), plating out five replicates each of 1/20000 and 1/50000 dilutions onto modified peptone dextrose yeast extract agar (PAPAVISAS and DAVEY, 1959). In the leaf and litter samples, surface sterilisation using 0.1% silver nitrate solution was carried out, followed by direct plating of whole leaves or fragments (HAYES, 1965a, b, 1973).

Results and Discussion

Although in both 1972 and 1973 a large percentage of the isolates were found to be sterile, the most striking difference between the years is the comparative absence of *Penicillium* and *Aspergillus* in 1973 (Table 1). This may well be due to the difference between the two growing seasons, since 1972 was unusually hot and dry, whereas 1973 was cool and moist. KOUYEAS and BAILES (1969) were able to isolate *Mortierella* spp. at high water potentials, but observed that species of *Mucor*, *Penicillium* and *Aspergillus* all occurred with increasing frequency as conditions became drier. It should also be remembered that since the *Aspergillus* and *Penicillium* are fast growing on most culture media, slower growing organisms tend to be swamped, and their importance is generally underestimated. The characteristic tundra genus *Chrysosporium* was comparatively rare at Stordalen in 1972, whereas in 1973 it was dominant in the moister Stordalen subtype. On the drier subtype, *Chrysosporium* was largely replaced by *Oidiodendron* sp., but *Penicillium* (principally *P. frequentans* Westling and *P. waksmani* Zaleski) still occupied a subordinate position.

Table 1. Fungi comprising 80% of isolates from Stordalen peat (0–5 cm)

Isolate	Hummocks						Small depressions					
	1972			1973			1972			1973		
	June	July	Aug.	June	July	Aug.	June	July	Aug.	June	July	Aug.
Mortierella sp.					10	75				17		
Aspergillus niger			6						27			
Chrysosporium										15	20	22
Oidiodendron sp.				30	15	28				30	3	7
Penicillium decumbens	20							4				
Penicillium frequentans				9			9			7		
Penicillium implicatum										3		
Penicillium janthinellum							59		25			
Penicillium odoratum		18										
Penicillium spinulosum			73									
Penicillium waksmani										5		20
Yeasts					15							
Sterile mycelia	69	68	5	40	35	35	14	69	30	3	55	31
Uncolonised								13				

Table 2. Fungi comprising 80% of isolates from Njulla (0–5 cm)

Isolate	1972			1973
	June	July	Aug.	July
Mortierella ramanniana	13			15
Chrysosporium sp.				12
Humicola sp.				7
Penicillium brevicompactum	3			
Penicillium canescens	4			8
Penicillium citrinum	2			
Penicillium chrysogenum	2			
Penicillium cyclopium	5		49	
Penicillium crustosum	7			
Penicillium decumbens	2			
Penicillium frequentans	2			3
Penicillium janthinellum	2	17		
Penicillium implicatum		25		
Penicillium roquefortii			19	
Penicillium rugulosum	2			
Penicillium spinulosum	3			
Penicillium stoloniferum	6			
Sterile mycelia	20	36	13	38
Uncolonised	3			

On Njulla, although sterile forms were again common in 1973, the number of isolates was greater compared with Stordalen (Table 2). *Chrysosporium* sp. was a regular component of the mycoflora in 1973, but was rare in 1972, while *Humicola* sp. was not recorded in 1972, but was common on Njulla in 1973. *Humicola grisea* Traaen and *Penicillium rugulosum* Thom (also found on Njulla) were reported by WARCUP (1951) to be virtually restricted to shallow alkaline grassland soils. *Chrysosporium* and *Humicola* occur regularly on Devon Island, northern Canada (WIDDEN et al., 1972).

In both years, a high percentage of the leaves and litter samples were found to be completely sterile. It may well be that the sterilisation procedure, designed to eliminate fungal spores merely present on the outer surface of the leaf, also killed the organisms active within. Different categories of leaf material might be differently affected by the sterilant according to their water repellant properties; or else small changes in experimental procedure may have had wide effects. However, in view of the results obtained with the same technique in the past (WARD, 1952; HAYES, 1965a, b) this seems somewhat unlikely, and on balance it seems probable that many samples were indeed sterile. Interesting results did emerge from this investigation, notably the apparent succession of *Dryas* in the different years (Table 3). On *Rubus* litter, *Penicillium frequentans* was found predominantly on brown litter, whereas *P. waksmani* was found on black. Furthermore, *P. frequentans* also appeared to be the characteristic species on *Eriophorum* standing dead, and also of *Empetrum* (the second species examined on the drier mire sub-type). *Penicillium waksmani* was isolated from *Eriophorum* litter lying in depressions, and this agrees with the known distribution of these *Penicillium* among hum-

Table 3. Fungi occurring on leaves of different plant species at Stordalen and Njulla 1972–1973

	Rubus chamaemorus				*Empetrum hermaphroditum*				*Eriophorum vaginatum*				*Dryas octopetala*		
	1972		1973		1972		1973		1972		1973		1972		1973
	June	Aug.	June	Aug.	June	Aug.	June	Aug.	June	Aug.	June	Aug.	June	Aug.	July
Chysosporium sp.	1						1								
Cladosporium herbarum	11									2			4		7
Cylindrocarpon candidum	2														
Penicillium cyclopium						4				4					
P. decumbens	2	21			1	67				50				2	
P. frequentans				1				24				4	2		
P. janthinellum														1	
P. roquefortii														34	
P. thomii													1		
P. waksmani								1				1			
Other Penicillia						2	6	6	1	3		4	3	15	
Phoma exigua					7								17	31	
Scytalidium sp.							14								17
Other microfungi	2				2				1				4		1
Yeasts	4	20	1		1	16		1	4	40			9	31	
Sterile mycelia	16	31	2	3	7		33	2	7			48	39	16	105
Uncolonised		10	97	96	82		48	66	87	1	100	43	19		9

Table 4. Fungi occurring on litter of different plant species at Stordalen and Njulla 1972–1973 (%)

	Rubus chamaemorus (brown litter)				*Empetrum hermaphroditum*				*Eriophorum vaginatum*				*Dryas octopetala*		
	1972		*1973*		*1972*		*1973*		*1972*		*1973*		*1972*		*1973*
	June	Aug.	June	Aug.	June	Aug.	June	Aug.	June	Aug.	June	Aug.	June	Aug.	July
Chrysosporium sp.				6				4							62
Cladosporium herbarum													3		
Penicillium aurantio-candium														5	
P. canescens									39						2
P. cyclopium		1				2				2					
P. decumbens	46	37	2		1	36			10	26					
P. frequentans			10	11			16	42	2		6	1	2		
P. implicatum													39	1	2
P. janthinellum													34	1	
P. thomii		1				3		6					2		
P. waksmani				5			1	1				6			
Other Penicillia	9	3	12	8		6	3		3	6			7	60	7
Phoma exigua	41	1			25	6				19					
Scytalidium sp.							31				51				
Trichoderma viride									2						
Other microfungi	2		1		4	1	16		1	26			1		4
Yeasts		34	58	3		10	23			3	8				
Sterile mycelia	5	9	89	22	38	16	21	43	35	8	40	53	4	31	20
Uncolonised	7			49		57	8	7		13	6	32	12	1	7

mocks and depressions of peats. Mention should also be made of *P. odoratum* Christensen and Backus at Stordalen, a characteristic species of peat bog mycofloras (CHRISTENSEN and BACKUS, 1961) (Table 4).

There are again marked differences between the two years, mainly noticeable is the absence of *Penicillium decumbens* Thom and *Phoma exigua* Desm. from leaves and litter of the mire in 1973, and their replacement by *Penicillium frequentans* and *Scytalidium* sp. On Njulla in 1973, the dominant forms were *Cladosporium*, *Chrysosporium* and *Scytalidium*, in direct contrast to the previous season when *Penicillium implicatum* Biourge, *P. janthinellum* Biourge and *Phoma exigua* were dominant. *Trichoderma viride* Pers. ex Fr. was observed on the mire on *Eriophorum* and on black *Rubus* litter.

In general terms, the populations of microfungi from the two sites compare reasonably well with those found by WIDDEN et al. (1972) in Arctic Canada. Their data show that up to 20% of the isolates were sterile forms, while *Cylindrocarpon* sp. and *Chrysosporium pannorum* (Link) Hughes were all relatively common. In contrast, *Penicillium* spp. tended to be rather uncommon, and *Trichoderma* spp. were always absent. However, FLANAGAN (Personal communication) has since confirmed that the Devon Island situation may be unique, since in Alaskan soils *Penicillium* and *Trichoderma* are regularly present as components of the soil mycoflora. The occurrence of *Trichoderma viride* at Stordalen as an occasional isolate from *Eriophorum vaginatum* and black *Rubus chamaemorus* litter, and as a regular member of the July *Carex rostrata* sample, contrasts sharply with the findings from some of the other tundra sites, and also with HUDSON'S (1968) assertion that *Penicillium* are rare and *Trichoderma* is typically absent from tundra soil mycofloras. In this connection, it is significant that *Trichoderma* was not isolated from the Njulla site in either year.

The contrast in fungal populations found in the two years 1972 and 1973 illustrates the danger of making firm conclusions based on only one years sampling. Only if it is possible to contrast results obtained in several different growing seasons with different weather conditions is it likely that a meaningful assessment of the soil mycoflora will be obtained.

References

BERGMAN, H., JONSSON, S., ROSSWALL, T., SONESSON, M.: The Swedish Tundra Biome Project: A presentation. Swedish IBP Tundra Biome Project Tech. Rep. **1**, 1–19 (1970).

CAMPBELL, M. E.: An investigation of the Mucorales in the soil. Trans. Roy. Soc. Edinb. **59**, 411–436 (1938).

CHRISTENSEN, M., BACKUS, M. P.: New or noteworthy Penicillia from Wisconsin soils. Mycologia **53**, 451–463 (1961).

DALE, E.: On the fungi of the soil. Ann. Mycol. Berl. **10**, 452–477 (1912).

ELLIS, M.: Fungi isolated from pinewood soil. Trans. Brit. Mycol. Soc. **24**, 87–97 (1939).

FRANKLAND, J.: Succession of microfungi on decaying petioles of *Pteridium aquilinum*. J. Ecology **54**, 41–65 (1966).

FRANKLAND, J.: Fungal decomposition of bracken petioles. J. Ecology **57**, 25–36 (1969).

GREMMEN, J.: Microfungi decomposing organic remains of pines. Fungus **27**, 34–42 (1957).

HAGEM, O.: Untersuchungen über Norwegische Mucorineen, I. Skr. Vidensk. Selsk., I Mat. Nat. Kl. **7**, 1–50 (1907).

HAGEM, O.: Untersuchungen über Norwegische Mucorineen, II. Skr. Vidensk. Selsk., I Mat. Nat. Kl. **10**, 1–152 (1910a).

HAGEM, O.: Neue Untersuchungen über Norwegische Mucorineen. Ann. Mycol. Berl. **8**, 265–286 (1910b).

HANSSEN, J. F., GOKSØYR, J.: Biomass and production of soil and litter fungi at Scandinavian tundra sites. In: WIELGOLASKI, F. E. (Ed.): Fennoscandian tundra ecosystems, Part 1: Plants and microorganisms, pp. 239–243. Berlin-Heidelberg-New York: Springer 1975.

HAYES, A. J.: Studies on the decomposition of coniferous litter. II Changes in external features and succession of microfungi. J. Soil Sci. **16**, 242–257 (1965a).

HAYES, A. J.: Some microfungi from Scots pine litter. Trans. Brit. Mycol. Soc. **48**, 179–185 (1965b).

HAYES, A. J.: Studies on the microfungi occurring at Stordalen and Njulla 1972. Swedish IBP Tundra Biome Project. Tech. Rep. **15**, 1–22 (1973).

HERING, T.: Succession of fungi in the litter of a Lake District oakwood. Trans. Brit. Mycol. Soc. **49**, 500–509 (1965).

HERING, T.: Fungal decomposition of oak litter. Trans. Brit. Mycol. Soc. **50**, 267–273 (1967).

HUDSON, H.: The ecology of fungi on plant remains above the ground. New Phytologist **67**, 837–874 (1968).

JENSEN, V.: Studies on the microflora of the Danish beech forest soils. I. The dilution plate count technique for the enumeration of bacteria and fungi in soil. Zbl. Bakt. II **116**, 13–32 (1962).

KENDRICK, W. B., BURGES, A.: Biological aspects of the decay of *Pinus sylvestris* leaf litter. Nova Hedwigia **4**, 313–342 (1962).

KOUYEAS, V., BAILES, C.: Influence of moisture on the restoration of mycostasis in air dried soils. Ann. Inst. Phytopathol. Benaki **8**, 123–144 (1969).

NIELSEN, N.: Fungi isolated from soil and from excrements of arctic animals derived from Disko and North Greenland. Meddel. om Grønland **74**, 1–8 (1927).

PAPAVISAS, G. C., DAVEY, C. B.: Evaluation of various media and antimicrobial agents for the isolation of soil fungi. Soil Sci. **88**, 112–117 (1959).

RHEINBERG, P.: Microfungal studies at Stordalen and Njulla 1973. Swedish IBP Tundra Biome Project Tech. Rep. **17** (1975) In press.

TRAAEN, A. E.: Untersuchungen über Bodenpilze aus Norwegen. Nyt Mag. Naturvidensk. **52**, 10–120 (1914).

WARCUP, J. H.: The ecology of soil fungi. Trans. Brit. Mycol. Soc. **34**, 376–399 (1951).

WARD, G. M.: Studies on the succession of fungi in decomposing litter of forest soil. Ph.D. Thesis, University of Nottingham 1952.

WIDDEN, P., NEWELL, T., PARKINSON, D.: Decomposition and microbial populations of Truelove Lowland, Devon Island. In: BLISS, L. C. (Ed.): Devon Island IBP project High arctic ecosystem. Project report 1970 and 1971, pp. 341–358. Edmonton: University of Alberta, Dept. of Botany 1972.

Bacterial Populations of Some Fennoscandian Tundra Soils

M. CLARHOLM, V. LID-TORSVIK, and J. H. BAKER

Bacteria are found in substantial numbers at all the tundra sites investigated despite low pH values which favor the growth of fungi compared to bacteria. Low oxygen concentrations have an inhibiting effect on most fungi, while many groups of bacteria grow well under microaerophilic and anaerobic conditions, which are common in waterlogged milieus like peat. When plant production exceeds the rate of decomposition and removal, organic material accumulates, often in the form of peat. The presence of peat is thus an indication of low activity on the part of the soil microflora.

Most of the methods used for assessing microbial activity relate to total metabolism in soil, and the results are due to fungal, animal or plant-root activity as well as bacterial activity. The number of active bacterial cells in soil is an important parameter in assessing total soil metabolism, but neither plate nor direct counts can give exact information, as discussed by GOKSØYR (1975). The opinion of many soil microbiologists is that the plate count is a better way of measuring bacterial activity at any particular time in soil than the direct count (HOLM and JENSEN, 1972). It has been shown that only a small fraction of soil bacteria are metabolically active at any given time, and that a very large part of the population is in a resting state (CLARK and PAUL, 1970). If all the cells were active, the average activity of a single cell would be extremely low on the basis of direct counts.

One of the aims of the IBP tundra investigations is to study the energy flow of the ecosystems concerned (DAHL and GORE, 1968). When the energy locked up in bacteria is to be estimated, direct counts are probably the best basis for an estimate, since inactive cells also represent energy stored in the bacterial mass. Another advantage of basing estimates of bacterial mass on direct counts is that they then become comparable to fungal mass, which is usually estimated by microscopic examination.

Material and Methods

Ten soils from three areas in Fennoscandia were studied in varying degree: three forest podzols, three peaty soils, two dry, lichen-dominated soils, one eutrophic meadow and one soil only briefly free from snow each summer. Four of these sites are located near Kevo, Finland, one at Stordalen, Abisko, Sweden and the rest at Hardangervidda, Norway. Site descriptions and soil characteristics are found in SONESSON et al. (1975) for all the sites except the Kevo palsa, which is described by BAKER (1974).

Estimates of bacterial numbers were based on plate counts and direct counts. The latter were made on specially prepared slides (CLARHOLM, 1974) or on black membrane filters (Oxoid) (LID-TORSVIK and HANSSEN, 1972). The samples for direct counting were stained with acridine orange and counted in incident or transmitted UV-light. The numbers of anaerobs at Stordalen were determined by the roll tube method (SVENSSON, unpublished). The shake tube technique used at Hardangervidda included both facultatively and strictly anaerobic bacteria.

Generation times of bacteria were estimated by the method of PARINKINA (1973). Periodic increases in numbers and biomass of bacteria, as observed by direct counting, were used for calculating generation time and production. For the biomass calculations, the sizes of the first 25–30 cells were determined in each sample and divided into six size-classes. By multiplying the proportion of each size-class by its volume, the cell density and the determined number of bacteria, six "sub-biomass" units were obtained, and addition gave the total biomass in fresh weight. The density of the bacterial cell was taken as unity, dry weight as 25% of fresh weight (CLARHOLM, 1974), and energy content as 5000 calories per gram dry weight (LID-TORSVIK and HANSSEN, 1972). The number of generations over the season at Stordalen (*viz.* when the mean, daily temperature at -5 cm was above 0° C; ca. 4 months) was calculated by using average generation times for 5-day periods, with temperature as the driving force and a $Q_{10}=2$. The experiments on generation times were, for ease of access, carried out at a mire very similar to the Stordalen mire, but near the Scientific Station at Abisko.

At Kevo the amounts of two enzymes, dehydrogenase and β-fructo-furanosidase (invertase), were determined at three depths. Invertase was measured by the method of ROSS (1965), and the method of CASIDA (CASIDA et al., 1964) was used, slightly modified, for determining dehydrogenase activity.

For estimation of total metabolic activity, the oxygen uptake of the organic A_0 horizons of the Finnish soils was measured in the laboratory in a Dixon respirometer at 10° C.

As part of a characterization of tundra bacteria, physiological classification was carried out by a multi-point technique on 100 strains each from the wet meadow at Hardangervidda and the Stordalen mire (ROSSWALL and CLARHOLM, 1974).

Results and Discussion

The number of aerobic and anaerobic bacteria at Kevo (BAKER, 1974), Stordalen, Abisko (CLARHOLM, 1974 and Hardangervidda (LID, 1971; LID-TORSVIK and HANSSEN, 1972; and HANSSEN, LID-TORSVIK and TORSVIK, 1973) are given in Tables 1 and 2.

At sites with organic soils, variations with depth were almost the same whether the numbers of bacteria were expressed per gram dry weight or per cm^3 of soil or per gram organic matter. At sites with developed soil horizons, variation with depth changed, depending on differences in content of organic matter and bulk densities.

At Kevo and Hardangervidda, plate counts as expressed per gram dry weight or per cm^3 of soil were highest in the surface zones, and decreased markedly from the A_{0-1} to the A_2 horizons. At Stordalen neither plate counts of aerobic bacteria nor the numbers of anaerobs changed much with depth compared with the other sites.

Table 1. Mean values and % standard deviations of aerobic and anaerobic bacterial plate counts expressed per gram dry weight of soil and per gram organic matter. The figures from Hardangervidda are based on results from 3 or 4 seasons, the ones from Kevo are from 1970, and the values for the aerobic and anaerobic bacteria at Stordalen are from 1973 and 1974 respectively

Site	Horizon or depth in cm	Aerobic counts			Anaerobic counts		
		per g d.w. $\times 10^{-4}$	% SD	per g O.M. $\times 10^{-4}$	per g d.w. $\times 10^{-4}$	% SD	per g O.M. $\times 10^{-4}$
Kevo							
Pine forest	A_0	723		8100			
	A_2	2.3		468			
	B_2	4.4		147			
Birch forest	A_0	226		410			
	A_2	1.7		95			
	B_2	6.0		206			
Subalpine heath	A_0	225		450			
	B_2	9.1		71			
Palsa	0–10	62		65			
	25–35	8.3		8.8			
	50–60	3.9		4.2			
Hardangervidda							
Wet meadow	0–5	6760	96	8360	170	96	210
	15	120	71	150	24	101	29
	30	20	150	24	7	132	8
Dry meadow	0–2 A_{0-1}	34300	150	42850	1300	86	1600
	5–7 A_{0-1}	1875	175	2340	410	235	510
	10–15 A_2	215	130	5370	15	115	375
	20 B	31	87	1575	7	180	340
	30 B	90	120	4485	0.5	70	25
Lichen heath	0–1 A_{0-1}	170	90	600	5	90	17
	5 A_2	13	60	480	0.7	62	26
	15 B_1	6	77	315	0.3	79	16
Birch forest	0–2 A_{0-1}	4180	135	5330	280	107	356
	8–10 A_2	145	80	10360	32	195	2300
	20 B	150	88	2300	4	180	66
Snow bed	0–2 A_{0-1}	396	80	970	26	120	65
	10–15 $(A_2)+B$	50	122	2600	6	125	310
	20 B	10	106	537	4	108	200
Stordalen							
Mire	1–3	11900	77	12200	27	35	28
	3–5	6100	67	6250	—	—	—
	5–7	4200	62	4300	76	10	78
	10	2600		2665	53	16	55
	20	1700		1750	102	33	105

Table 2. Direct counts of bacteria in numbers per gram dry weight of soil and per gram organic matter. Mean values and % standard deviation are given

Site	Horizon or depth in cm	per g d.w. $\times 10^{-9}$	% S.D.	per g O.M. $\times 10^{-9}$
Hardangervidda				
Wet meadow	0–5	205	58	253
	15	115	105	140
	30	110	—	134
Dry meadow	0–2 A_{0-1}	86	80	107
	5–7 A_{0-1}	45	79	56
	10–15 A_2	20	—	500
	20 B	18	102	875
	30 B	10	47	475
Lichen heath	0–1 A_{0-1}	33	81	117
	5 A_2	15	47	556
	15 B_1	15	90	2900
Birch forest	0–2 A_{0-1}	106	92	360
	8–10 A_2	15	47	1070
	20 B	20	—	308
Snow bed	0–2 A_{0-1}	37	50	90
	10–15 $(A_2)+B$	20	—	1050
	20 B	10	—	525
Stordalen				
Mire	1–3	31	39	32
	3–5	20	15	21
	5–7	21	10	22
	10	16		16
	20	6		6

When the results were expressed per gram organic matter, the diminution in number of bacteria with depth in mineral soil was much smaller than when expressed per gram dry weight of soil. At sites like the birch forest and snow bed, the number of bacteria in the A_{0-1} horizons was even lower than in the A_2 horizons.

In the surface zones, the direct counts per gram dry soil were 10^2–10^5 times greater than the corresponding plate counts, and about 10^6 times greater further down (Tables 1 and 2), variations with depth at the same site being mostly small. When the direct counts were expressed per gram organic matter the figures were lowest in the surface zones, except for the wet meadow and the mire at Stordalen.

To explain some of the differences in results between plate counts and direct counts, the number of bacteria per particle after homogenization was estimated in soil samples from the Hardangervidda sites. In 40 different counts, about 30% of the bacteria were associated with soil particles, the number of bacteria ranging from 1–10. Complete homogenization can at most be expected to double the plate counts.

Comparison of results between countries is difficult, because the methods used were not the same. The differences estimated by plate counts may depend on

Table 3. μl oxygen evolved and amount of invertase and dehydrogenase at three depths in four soils near Kevo. Invertase expressed as mg glucose produced in 24 hrs, dehydrogenase expressed as micromoles formazan produced in 24 hrs, O.M. = organic matter. No dehydrogenase measurements were made on the palsa (From BAKER, 1974)

Site	Horizon or depth	Invertase		Dehydrogenase		μl O_2 evolved	
		g^{-1} soil	g^{-1} O.M.	g^{-1} soil	g^{-1} O.M.	g^{-1} soil	g^{-1} O.M.
Pine-forest	A_0	140	1570	1.70	19.1	8.1	79.7
	A_2	6.3	1310	0.205	42.6	—	—
	B_2	8.4	280	0.144	4.75	—	—
Birch-forest	A_0	188	290	5.22	8.05	27.3	42.0
	A_2	5.8	308	0.228	12.7	—	—
	B_2	2.9	96.7	0.083	2.86	—	—
Subalpine heath	A_0	112	224	3.84	7.68	5.6	11.2
	B_2	7.4	252	0.281	9.55	—	—
Palsa	0–10 cm	103	108	—	—	—	—
	25–35 cm	11.8	12.6	—	—	—	—
	50–60 cm	25.0	27.2	—	—	—	—

differences in methods and techniques rather than on differences between sites, but any variations found within a country can be considered real. Thus, the Hardangervidda sites, in order of decreasing plate counts, are: the eutrophic sites (dry and wet meadows), the oligotrophic birch forest site, and lastly the two remaining oligotrophic sites. The high number at the birch forest site is probably caused by the higher mean temperature at this site. At the lichen heath and snow bed, the plate counts were one to two orders of magnitude less than at the eutrophic sites. Differences between the Norwegian sites were much smaller when direct counts were compared (Table 2).

Peaks in plate counts for the surface layers of some tundra soils at Signy Island and Point Barrow were recorded in association with rapid thaw (BAKER, 1970b; BOYD, 1958). No such peaks were observed at Hardangervidda either in plate counts or in direct counts, not even in samples taken immediately after snow melt. In samples taken from the surface layers of frozen soil under the snow in May, plate counts and direct counts were mostly of the same order of magnitude as for samples taken later in the summer.

In general, the number of bacteria obtained from viable counts from the Finnish tundra soils was similar to that found in other investigations at comparable latitudes. BOYD'S (1958) site 1 at Point Barrow in Alaska approximates to the present subalpine heath site, and IVARSON'S (1965) site 1 in the Canadian Northwest Territories appears to be similar to the palsa. In each case, the surface horizons of the soils have yielded similar results. However, the numbers of bacteria further down the palsa do seem to be low compared with the results of IVARSON (1965) and BAKER (1970a). The results from the A_0 horizons of the podzols are only slightly lower than those from forest podzols in less severe climates (JAGNOW, 1971).

The enzyme determinations made at Kevo (Table 3) have been expressed on the basis of organic matter, as well as per gram soil. The highest activity was

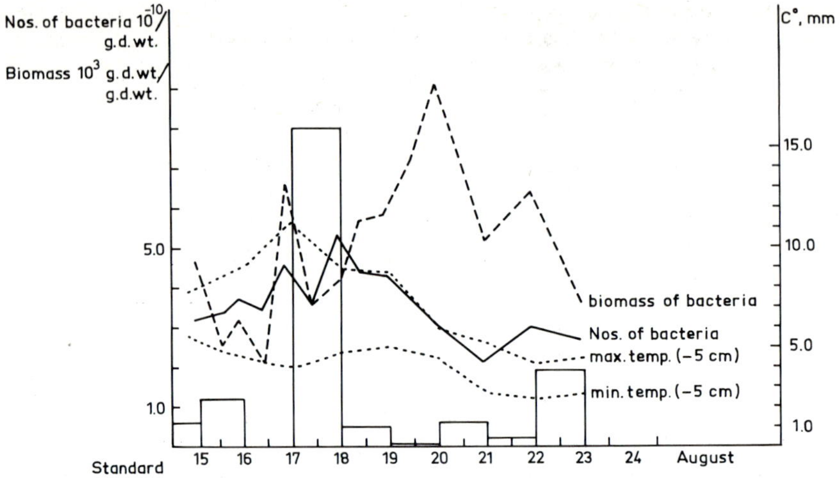

Standard deviation

Date	15	16			17		18		19		20	21	22	23
time	11	06	11	22	08	21	10	21	09	21	10	10	10	10
biomass	3.3	0.7	1.8	0.3	4.3	0.8	3.1	0.6	1.9	2.1	3.3	2.8	3.1	2.4
number	0.8	0.4	1.3	0.6	1.1	0.9	1.3	1.2	0.4	0.8	0.3	0.8	0.6	0.5

Fig. 1. Fluctuations in bacterial numbers (direct counts) and biomass in the Abisko mire from four replicates. Precipitation is given in columns, together with graphs of maximum and minimum temperatures (CLARHOLM, 1974)

always found in the A_0 horizon when expressed as per gram soil. If expressed as per gram organic matter, then the enzyme levels in the bleached A_2 horizon were almost equal to, or greater than, those in the A_0 horizon.

Both oxygen uptake measurements and enzyme determinations (Table 3) indicated that the birch forest soil was the most active, followed by pine forest, lichen heath and palsa. There did not seem to be a correlation between bacterial plate counts and the other activity measurements. KISS (1958) has shown that roots produce invertase, but the method used in the present investigation for enzyme determinations is unlikely to implicate roots. Oxygen is also required by roots as well as microfungi; the latter have been suggested to respire substantially more oxygen than bacteria in an agricultural soil (ANDERSON and DOMSCH, 1973).

PARINKINA (1973) observed by direct count that the number of bacteria in tundra soils fluctuated markedly within a few days. In the long-term experiment with daily sampling carried out at Abisko in August 1973 (Fig. 1), the observed fluctuations were found to be greater than those in an experiment made at Stordalen in July. Biomass values fluctuated more than bacterial numbers, with a peak after heavy rainfall associated with high maximum temperatures.

Accumulation of metabolic by-products has been advanced as a theory to explain observed sudden decreases in bacterial numbers (HUDYAKOV, 1957). The by-products may, when they reach a certain level, be poisonous to the bacteria, thus causing sudden mass death. In the ombrogenous peat at the Swedish Stordalen site, the lack of easily available nutrients is an important limiting factor for

Table 4. Bacterial biomass, number of generations and energy flow through the bacterial population per season at the Stordalen mire

Depth	Biomass g d.w. m^{-2}	Number of generations	Kcal produced per m^2
0– 5	7.0	9.2	325
5–15	11.0	7.7	421
15–25	2.8	4	54
25–35	1.5	3	23
0–35	22.3		823

bacterial growth. The input of fresh nutrients comes from decomposing litter, faeces and from root exudates. Nutrients leached from the surface litter would already have been utilized before reaching the five-centimetre depth where the samples were taken. Hence the peat itself—although locally enriched with exudates—is the main source of energy and nutrients for the maintenance and reproduction requirements of microorganisms. Only a small fraction of the bacteria living in the peat can decompose complex substrates, the rest having to metabolize catabolic products from the first stages of decomposition affected by other microbes. Large amounts of metabolic by-products are thus not allowed to accumulate.

The mean daily production of bacteria at Stordalen was $0.2–0.5 \times 10^{-3}$ g d.w. per g d.w. peat in the 0–5 cm layer, and over the whole season 140 g d.w. of bacteria were produced per m^2 down to 35 cm. In Table 4, biomass, number of generations and energy flow through the bacterial population per season are listed for the Stordalen mire. The biomass of the comparable wet meadow at Hardangervidda was 940 g d.w. of bacteria down to 35 cm depth. This high figure results from the numbers of bacteria being more than an order of magnitude higher than at Stordalen, and a 2.5-fold higher bulk density. The biomass of bacteria of the other Norwegian sites was estimated to 730 g d.w. m^{-2} in the birch forest, 630 g d.w. in the dry meadow and 550 and 530 g d.w. respectively, at the lichen heath and snow bed (down to 20 cm only for the last site).

The average generation time at five centimetres below the surface was determined for the periods when an increase was registered. It was 37.4 ± 4.7 hrs ($n = 16$) during the long-term experiments, and 9.4 ± 0.4 hrs ($n = 14$) when estimated from the short-term experiment with sampling every three hours (Fig. 2). The shorter generation times found with more frequent sampling were due to the larger number of fluctuations detected. The more varied generation times obtained in the long-term experiments may be an expression of the more varied abiotic conditions prevailing during a longer period of time.

Differences between areas with a general paucity of nutrients and those enriched by root exudates—which enhance microbial activity with a consequent rise in biomass—become accentuated later in the season at Stordalen, when all soil activities are heightened. This situation could create an exaggerated difference between parallel samples, thus leading to a false impression of fluctuations. It is therefore difficult to state whether or not the recorded fluctuations in the present

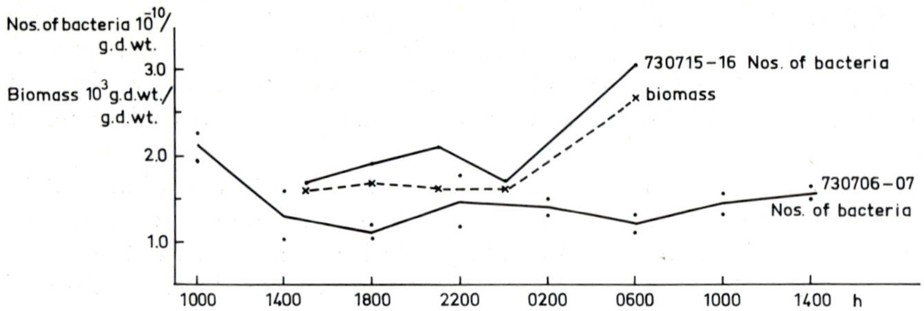

Standard deviation 730715–16	number	0.42	0.15	0.53	0.27	0.35
	biomass	0.44	0.26	0.66	0.48	0.50

Fig. 2. Two short-term experiments with frequent sampling. *Stordalen mire:* number of bacteria determined (730706-07). *Abisko mire:* both bacterial number and biomass determined (730715-16) (CLARHOLM, 1974)

study give a true picture of rapid changes in bacterial numbers within the whole mire.

At Stordalen, growth rates of 3.1 generations per month in July and 4.1 generations in August were determined with temperature as the driving force. Because the moisture (% dry weight) was constantly high (700 ± 33%), it was considered to have minor importance in this case. In investigations on the Russian tundra, PARINKINA (1973), using the same technique, obtained 2.8 generations per month for a polygonal bog.

With 720 hrs in a month, 3.1 and 4.1 generations per month give an average generation time of 200 and 180 hrs respectively, compared with 260 hrs calculated for the Russian bog. This is lower than, but of the same order of magnitude as, the figures that MORRILL and DAWSON (1962) obtained in their experiment with *Nitrosomonas* and *Nitrobacter* in an agricultural soil, where they recorded 21–100 hrs generation time. The shorter generation times are easily explained by the more favourable conditions prevailing, e.g. higher pH value, higher temperature and easily available nutrients in greater supply. Growth rate determinations for bacteria in soil by energy input calculations were done for grassland by BABIUK and PAUL (1970), and for deciduous forest by GRAY and WILLIAMS (1971); they obtained generation times of 1200 and 16 hrs respectively. GRAY and WILLIAMS also made calculations based on CO_2 evolution from an agricultural soil, and they obtained a 29-hour generation time; in both their calculations, they did not take fungi into consideration, and the figures must hence be overestimates—only of theoretical interest for setting the upper limits on possible growth rates.

The physiological tests showed an adaptation on the part of the Norwegian strains to the cold winters and the cool summers at Hardangervidda, with 84% growing at 2° C, and only 5% at 37° C. At Stordalen, Abisko with less severe winters and warm summers, only 20% grew at 2° C, but 70% at 37° C. The Hardangervidda strains were more tolerant of low pH; at pH 5.5, 44% of the population grew out, compared with only 5% for Stordalen. Very low biopolymer degrading capacity—particularly for more resistant polysaccharides like chitin and cellulose—were recorded for the Norwegian strains. In hydrolysis tests on nine different compounds, the Stordalen strains showed twice as high a capacity for hydrolysis as the Hardangervidda strains, but the latter had a fivefold greater frequency for utilization of simple compounds like sodium salts of organic acids for growth.

Tests with selective media and antibiotics supported indications from other tests that the bacterial population in the Hardangervidda wet meadow was mainly built up of *Pseudomonas* spp. The Swedish Stordalen population was more complex with *Bacillus* spp. as the predominant group, but *Pseudomonas*-like bacteria were also present.

References

ANDERSON, J. P. E., DOMSCH, K. H.: Quantification of bacterial and fungal contributions to soil respiration. Arch. Mikrobiol. **93**, 113–127 (1973).

BABIUK, L. A., PAUL, E. A.: The use of fluorescein isothiocyanate in the determination of the bacterial biomass of grassland soil. Can. J. Microbiol. **16**, 57–62 (1970).

BAKER, J. H.: Quantitative study of yeasts and bacteria in a Signy Island peat. Br. Antarct. Surv. Bull. **23**, 51–55 (1970a).

BAKER, J. H.: Yeast, moulds and bacteria from an acid peat on Signy Island. In: HOLDGATE, M. W. (Ed.): Antarctic Ecology, pp. 717–722. New York-London: Academic Press 1970b.

BAKER, J. H.: A comparison of the microbiology of four soils in Finnish Lapland. Oikos **25**, 209–215 (1974).

BOYD, W. J.: Microbial studies of arctic soils. Ecology **39**, 332–336 (1958).

CASIDA, L. E., KLEIN, D. A., SANTORO, T.: Soil dehydrogenase activity. Soil Sci. **98**, 371–376 (1964).

CLARHOLM, M.: Direct counts of bacteria in tundra peat for estimating generation time and biomass production. Progress Report 1973. Swedish IBP Tundra Biome Project Tech. Rep. **16**, 43–57 (1974).

CLARK, F. E., PAUL, E. A.: The microflora of grasslands. Advan. Agron. **22**, 375–435 (1970).

DAHL, E., GORE, A. J. P.: Proceedings Working meeting on analysis of ecosystems; Tundra zone. Ustaoset, Norway. Oslo: Norsk IBP 1968. Mimeographed.

GOKSØYR, J.: Decomposition, microbiology and ecosystem analysis. In: WIELGOLASKI, F. E. (Ed.): Fennoscandian tundra ecosystems, Part 1: Plants and microorganisms, pp. 230–238. Berlin-Heidelberg-New York: Springer 1975.

GRAY, T. R. G., WILLIAMS, S. T.: Microbial productivity in soil. Symp. Soc. Gen. Microbiol. **21**, 255–286 (1971).

HANSSEN, J. F., LID-TORSVIK, V., TORSVIK, T.: Microbiology. In: Annual Report, pp. 80–92. Oslo: Norsk IBP 1973 (in Norwegian).

HOLM, E., JENSEN, V.: Aerobic chemoorganotrophic bacteria of a Danish beech forest. Oikos **23**, 248–260 (1972).

HUDYAKOV, J. P.: Methods for investigating soil toxicity. Symposium: Méthodes d'Études Microbiologiques du Sol. V. 5, 13 pp. (1957).

IVARSON, K. C.: The microbiology of some permafrost soils in the MacKenzie Valley, N.W.T. Arctic **18**, 256–260 (1965).

JAGNOW, G.: Seasonal amounts of fungal mycelium, numbers of aerobic bacteria and bacterial spores in a beech and spruce forest soil of the Solling. In: Organismes du Sol et Production Primaire. IV. Colloq. pedobiol. pp. 303–311. Dijon: I.N.R.A. Publ. 71–7, 1971.

KISS, S.: Untersuchungen über die Produktion von Saccarase im Boden. Z. Pfl. Ernähr. Düng. Bodenk. **18**, 117–125 (1958).

LID, V.: Bacterial counting. In: Annual Report 1970, pp. 20–25. Oslo: Norsk IBP 1971 (in Norwegian).

LID-TORSVIK, V., HANSSEN, J. F.: Microbiology. In: Annual Report 1971, pp. 24–34. Oslo: Norsk IBP 1972 (in Norwegian).

MORRILL, L. G., DAWSON, J. E.: Growth rates of nitrifying chemoautotrophs in soil. J. Bacteriol. **83**, 205–206 (1962).

PARINKINA, O. M.: Determination of bacterial growth rates in tundra soils. In: ROSSWALL, T. (Ed.): Modern Methods in the Study of Microbial Ecology. Bull. Ecol. Res. Comm. (Stockholm) **17**, 303–309 (1973).

ROSS, D. J.: A seasonal study of oxygen uptake of some pasture soils and activities of enzymes hydrolysing sucrose and starch. J. Soil Sci. **16**, 73–85 (1965).

ROSSWALL, T., CLARHOLM, M.: Characteristics of tundra bacterial populations and a comparison with populations from forest and grassland soils. In: HOLDING, A. J., HEAL, O. W., MACLEAN, S. F., FLANAGAN, P. W. (Eds.): Soil organisms and decomposition in tundra, pp. 93–108. Stockholm: IBP Tundra Biome Steering Committee 1974.

SONESSON, M., WIELGOLASKI, F. E., KALLIO, P.: Description of Fennoscandian tundra ecosystems. In: WIELGOLASKI, F. E. (Ed.): Fennoscandian tundra ecosystems, Part 1: Plants and microorganisms, pp. 3–28. Berlin-Heidelberg-New York: Springer 1975.

Comparisons of Decomposition Rates Measured by Means of Cellulose

B. BERG, L. KÄRENLAMPI, and A. K. VEUM

Cellulose has been used as a model substance in decomposition studies in the IBP tundra projects (ROSSWALL, 1974), and should prove particularly suitable for use in comparative studies. It is free of leaching associated with litter, a factor which can cause difficulties in determining decomposition, and large numbers of samples are relatively simple to prepare and analyse for weight loss. Different plant species produce different types of litter; it is difficult to compare their decomposition rates, because the microflora has probably adapted itself to the specific litter types and their various inhibiting substances. The use of cellulose affords a general and non-specific compound for decomposition experiments. Practically all plant material contains cellulose in varying amounts. Grasses contain a few per cent, and wood as much as about 40%. Organic compounds of Norwegian alpine plants are given by SKRE et al. (1975).

The property of degrading cellulose is restricted to relatively few groups of organisms, viz. bacteria, fungi and actinomycetes, which are either free-living or existing in symbiotic relationships in the guts of various soil animals. Besides, some protozoa and snails form cellulose-degrading enzymes. BERG et al. (1972a, b) have shown that different cellulose preparations are attacked and degraded to differing extents by the same organisms, depending on the source of the material and the properties it has acquired by various treatments. In order to obtain reliable results in an experiment or a comparative study, it is therefore important to work with a standardised type of cellulose.

The most common method of using cellulose in model studies appears to be the one where cellulose pieces are buried or placed on the soil surface and weight loss determined after a certain time. Another common method consists in measuring the tensile strength of cotton strips buried in soil.

Materials and Methods

Two cellulose preparations were used in the present studies: A. Dried sheet-cellulose containing 100% glucan and 0.04% ash. The experimental procedure with this preparation has been described by ROSSWALL and VEUM (1971), BERG and ROSSWALL (1973), and ROSSWALL (1974). B. Never-dried, very porous cellulose preparation containing: glucan 85%, xylan 10%, mannan 5%, ash 0.2%. The preparation of samples of this material, their burial in the peat, their collection, as well as determination of weight loss, have been described by BERG and ROSSWALL (1972).

The effect of nitrogen and phosphorus additions on cellulose decomposition was studied by enriching cellulose samples with different concentrations of protein, urea, ammonium chloride, sodium nitrate and potassium dihydrogen phosphate (0.1, 0.5 and 1% N and P of d.w.). The samples were placed randomly at a depth of 5 cm at the Stordalen site (Sweden) in May 1973, collected in September 1973, and analyzed for weight loss ROSSWALL et al., 1974).

Results and Discussion

Corresponding investigations on rates of cellulose decomposition showed a low rate at most of the Fennoscandian IBP tundra sites at Hardangervidda (Norway), Kevo (Finland), and Stordalen (Sweden). After one year, dry sheet-cellulose was decomposed by $3 \pm 0.3\%$ in the mire at Stordalen (ROSSWALL et al., 1974), and also in the wet meadow at Hardangervidda. Both sites have similar soil characteristics, i.e. organic content and moisture, whereas the nitrogen and phosphorus contents are about twice as high at Hardangervidda (HINNERI et al., 1975) and the temperature conditions more favorable, with a higher mean soil temperature (SONESSON et al., 1975). At Stordalen the low precipitation, together with frequent winds and sunshine, often resulted in very dry cellulose samples.

The birch forests at Kevo and Hardangervidda had the highest decomposition values of the sites, with 34 and 85% respectively. In both cases the samples were placed on the surface; the soil characteristics are similar, apart from the comparatively high phosphorus content at Hardangervidda (HINNERI et al., 1975). The valley with the Hardangervidda birch forest is moist and often foggy, which could explain the decomposition being higher than in the Kevo birch forest.

The dry and the wet meadow sites at Hardangervidda have a very similar climate, and the extent of decomposition on the surface is respectively 6 and 3% annually. This difference might be explained by the differences in soil characteristics. Comparison between these two sites and the Hardangervidda birch site showed that decomposition at the latter was greater (85%). A considerable part (62%) was decomposed during the cold period from September to June. The Hardangervidda lichen heath also gave low values, with less than 1% decomposed annually. Both the annual temperature and moisture were lower than at the meadow sites (SONESSON et al., 1975), and it is likely that evaporation caused by wind and direct sunlight dries out the exposed cellulose pieces.

It is to be expected that differences in vegetation should affect the rates of decomposition and general microbial activity. The latter could be expected to vary with changing plant communities and along gradients of moisture, temperature, and redox-potential. It is therefore important to study the effects caused by changes in abiotic factors when determining decomposition in large areas. Such studies have been carried out at the Fennoscandian tundra sites to explain the large differences in decomposition often obtained among parallel samples.

Decomposition rates were found to differ widely in the case of two Kevo sites—the pine and birch forests—both within each site and between them (Table 1). Both sites are classed as minerotrophic and have almost the same pH (SONESSON et al., 1975), but there is about ten times as much nitrogen in the birch forest soil as in that of the pine forest, and about twenty times as much phosphorus (HINNERI et al., 1975). These soil characteristics may partly explain the higher decomposition in the birch stand, and nitrogen may be limiting in the pine forest.

Both stands have the same precipitation, but the evaporation from the birch stand is less and the soil has a higher moisture content due to a luxuriant vascular vegetation, whereas the sparse vegetation in the pine stand enhances evaporation (KÄRENLAMPI, 1972). At leaf fall in the birch forest the cellulose pieces were embedded in wet leaves, which conserved their moisture. The difference in moisture between the two stands may partly explain the difference in decomposition. Comparison of the data obtained in the pine forest at selected points with differing canopy densities shows that decomposition increased with decreasing canopy density. It is possible that the low decomposition under the dense canopies depends on smaller precipitation, but inhibiting substances from the needles might also play a part. In the birch stand no such correlation between canopy density and decomposition is observed (Table 1). The nearby alpine lichen heath site at Kevo was somewhat colder (SONESSON et al., 1975) and drier than the other sites. Precipitation and soil characteristics with regard to pH, nitrogen and phosphorus contents were similar to those of the other sites. Decomposition values were somewhat lower on the average than in the pine forest (Table 1) but varied consid-

Table 1. Decomposition of cellulose samples on the litter layer surface at three different sites at Kevo. The samples were in the field for 401 days in the period 1972-06-10 to 1973-07-16

Site	Habitat description	Decomposition %	
Pine forest	Lichen vegetation, no covering tree canopy	18	
	Dense *Empetrum* vegetation under pine canopy	14	
	Litter layer, needles, no vegetation, under dense pine canopy	7	
	Mean value		13
Birch forest	*Pleurozium*, partly covered with sparse canopy	34	
	Dry place with lichens and some mosses. No birch canopy	34	
	Moist place. Mosses and dwarf shrubs	33	
	Mean value		34
Mountain site	Open heath with lichens	6	
	Open heath with sparse *Empetrum* vegetation	6	
	Open heath, under dense *Empetrum* vegetation	13	
	Open heath, under *Calamagrostis* vegetation	14	
	Mean value		10

erably within the site, and it appears that the covers of *Empetrum* and *Calamagrostis* play an important role—possibly by conserving moisture on the otherwise open and windy site.

The differences in decomposition between the pine and birch stands are also pronounced at lower depths. In one degradation experiment lasting 365 days, cellulose disappeared by 23, 41, 39, and 38% at depths of 1, 3, 5, and 7 cm, respectively, in the birch forest, whereas the corresponding figures for the pine forest were 8, 21, 17 and 18%, respectively (LÄHDE, 1974).

Dependence of decomposition on depth in peat was investigated in a long-term experiment with never-dried pulp cellulose at Stordalen (BERG and ROSSWALL, 1972, 1973; ROSSWALL et al., 1974). Cellulose was placed in litter-bags and buried at 2, 10, and 20 cm depth in the peat at ten randomly selected points on a limited area of the mire (BERGMAN et al., 1970). Although this type of cellulose is comparatively easily decomposed (BERG et al., 1972a), decomposition in the field was slow. A notable variation in decomposition with depth was, however, observed (Table 2). As the pH in the mire is low (about 4), bacterial activity can be expected to be low compared with fungal activity. The peat is moist and conditions rapidly become anaerobic down the peat profile, thus lowering fungal activity. There seems to be a boundary between aerobic and anaerobic conditions at about 10 cm depth (ROSSWALL et al., 1974), which possibly explains the relatively large differences in decomposition between the 10 and 20 cm levels.

Although the samples were buried in the middle of the growing season and ought to have been quickly attacked, the decomposition values show that only a small percentage was decomposed after a year. Some time is probably needed for infection. The samples collected after two years show a relatively large increase in decomposition as compared with those collected after one year. In an experiment, wet cellulose pieces were laid out at 10 cm depth and their decomposition compared with that of the corresponding dry sheets. There was a markedly higher decomposition in the case of the wet pieces −9% compared with 2% (BERG and ROSSWALL, 1973). It is possible that moisture increases the infection rate and thus influences decomposition. Within the area investigated, large differences were noted among the samples placed at the same depth for the same time interval (Table 2). No correlation could, however, be found between the different degrees of decomposition and the number of fungi, bacteria or soil animals. Variations in decomposition also appeared between samples at the same point, e.g. point C 57 (Table 2), where the decomposition value at 2 cm was 98% for two years and 4% for three years. Attempts were made to relate decomposition to wet or dry areas, but were not successful. The location of the various plant communities in the area was dependent on moisture conditions (MADSEN and WIDELL, 1974). Decomposition could not therefore be correlated with the plant communities.

A transection from a lake up to a dry area in the mire was investigated with the cotton-strip method at Stordalen (ROSSWALL et al., 1974) and significant differences—probably dependent on moisture—were noted. The most decomposed cotton strips were those placed in the water line of the lake among *Sphagnum lindbergii* and *Eriophorum angustifolium* in a minerotrophic area. Those less decomposed were placed at increasing levels from the lake in ombrotrophic peat with less moisture.

Table 2. Per cent weight loss ± S.E. from pulp cellulose buried for one, two and three years at three different depths in the peat profile, $n = 10$. (From ROSSWALL et al., 1974)

Depth	2 cm			10 cm			20 cm			Site mean value		
Small quadrat	1971	1972	1973	1971	1972	1973	1971	1972	1973	1971	1972	1973
C 15	3	17	90	3	25	65	(+200)	0	0	3	14	52
C 24	6	19	54	1	31	35	2	5	0	3	18	30
C 42	11	6	72	6	7	45	5	25	4	7	13	40
C 43	5	13	16	6	15	33	7	11	20	6	13	23
C 45	10	64	48	7	10	18	6	11	19	8	28	28
C 57	9	98	4	2	50	23	8	11	4	6	53	10
C 58	5	17	46	4	17	13	(+8)	0	6	3	11	22
C 77	10	85	100	11	53	90	8	0	24	10	46	71
C 82	11	14	99	3	8	98	8	6	24	7	9	74
C 96	11	34	39	10	14	27	6	3	5	9	17	24
Mean value 1971	8±1			5±1			6±1			6±1		
± S.E. 1972		37±10			23±5			7±3			22±5	
1973			57±10			45±9			11±3			37±7

Table 3. Effect of addition of different N and P-sources on decomposition of cellulose placed at 5 cm depth in the peat at Stordalen during the period 1973-05-23 to 09-20. * indicate pairs of values significantly different at the 5% level, as tested by Student's t-test. (From ROSSWALL et al., 1974; BERG and ROSSWALL, 1973)

	casein-N %			urea-N %			NH_4-N %			NO_3-N %			PO_4-P %			Control
	0.1	0.5	1.0	0.1	0.5	1.0	0.1	0.5	1.0	0.1	0.5	1.0	0.1	0.5	1.0	
Decomposition %	9.0	7 1 *	9.4	9.4	10.3	7.5	11.9	11.7	10.5	11.8 *	7.8 *	11.9	2.9	3.7	3.8	2.0
S.E.	1.9	1.4	2.1	2.1	2.4	2.3	3.4	2.4	3.0	6.1	4.4	5.7	0.8	0.8	1.2	2.0
n	13	13	13	13	13	13	13	13	12	13	12	13	13	13	13	9

Fertilization of cellulose samples caused an increase in decomposition rates for all nitrogen sources and concentrations (Table 3). Phosphorus did not influence the decomposition, but this could be due to the fact that many microfungi are inhibited by soluble phosphates (ROSSWALL et al., 1974). The data indicate a lack of available nitrogen for cellulose decomposition in the Stordalen peat. The scarcity of available nitrogen to the microorganisms in tundra areas is also pointed out by GOKSØYR (1975). KONG and DOMMERGUES (1970) studied cellulose decomposition in an acid peat, and found that there was a deficiency in mineral nitrogen but there were also indications of lack of available phosphorus. In addition to these two factors, they stated that low pH and absence of calcium ions had a retarding effect on the mineralisation of cellulose.

Acknowledgements. We wish to thank Dr. O.W. HEAL for his kind help in supplying the cotton strips and testing them for tensile strength. We also wish to thank Dr. T. ROSSWALL for analysing the Kevo cellulose samples. The cellulose preparations used at the Fennoscandian tundra sites were gifts from Borregaard A/S, Sarpsborg, Norway and Mo & Domsjö AB, Örnsköldsvik, Sweden. We are grateful to Mr. R. KUMAR for the linguistic revision of the manuscript.

References

BERG, B., v. HOFSTEN, B., PETTERSSON, G.: Growth and cellulose degradation by *Cellvibrio fulvus*. J. Appl. Bacteriol. **35**, 201–214 (1972a).

BERG, B., v. HOFSTEN, B., PETTERSSON, G.: Electron microscopic observations on the degradation of cellulose fibres by *Cellvibrio fulvus* and *Sporocytophaga myxococcoides*. J. Appl. Bacteriol. **35**, 215–219 (1972b).

BERG, B., ROSSWALL, T.: Microbiology and decomposition studies. Cellulose decomposition. Progress Report 1971. Swedish IBP Tundra Biome Project Tech. Rep. **9**, 27–31 (1972).

BERG, B., ROSSWALL, T.: Studies in cellulose decomposition in a mire at Stordalen. Progress Report 1972. Swedish IBP Tundra Biome Project Tech. Rep. **14**, 134–141 (1973).

BERGMAN, H., JONSSON, S., ROSSWALL, T., SONESSON, M.: The Swedish Tundra Biome Project. A presentation. Swedish IBP Tundra Biome Project Tech. Rep. **1**, 1–13 (1970).

GOKSØYR, J.: Decomposition, microbiology and ecosystem analysis. In: WIELGOLASKI, F. E. (Ed.): Fennoscandian tundra ecosystems, Part 1: Plants and microorganisms, pp. 230–238. Berlin-Heidelberg-New York: Springer 1975.

HINNERI, S., SONESSON, M., VEUM, A. K.: Soils of Fennoscandian IBP tundra ecosystems. In: WIELGOLASKI, F. E. (Ed.): Fennoscandian tundra ecosystems, Part 1: Plants and microorganisms, pp. 31–40. Berlin-Heidelberg-New York: Springer 1975.

KÄRENLAMPI, L.: Comparisons between the microclimates at the Kevo ecosystem study sites and the Kevo Meteorological Station. Rep. Kevo Subarctic Res. Stat. **9**, 50–65 (1972).

KONG, K. T., DOMMERGUES, Y.: Limitation de la cellulolyse dans les sols organique. I. Etude respirometrique. Rev. Ecol. Biol. Sol **7**, 441–456 (1970).

LÄHDE, E.: Rate of decomposition of cellulose in forest soils in various parts of the Nordic countries. Rep. Kevo Subarctic Res. Stat. **11**, 72–78 (1974).

MADSEN, J.-L., WIDELL, S.: A vegetation map of the Stordalen site. Progress Report 1973. Swedish IBP Tundra Biome Project Tech. Rep. **16**, 3–15 (1974).

ROSSWALL, T.: Cellulose decomposition on the tundra. In: HOLDING, A. J., HEAL, O. W., MACLEAN, S. F., FLANAGAN, P. W. (Eds.): Soil Organisms and Decomposition in Tundra, pp. 325–340. Stockholm: IBP Tundra Biome Steering Committee 1974.

ROSSWALL, T., BERG, B., LUNDKVIST, H.: Use of cellulose in field decomposition experiments. Progress Report 1973. Swedish IBP Tundra Biome Project Tech. Rep. **16**, 109–121 (1974).

ROSSWALL, T., VEUM, A. K.: Notes on the use of paper pulp for cellulose decomposition determinations. IBP Swedish Tundra Biome Project Tech. Rep. **6**, 7–8 (1971).

SKRE, O., BERG, A., WIELGOLASKI, F. E.: Organic compounds in alpine plants. In: WIELGOLASKI, F. E. (Ed.): Fennoscandian tundra ecosystems, Part 1: Plants and microorganisms, pp. 339–350. Berlin-Heidelberg-New York: Springer 1975.

SONESSON, M., WIELGOLASKI, F. E., KALLIO, P.: Description of Fennoscandian tundra ecosystems. In: WIELGOLASKI, F. E. (Ed.): Fennoscandian tundra ecosystems, Part 1: Plants and microorganisms, pp. 3–28. Berlin-Heidelberg-New York: Springer 1975.

Plant Litter Decomposition at Fennoscandian Tundra Sites

T. ROSSWALL, A. K. VEUM, and L. KÄRENLAMPI

Plant litter decomposition is one of the most important processes in any ecosystem, yet little attention has hitherto been given to the study of it in tundra areas. Very scanty information existed before IBP—DOUGLAS and TEDROW (1959) had, for example, investigated the decomposition of organic matter in arctic soils close to Barrow in Alaska by the use of respirometry. This dearth of investigations is the more surprising, as many tundra ecosystems accumulate organic matter, which is caused by low rates of decomposition (OLSON, 1963). Large amounts of nutrients are locked up in this organic matter, and are not available to the plants until mineralized by microorganisms.

The rate of litter decomposition has been studied in many IBP tundra projects by the use of litter bag techniques (HEAL and FRENCH, 1974), and also by measuring the respiration of plant litter (FLANAGAN and VEUM, 1974). Faunal and microbial litter decomposition studies have been neglected, but BOOTH (1972) followed the decomposition of musk-ox dung on Devon Island in Canada. PARINKINA (1973) and CLARHOLM et al. (1975) presented data from which it can be concluded that bacterial cells decompose rapidly in tundra soils. They observed fluctuations in numbers of bacteria over short periods of time, and the rapid decreases suggest that the dead cells did not remain intact in the soil over long periods.

The present paper examines results from plant litter decomposition experiments at the Fennoscandian IBP Tundra sites (SONESSON et al., 1975).

Materials and Methods

Plant litter decomposition was, whenever possible, investigated by using individually tagged samples, e.g. woody roots at Hardangervidda, Norway, *Betula nana* twigs and *Rubus chamaemorus* leaves at Stordalen, Sweden and birch wood at Kevo, Finland, which were placed in the field for various lengths of time. In all other instances the litter bag method (BOCOCK and GILBERT, 1957) was used.

For the investigations at Hardangervidda, large bags (25 × 25 cm) with 0.5 mm mesh size were used. These were placed horizontally at various places in the profile or, for roots, vertically at 0–10 cm. A similar procedure was used for the investigations at Kevo and Stordalen, but with smaller bags (7 × 7 cm) with 1 mm mesh size. At both Kevo and Stordalen the bags were put on the litter surface, with the exception of the moss samples at Stordalen. These were placed vertically in the moss carpet peat profile at 0.5–5.5 cm in the habitats characteristic for the species in question (SONESSON, 1973).

At Hardangervidda, green vegetation was collected either by species or as "site vegetation" where the total aboveground plant biomass was collected by clipping. At Kevo, pine needles and birch leaves were collected from litter traps and wood, 1.5–3.0 cm in diameter, was collected fresh and cut into pieces 10 cm in length. *Rubus chamaemorus* was collected at Stordalen immediately after litter fall, and the other species of vascular plants were collected by harvesting in mid-September. The litter was not dried, but dry weight determinations were made on separate subsamples, dried at 105° C to constant weight. The bryophytes at Stordalen were cut 2 cm below the apices, and the dead parts were dried at 105° C to ensure that no live material was included (SONESSON, 1973).

After collection of the litter bags from the field, the material was dried to constant weight at 105° C. In many instances the material was analysed for organic and mineral constituents, but these results are not included in the present paper.

Rates of decomposition of litter were also determined on samples from the Stordalen site by laboratory respiration measurements in a Gilson respirometer (ROSSWALL, 1974b).

Results and Discussion

Intersite Comparison of Plant Litter Decomposition. Decomposition values for different types of litter at the nine sites investigated are presented in Table 1. In many cases decomposition was followed for three years, and to make the data sets comparable, missing data were estimated from the measured weight loss by using the fractional loss rate (k), assuming that decomposition follows a negative exponential function (JENNY et al., 1949).

Decomposition of plant litter as measured by weight loss is a complex process, the result of both biotic and abiotic processes, *viz.* mineralization, comminution and leaching. With the techniques used, the importance of these components cannot be elucidated, and this should be kept in mind when trying to interpret the results (ROSSWALL, 1971; GOKSØYR, 1975).

There is no single plant species, the decomposition of which was followed at all the investigated sites. It is still possible, however, to make some comparisons between countries from the data in Table 1. Birch leaves were studied both at Kevo *(Betula tortuosa)* and Stordalen *(Betula nana)*, and they showed a first-year weight loss of 27 and 21% respectively. Climatic conditions at the two sites are similar with regard to summer temperatures (SONESSON et al., 1975), although precipitation at Kevo is somewhat higher (SKARTVEIT et al., 1975).

Lichen decomposition was studied at both Hardangervidda and Kevo. *Cladonia alpestris* showed a first-year weight loss of 5.6% at Kevo, while mixed lichens at the Hardangervidda lichen heath lost 4.9%.

Graminid decomposition was studied at Hardangervidda and Stordalen. The Hardangervidda wet meadow is similar to the Stordalen site—*Carex nigra* lost

Table 1. Decomposition rates of plant litter at Fennoscandian tundra sites. Measured values are given together with calculated values (in italics) for missing years, estimated from the measured ones, assuming an exponential decomposition rate. S.E. is given and number of samples (*n*). Stordalen data from ROSSWALL (1973, 1974b) and SONESSON (unpubl.)

Site	Litter	One year loss %	Two years loss %	Three years loss %	Comments, year of experiment
Kevo	– Pine Forest				
	Pinus sylvestris – wood	(5.0)	*6.9*	*8.3*	69/73 Litter surface[a]
	Pinus sylvestris – needles	18	*33*	*45*	68/69, 69/70 Litter surface[b]
	Betula tortuosa – wood	(4.0)	*5.6*	*7.1*	69/73 Litter surface[a]
	Cladonia alpestris	5.6 ± 0.7 (3)	16.5 ± 2.8 (3)	27.9 ± 3.8 (3)	69/72 In plastic boxes
	– Birch Forest				
	Betula tortuosa – wood	(4.5)	*6.8*	*9.1*	69/73 Litter surface[a]
	Betula tortuosa – leaves	27	*47*	*61*	68/69, 69/70 Litter surface[b]
	– Lichen Heath				
	Betula tortuosa – wood	(5.0)	*6.9*	*8.3*	69/73 Litter surface[a]
Hardangervidda	– Lichen Heath				
	Barley straw	10.3 ± 0.8 (3)	16.6 ± 0.2 (3)	20.7 ± 0.6 (6)	69/72 L/A_0
	Festuca ovina	16.0 ± 2.0 (2)	*29.4*	*40.7*	71/72 Litter surface
	Juncus trifidus	6.7 ± 2.7 (2)	*13.0*	*18.8*	71/72 Litter surface
	Carex bigelowii	20.0 ± 1.2 (2)	*36.0*	*48.8*	71/72 In litter layer
	Lichens	4.9 ± 1.2 (3)	13.2 ± 5.2 (3)	17.5 ± 2.4 (6)	69/72 L/A_0
	Woody roots	6.2 ± 0.8 (9)	*12.0*	*17.5*	72/73 Vertically 0–10 cm
	Site vegetation – vascular plants	12.3 ± 2.3 (3)	1.4 ± 12.1 (3)	15.0 ± 4.8 (6)	69/72 L/A_0
	– Dry Meadow				
	Barley straw	24.8 ± 1.8 (3)	39.1 ± 3.4 (3)	53.1 ± 0.9 (6)	69/72 L/A_0
	Dryas octopetala	5.1 ± 0.8 (4)	*9.9*	*14.5*	71/72 In litter layer
	Mosses	10.5 ± 7.4 (3)	*19.9*	*28.3*	71/72 In litter layer
	Monocot. roots	29.2 ± 0.5 (3)	*49.9*	*64.5*	72/73 Vertically 0–10 cm
	Dicot. roots	19.8 ± 2.0 (10)	*35.7*	*48.4*	72/73 Vertically 0–10 cm
	Woody roots	32.2 ± 3.3 (6)	*54.0*	*68.8*	72/73 Vertically 0–10 cm
	Site vegetation (expt. *D. octopetala*)	38.0 ± 4.7 (3)	36.3 ± 4.4 (3)	58.1 ± 8.4 (6)	69/72 L/A_0

	– Wet Meadow					
	Barley straw	19.5 ± 2.1 (3)	26.0 ± 0.8 (3)	40.9 ± 2.2 (6)	69/72	L/A_0
	Carex nigra	42.0 ± 1.8 (4)	*66.4*	*80.5*	71/72	In litter layer
	Salix spp. – leaves-shoots	28.3 ± 2.2 (4)	*48.6*	*63.1*	71/72	In litter layer
	Carex nigra roots	21.5 ± 0.9 (4)	*38.4*	*51.6*	71/72	Vertically 0–10 cm
	Dicot. roots	29.6 ± 1.7 (10)	*50.5*	*65.1*	72/73	Vertically 0–10 cm
	Woody roots	21.7 ± 0.8 (10)	*28.7*	*52.0*	72/73	Vertically 0–10 cm
	Site vegetation (expt. bryophytes)	32.8 ± 5.7 (3)	45.9 ± 5.4 (3)	69.3 ± 4.6 (6)	69/72	L/A_0
	– Birch Forest					
	Barley straw	19.1 ± 1.0 (4)	44.4 ± 1.7 (4)	61.8 ± 4.6 (7)	69/72	L/A_0
	Site vegetation	32.7 ± 0.9 (4)	*54.7*	*69.5*	71/72	Litter surface
	– Snow Bed					
	Barley straw	23.3 ± 1.2 (3)	28.5 ± 2.1 (3)	36.2 ± 0.5 (8)	70/72	Litter surface[c]
	Site vegetation	38.0 ± 4.7 (3)	*61.6*	*76.2*	71/72	Litter surface
Stordalen						
	Rubus chamaemorus	23.5 ± 2.8 (10)	36.9 ± 3.6 (10)	47.5 ± 3.7 (10)	70/73	Litter surface
	Rubus chamaemorus	16.4 ± 3.9 (8)	21.1 ± 3.2 (11)		71/73	Litter surface
	Rubus chamaemorus	28.4 ± 1.8 (10)			72/73	Litter surface
	Empetrum hermaphroditum	5.7 ± 0.3 (10)	19.8 ± 1.7 (7)	26.2 ± 2.3 (8)	70/73	Litter surface
	Andromeda polifolia	15.0 ± 0.9 (9)	24.7 ± 1.4 (7)		71/73	Litter surface
	Betula nana leaves	20.9 ± 1.2 (11)	28.2 ± 1.1 (7)		71/73	Litter surface
	Betula nana twigs	7.8 ± 0.5 (9)	15.5 ± 0.9 (10)		71/73	Litter surface
	Eriophorum vaginatum	37.0 ± 1.0 (7)			72/73	Litter surface
	Sphagnum fuscum	5.4 (20)			71/74	Vertically 0.5–5.5 cm
	Sphagnum balticum	4.3 (10)	7.8 (9)		71/73	Litter surface Vertically 0.5–5.5 cm
	Sphagnum lindbergii	7.1 (10)	19.3 (11)		72/74	Litter surface Vertically 0.5–5.5 cm
	Dicranum elongatum	7.1 (20)			72/74	Litter surface Vertically 0.5–5.5 cm
	Drepanocladus schulzei	4.7 (10)			73/74	Litter surface Vertically 0.5–5.5 cm

[a] Measured for four years and over first summer (in brackets).
[b] Mean for two first-year observations.
[c] Experiment started in spring 1970.

42% in the first year in the wet meadow, and *Eriophorum vaginatum* lost 37% at Stordalen. In both instances, graminids had the largest weight losses of any of the litter types studied. There did not seem to be large differences between rates of decomposition of litter in the Hardangervidda wet meadow and Stordalen mire. The Hardangervidda wet meadow is eutrophic, while the Stordalen mire is oligotrophic (SONESSON et al., 1975); the data in Table 1 do not indicate that the nutrient status of the site influenced decomposition to any large extent, the phosphorus content in the Hardangervidda wet meadow is twice that of the Stordalen site (HINNERI et al., 1975), climatic conditions thus seem to be the major factors regulating decomposition rates. HEAL and FRENCH (1974), examining a wider variety of tundra sites, found that the nutrient status of the soil, especially phosphorus content, was correlated with rates of plant litter decomposition. This is probably to a large extent due to changes in litter quality, and not to a direct effect of soil phosphorus *per se*. The concentrations of soil nutrients will more markedly influence the rates of root litter and soil organic matter decomposition, and differences should be reflected in soil respiration measurements.

It is difficult to compare the different sites, as the same types of litter were not used. Even if the same plant species had occurred at all the sites investigated, the nutrient content of the litter would vary depending on the nutrient status of the site, and possibly also as a result of ecotypic differences in the species. For this reason, cellulose was used as a model substance for decomposition experiments, making direct comparison between sites easier (BERG et al., 1975; ROSSWALL, 1974a).

At Hardangervidda, barley straw was used as standard litter and put out at all five sites in a comparable manner, and the weight loss followed for three years. The results showed marked differences between sites (Fig. 1), with decomposition rates decreasing in the order: birch forest > dry meadow > wet meadow > snow

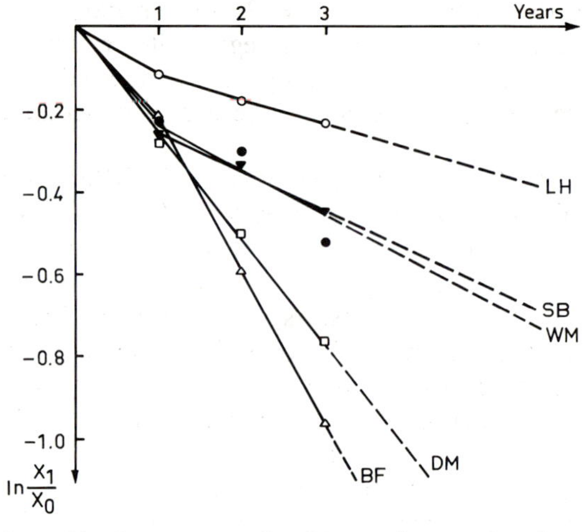

Fig. 1. Decomposition of barley straw at five Norwegian tundra sites. X_1 is the remaining weight after 1, 2, or 3 years, and X_0 is the weight at time zero. The plotting assumed a negative exponential loss rate after year one. *BF* birch forest, *DM* dry meadow, *WM* wet meadow, *SB* snow bed, *LH* lichen heath

Table 2. Comparison between primary production, decomposition and amounts of microorganisms at five Norwegian tundra sites. Data for primary production and respiration from SVENSSON et al. (1975), cellulose decomposition from BERG et al. (1975), bacteria from CLARHOLM et al. (1975) and fungi from HANSSEN and GOKSØYR (1975)

Site	Cellulose decomposition 1st year loss, %	Barley straw decomposition 1st year loss, %	Respiration $gC \cdot m^{-2} \cdot yr^{-1}$	Primary production $gC \cdot m^{-2} \cdot yr^{-1}$	Bacteria (direct count) $\cdot 10^9 \cdot g^{-1}$ soil dry weight top layer	Bacteria (plate count) $\cdot 10^7 \cdot g^{-1}$ soil dry weight top layer	Fungal hyphae $m \cdot g^{-1}$ soil dry weight top layer
Lichen heath (LH)	1	20.7	94	114	33	0.2	730
Snow bed (SB)	6	36.2	177	n.d.	37	0.4	850
Wet meadow (WM)	3	40.9	253	351	205	6.8	6850
Dry meadow (DM)	6	53.1	457	225	86	34.3	2900
Birch forest (BF)	85	61.8	689	328	106	4.2	3600

bed > lichen heath, with the birch-forest barley straw losing weight three times as rapidly as the barley straw placed at the lichen heath. At the lichen heath low moisture probably limits decomposition rates. The snow bed site has a shorter growing season, but the soil moisture is more favorable than at the lichen heath. The moisture level in the wet meadow is probably overoptimal for decomposition, and the soil temperature (−5 cm) is lower (annual mean 2.5° C) as compared with the dry meadow (annual mean 2.7° C). The rapid decomposition in the birch forest probably depends on several factors: the temperature is higher than at any other site investigated, soil moisture conditions in the A_0 layer are frequently nonlimiting, and there is also often high air humiditiy with fog. High phosphorus concentrations possibly also influenced rates of decomposition. SVENSSON et al. (1975), investigating soil respiration (Table 2), also found the highest activity at the birch forest and the lowest at the lichen heath. Primary production in the wet meadow is higher than at the other Norwegian sites referred to in the present paper, but due to the slow rates of decomposition reflected by both litter weight loss and soil respiration measurements, there is a slow accumulation of organic matter (SVENSSON et al., 1975).

Decomposers, both fungi (hyphal lengths) and bacteria (direct microscopic counts), are more abundant in the wet meadow (HANSSEN and GOKSØYR, 1975; CLARHOLM et al., 1975), with nearly twice the amount in the wet meadow compared with the birch forest. There is thus a good agreement between plant litter decomposition and soil respiration results, while biomass of microorganisms does not necessarily reflect the speed at which organic matter is decomposed.

A comparison can also be made between the different Kevo sites, where birch wood decomposition was studied over four years. Surprisingly, small variation between the sites was found, with 8.6% weight loss at the pine site under the canopy after four years, 12.1% at the same site but in an open place, 11.3% at the birch forest site and 12.6% at the lichen heath. These results did not indicate that edaphic factors were the major ones influencing decomposition rates, as e.g. the phosphorus content varied greatly between the sites, being 6, 140, and 65 ppm respectively in the three vegetation types.

Table 3. Decomposition of plant litter at Hardangervidda corrected for site differences by the use of the barley straw data (Table 1), and transformed to the abiotic conditions prevailing at the lichen heath site

Species	Decomposition, % 1st year	Site	Year
Carex nigra	22.2	WM	71/72
Carex bigelowii	20.0	LH	71/72
Festuca ovina	16.0	LH	71/72
Salix spp.	14.9	WM	71/72
Juncus trifidus	6.7	LH	71/72
Dryas octopetala	2.1	DM	71/72
Site vegetation	12.3	LH	69/70
Site vegetation	15.8	DM	69/70
Site vegetation	17.3	WM	69/70
Site vegetation	17.6	BF	71/72
Site vegetation	16.8	SB	71/72

Decomposition of Different Litter Types. Several different types of litter were investigated at the five Hardangervidda sites (Table 1). In order to be able to compare the rates obtained at different sites, the barley straw decomposition rates were used to correct the other data to a "standard site" (lichen heath) (Table 3). Graminids had the highest relative as well as absolute decomposition rates (*Carex nigra, C. bigelowii* and *Festuca ovina*), whereas *Dryas octopetala* decomposed very slowly with a first year relative weight loss of only 2%. The small *Dryas* decomposition is more surprising, as the material was collected green and a first leaching and decomposition of easily decomposable substrates would have been expected.

Mixed site vegetation decomposition was also studied at Hardangervidda. If the climatic differences between sites are disregarded by transformation of the data to the lichen heath conditions, there were no large differences in decomposition (Table 3). The birch forest (BF) and wet meadow (WM) showed the largest and very similar first-year weight losses for site litter (17.6 and 17.3%), although the sites varied with regard to the soil nutrients in the upper horizons (P-A_1 19.4 ppm for BF and 2.2 for WM, total-N 1.48% for BF and 2.09 for WM; HINNERI et al., 1975). The lichen heath vegetation decomposed at the lowest rate, and the nutrient concentrations at this site were low (P-A_1 2.5 ppm, N 0.8%; HINNERI et al., 1975). The snow bed and dry meadow occupy intermediate positions with regard to decomposition rates of site vegetation, if the effects of local climate are disregarded.

At Stordalen, *Eriophorum vaginatum* decomposed by 37% during the first year, thus having the fastest loss rate of all species studied (Table 1). The decomposition of *E. vaginatum* was also investigated on other IBP tundra sites; first-year weight loss at Moor House (UK) was 24% and at Glenamoy (Ireland) 20% (HEAL and FRENCH, 1974). Decomposition of *Eriophorum* spp. has also been estimated on the standing dead fraction, giving first-year weight losses of 28% for *E. angustifolium* at Point Barrow (USA) as compared to 27% in the Hardangervidda wet meadow and 25% for *E. vaginatum* at Eagle Summit (USA) (HEAL and FRENCH, 1974).

First year decomposition values for *Rubus chamaemorus* litter varied between 16 and 28%. This variation could not be explained by differences in concentrations of easily decomposable and leachable substances present in the litter, caused by collection in different years and with varying extents of leaching preceding collection. When the three-year data (1970/73) (Table 1) were plotted in the same manner as the barley straw data in Fig. 1, the line did not pass through the origin, and the intercept indicated a too rapid loss rate during the first year for the process to be exponential. The weight loss was calculated to be 9% too high during the first year, and this could be attributable to leaching and decomposition of easily metabolisable substrates. The small weight loss for the first year 1971/72 was, however, also followed by a small second-year weight loss as compared to the second-year weight loss for the material from 1970.

For *Empetrum hermaphroditum*, there was only a very small first-year loss followed by considerable decomposition during the second and third years. The *Empetrum* material used was harvested in September and consisted of the above-ground part of the plants, and no attempt was made to separate material into new and old leaves, etc.

Different species of *Sphagnum* occur at the Stordalen site, depending on the habitat (SONESSON, 1973). *Sphagnum* decomposition was studied in these different habitats, and the variation in rates of decomposition presented in Table 1 were found to depend both on habitat and species. *S. lindbergii* grows in very wet, minerotrophic habitats, while *S. fuscum* and *Dicranum elongatum* grow in dry, ombrotrophic ones. It seemed as if the plant species had a greater influence on decomposition rate than habitat; *S. fuscum* decomposed slowly in the dry, nutrient-poor habitat, but *D. elongatum*, in the same habitat, decomposed at the same rate as the *S. lindbergii* in wet, nutrient-richer environment. *Sphagnum* species are known to decompose slowly, and they are often the main constituents of peat. An inhibitory substance, sphagnol (CZAPEK, 1899), isolated from *Sphagnum* spp. was shown to inhibit bacteria, and has recently been reisolated (ENGMANN, 1972) and different components identified (RUDOLPH, 1972). Also other, water soluble, inhibitory substances have been isolated from *Sphagnum* peat (RAMAUT, 1953), and this could explain the slow decomposition of the mosses which is one important factor causing peat accumulation.

It is possible that the more rapid weight loss of *Dicranum elongatum* than *Sphagnum fuscum* at the dry site was a result of the smaller *D. elongatum* leaves falling out of the bags. CLYMO (1965) found the largest weight losses from *Sphagnum* spp. with small leaves, but he showed that this was not due to leaves lost by handling or by water movement through the litter bags. BAKER (1972) found good agreement between decomposition of the moss *Chorisodontium aciphyllum* in Antarctica whether measured by litter bag, weight per length or bulk density of accumulating peat, thus supporting the validity of using the litter bag method with mosses.

Root decomposition was studied at three of the Hardangervidda sites (Table 1). Rates of decomposition were high at both the wet and dry meadow sites for all root types studied. At the lichen heath, decomposition was small for the first year, probably due to the low moisture content of the soil in which the root samples were buried. Belowground parts of *Carex nigra* decomposed at half the rate of the aboveground parts at the wet meadow, probably due to too high moisture content, and contributed to the accumulation of organic matter.

Influence of Temperature and Moisture on Litter Decomposition. In an attempt to determine the influence of major climatic variables on rates of decomposition, KÄRENLAMPI (1971) observed birch leaf and pine needle decomposition at 2–3 week intervals during the summer of 1970. By the use of various linear regression equations, he found that there was a highly significant positive correlation between the rate of decomposition of birch leaves and long periods of drought; $r = 0.99$, $p < 0.1\%$ with no days < 0.1 mm rainfall. The correlation coefficient for pine needles was 0.78 ($p < 10\%$). Mean daily rainfall, percentage of days with rainfall or mean temperature during the observation period did not give significant correlations with decomposition. Both Kevo and Stordalen have low precipitation, with high summer temperatures. Moisture is probably the main abiotic factor limiting the decomposition process, in the litter layer at any rate.

A simulation model for plant litter decomposition has been developed within the Tundra Biome International Decomposition Work Group (BUNNELL and DOWDING, 1974). This model was used on data for *Rubus chamaemorus* litter decomposition at Stordalen, Sweden (LOHAMMAR and ROSSWALL, 1974), and it predicted no decomposition during July and August, due to low precipitation. This does not agree with field observations; although decomposition is very slow in summer, when the temperature is optimal. Decomposition is most rapid during September, when moisture increases (ROSSWALL, 1973). Precipitation also influences leaching rates, and it is probable that the first rapid weight loss in autumn after litter fall is due to leaching and decomposition of easily metabolisable compounds, e.g. soluble carbohydrates and amino acids. *Betula nana* and *Andromeda*

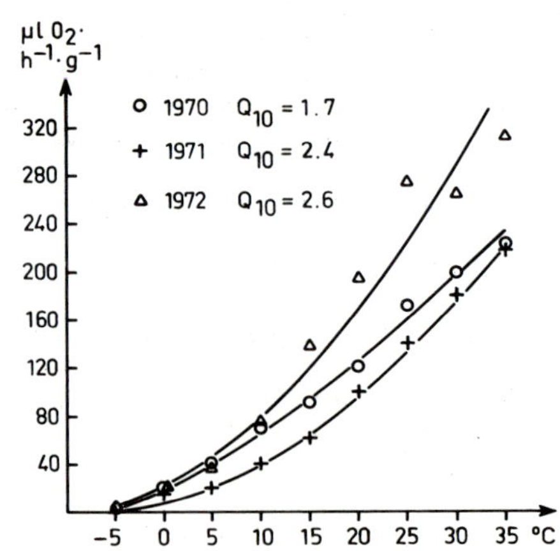

Fig. 2. Respiration (O_2-uptake, $\mu l\, O_2 \cdot h^{-1} \cdot g^{-1}$ d.w. at 760 mm Hg and 0° C) of *Rubus chamaemorus* litter collected at the Stordalen site in September 1973. All data points are means for three separate samples. Litter moisture was adjusted to ca. 300% of d.w. Data plotted from ROSSWALL (1974b)

polifolia lost 50 and 40% respectively of the first-year weight loss during September-May at Stordalen (ROSSWALL, 1973). This is in agreement with the data presented by WITKAMP (1966), who suggested that 40% of the first-year weight loss from deciduous leaves was independent of microbial activity and could mostly be attributed to leaching.

The effect of temperature on decomposition of *Rubus chamaemorus* litter at Stordalen was estimated by respiration measurements (ROSSWALL, 1974b). These were made on litter which had been decomposing for one, two and three years (Fig. 2, cf. Table 1). The decomposition rate decreased in the order 1972, 1970, and 1971 litter, which is in agreement with the weight loss results presented in Table 1. Q_{10}(10–20° C) was 2.4 for the 1971 and 2.6 for the 1972 material, while it was lower (1.7) for the oldest leaves. These Q_{10} values are low compared with the values for soil respiration given by SVENSSON et al. (1975).

References

BAKER, J. H.: The rate of production and decomposition of *Chorisodontium aciphyllum* (Hook. f. & Wils.). Broth. Br. Antarct. Surv. Bull. **27**, 123–129 (1972).

BERG, B., KÄRENLAMPI, L., VEUM, A. K.: Comparison of decomposition rates measured by means of cellulose. In: WIELGOLASKI, F. E. (Ed.): Fennoscandian tundra ecosystems, Part 1: Plants and microorganisms, pp. 261–267. Berlin-Heidelberg-New York: Springer 1975.

BOCOCK, K. L., GILBERT, O. J. W.: The disappearance of leaf litter under different woodland conditions. Plant and Soil **9**, 179–185 (1957).

BOOTH, T.: Studies on dung decomposition. In: BLISS, L. C. (Ed.): Devon Island IBP project High arctic ecosystem. Project report 1970 and 1971, pp. 359–368. Edmonton: University of Alberta, Dept. of Botany 1972.

BUNNELL, F., DOWDING, P.: Abisko model—A generalised decomposition model for comparison between tundra sites. In: HOLDING, A. J., HEAL, O. W., MACLEAN, S. F., FLANAGAN, P. W. (Eds.): Soil Organisms and Decomposition in Tundra, pp. 227–248. Stockholm: IBP Tundra Biome Steering Committee 1974.

CLARHOLM, M., LID-TORSVIK, V., BAKER, J. H.: Bacterial populations of some Fennoscandian tundra soils. In: WIELGOLASKI, F. E. (Ed.): Fennoscandian tundra ecosystems, Part 1: Plants and microorganisms, pp. 251–260. Berlin-Heidelberg-New York: Springer 1975.

CLYMO, R. S.: Experiments on breakdown of *Sphagnum* in two bogs. Ecology **53**, 747–758 (1965).

CZAPEK, F.: Zur Chemie der Zellmembranen bei den Laub- und Lebermoosen. Flora **86**, 361–381 (1899).

DOUGLAS, L. A., TEDROW, J. C. F.: Organic matter decomposition rates in arctic soils. Soil Sci. **88**, 305–312 (1959).

ENGMANN, B.: Beitrag zur Klärung des Czapekschen Sphagnolbegriffes. Biochem. Physiol. Pflanzen **163**, 93–102 (1972).

FLANAGAN, P. W., VEUM, A. K.: The influence of temperature and moisture on decomposition rates on tundra. In: HOLDING, A. J., HEAL, O. W., MACLEAN, S. F., FLANAGAN, P. W. (Eds.): Soil Organisms and Decomposition in Tundra, pp. 249–278. Stockholm: IBP Tundra Biome Steering Committee 1974.

GOKSØYR, J.: Decomposition, microbiology and ecosystem analysis. In: WIELGOLASKI, F. E. (Ed.): Fennoscandian tundra ecosystems, Part 1: Plants and microorganisms, pp. 230–238. Berlin-Heidelberg-New York: Springer 1975.

HANSSEN, J. F., GOKSØYR, J.: Biomass and production of soil and litter fungi at Scandinavian tundra sites. In: WIELGOLASKI, F. E. (Ed.): Fennoscandian tundra ecosystems, Part 1: Plants and microorganisms, pp. 239–243. Berlin-Heidelberg-New York: Springer 1975.

HEAL, O. W., FRENCH, D. D.: Decomposition of organic matter on tundra. In: HOLDING, A. J., HEAL, O. W., MCLEAN, S. F., FLANAGAN, P. W. (Eds.): Soil Organisms and Decomposition in Tundra, pp. 279–310. Stockholm: IBP Tundra Biome Steering Committee 1974.

HINNERI, S., SONESSON, M., VEUM, A. K.: Soils of Fennoscandian IBP tundra ecosystems. In: WIELGOLASKI, F. E. (Ed.): Fennoscandian tundra ecosystems, Part 1: Plants and microorganisms, pp. 31–40. Berlin-Heidelberg-New York: Springer 1975.

JENNY, H., GESSEL, S. P., BINGHAM, F. T.: Comparative study of decomposition rates of organic matter in temperate and tropical regions. Soil Sci. **68**, 419–432 (1949).

KÄRENLAMPI, L.: Weight loss of leaf litter on forest soil surface in relation to weather at Kevo Station, Finnish Lapland. Rep. Kevo Subarctic Res. Stat. **8**, 101–103 (1971).

LOHAMMAR, T., ROSSWALL, T.: Use of the "Abisko model" for simulating *Rubus chamaemorus* litter decomposition. Progress Report 1973. Swedish IBP Tundra Biome Project Tech. Rep. **16**, 97–107 (1974).

OLSON, J. S.: Energy storage and the balance between producers and decomposers in ecological systems. Ecology **44**, 322–331 (1963).

PARINKINA, O. M.: Determination of bacterial growth rates in tundra soils. In: ROSSWALL, T. (Ed.): Modern methods in the study of microbial ecology. Bull. Ecol. Res. Comm. (Stockholm) **17**, 303–309 (1973).

RAMAUT, J.: Contribution à l'étude du pouvoir inhibiteur des extraits aqueux et alcoholiques de tourbe vis-à-vis d'un germe du sol. Bull. Soc. Roy. Sci. Liège **12**, 561–569 (1953).

ROSSWALL, T.: Decomposition and ecosystem research. In: ROSSWALL, T. (Ed.): Systems analysis in northern coniferous forests–IBP Workshop. Bull. Ecol. Res. Comm. (Stockholm) **14**, 149–157 (1971).

ROSSWALL, T.: Plant litter decomposition studies at the Swedish Tundra site. Progress Report 1972. Swedish IBP Tundra Biome Project Tech. Rep. **14**, 124–133 (1973).

ROSSWALL, T.: Cellulose decomposition studies on the tundra. In: HOLDING, A. J., HEAL, O. W., MACLEAN, S. F., FLANAGAN, P. W. (Eds.): Soil Organisms and Decomposition in Tundra, pp. 325–340. Stockholm: IBP Tundra Biome Steering Committee 1974a.

ROSSWALL, T.: Decomposition of plant litter at Stordalen. In: Progress Report 1973. Swedish IBP Tundra Biome Project Tech. Rep. **16**, 207–212 (1974b).

RUDOLPH, H.: Identifikation der Czapekschen Sphagnolkristalle. Biochem. Physiol. Planzen. **163**, 110–112 (1972).

SKARTVEIT, A., RYDÉN, B.-E., KÄRENLAMPI, L.: Climate and hydrology of some Fennoscandian tundra ecosystems. In: WIELGOLASKI, F. E. (Ed.): Fennoscandian tundra ecosystems, Part 1: Plants and microorganisms, pp. 41–53. Berlin-Heidelberg-New York: Springer 1975.

SONESSON, M.: Studies in production and turnover of bryophytes at Stordalen 1972. Progress Report 1972. Swedish IBP Tundra Biome Project Tech. Rep. **14**, 66–75 (1973).

SONESSON, M., WIELGOLASKI, F. E., KALLIO, P.: Description of Fennoscandian tundra ecosystems. In: WIELGOLASKI, F. E. (Ed.): Fennoscandian tundra ecosystems, Part 1: Plants and microorganisms, pp. 3–28. Berlin-Heidelberg-New York: Springer 1975.

SVENSSON, B. H., VEUM, A. K., KJELVIK, S.: Carbon losses from tundra soils. In: WIELGOLASKI, F. E. (Ed.): Fennoscandian tundra ecosystems, Part 1: Plants and microorganisms, pp. 279–286. Berlin-Heidelberg-New York: Springer 1975.

WITKAMP, M.: Decomposition of leaf litter in relation to environment, microflora and microbial respiration. Ecology **47**, 194–201 (1966).

Carbon Losses from Tundra Soils

B. H. SVENSSON[1], A. K. VEUM[2], and S. KJELVIK[3]

Biological activities in soils are often indirectly assessed by measuring gaseous carbon losses as CO_2. There is a fundamental difference between respiration in roots as compared with that in soil fauna and microorganisms: energy for root respiration is obtained from translocated photosynthesized carbon within the plant, while energy for soil fauna and microorganisms is obtained from mineralization of organic compounds in the soil. Conversion of soil carbon to CO_2 releases nutrients fixed in litter, thereby making them available to plants again. The carbon dioxide produced by microorganisms may therefore reflect the turnover of certain minerals in soil (WITKAMP, 1973). It is very difficult to separate root respiration from that of microorganisms and soil fauna when measuring evolution of carbon dioxide in the field; these figures can only be used as integrated indices for total metabolism in the soil (MACFADYEN, 1971). The use of respiration figures as indices of biological activities in soils may be valid for aerobic soils. Quite another situation prevails in anaerobic soils, where considerable amounts of mineralized carbon may be lost to the atmosphere as methane, and methane production has been observed by CLYMO and REDDAWAY (1971) and SVENSSON (1974) as playing an important role in very wet areas of oceanic and subarctic peat bogs.

Measurements of gaseous carbon losses have been made for both the Norwegian and the Swedish IBP tundra sites (SONESSON et al., 1975). The Norwegian studies were made on total carbon dioxide losses only (caused by litter and soil fauna, litter and soils microorganisms and root respiration), later called soil respiration, while the Swedish investigations also took methane production into account. The purpose of this paper is to compare carbon losses from the sites in relation to governing environmental factors, and to compare the losses with the amount of carbon fixed by primary production.

Materials and Methods

Gas collections were made in a similar manner at all sites, using cuvettes of different kinds inserted into the soil, and withdrawing samples for analysis (FLAN-

[1] Responsible for the investigation at the Stordalen site as well as for the statistical analyses and the manuscript.
[2] Responsible for investigations at Hardangervidda 1972.
[3] Responsible for investigations at Hardangervidda 1973.

AGAN and VEUM, 1974; SVENSSON, 1974). Gas sampling and analysis were performed in different ways: the Norwegian investigations were made by means of pump and ampoule equipment in the field (KJELVIK et al., 1975), while the analysis for the Swedish site was performed by gas chromatography on samples transported to the laboratory in plastic syringes (SVENSSON, 1974).

Results and Discussion

At the beginning of the season (1972) the soil respiration values for the sites, except birch forest, at Hardangervidda, Norway rose quickly to a maximum. In the birch forest there were no measurements at the beginning of the season, but a rapid increase was noted at the other sites after snow melt. There was a decrease in respiration at all the sites towards the end of the growing season. The mean temperatures of the litter layer and the temperature range for the different days of measurements are given in Table 1, as are the mean and range for the moisture. Since biological processes follow exponential functions in relation to temperature, the natural logarithms of the soil respiration figures were used in regression models with temperature as the independent variable for each site. The r-values varied from 0.80 to 0.90 (r_1, Table 1), and the intercepts for these models predicted respiration at 0° C, the same respiration rate being assumed for temperatures outside the range covered by measurements. The respiration above 0° C is in agreement with laboratory measurements on plant litter respiration (FLANAGAN and VEUM, 1974; ROSSWALL et al., 1975), and can be seen from Fig. 1, where the exponential graphs predicted by the regression lines for the different sites are shown.

By adding moisture as a second independent variable in the models, i.e.:

$$\ln(CO_2) = a + b_1x + b_2z + e$$

where x and z are the temperature and moisture variables respectively, it is assumed that soil respiration is influenced by temperature and moisture together in an exponential process. An increase was obtained in the r-values when moisture was added in a stepwise regression analysis (Table 1). The increase was not significant for the dry meadow, wet meadow and birch forest models, whereas it was significant for the driest site, viz. lichen heath and the snow bed in a relatively dry, but not in a moister year. The response of soil respiration in the Hardangervidda dry meadow to air temperature fluctuations at 5 cm above ground level on a cloudless day is shown in Fig. 2 ($r = 0.92$).

To compare the output of carbon from the Norwegian Hardangervidda and Swedish Stordalen sites due to gaseous losses to the atmosphere, with the input by primary production, the losses were summed over a season (Tables 2 and 3). For the Norwegian sites the time without snow cover was used as the respiration season, and for the Swedish ones, from the disappearance of snow until the upper layer of the peat became frozen. Estimation of soil respiration at the border of a melting snow patch at the Hardangervidda birch forest site in 1973 showed a carbon loss of 39 mg m^{-2} h^{-1} at a snow depth of 15 cm, 137 at the melting border and 259 in a spot that had been free from snow for three hours. This supports the

Table 1. The mean temperatures in the litter layer and soil moistures for the different days of measurement (*n*) are given together with their ranges for the sites at Hardangervidda. The constants (*a*) and the parameters (b_1, b_2) for the regressions of logarithmized respiration ($gC\,m^{-2}\,h^{-1}$) against temperature and moisture are also presented, together with the correlation coefficient for the models for the relation to temperature, r_1, and moisture, r_2, and r_{12} taking both variables into account. The marks on the parameter values show their significance, $^{+}$ indicating significant at the 1 and ‡ at the 5% level respectively, while ^{x} indicates no significance even at the 20% level. Q_{10} for the different site models are also given

	Mean temp. (°C)	Range	Mean moist. (% d.w.)	Range	*n*	*a*	b_1	b_2	r_1	r_2	r_{12}	Q_{10}
Lichen heath	9.8	1.6–19.5	40	21– 62	12	1.40	0.136^{+}	$0.017^{\ddagger}$	0.87	0.20	0.93	4.5
Dry meadow	11.5	5.0–20.3	175	95–467	11	3.81	0.106^{+}	0.00077^{x}	0.86	0.25	0.88	3.0
Wet meadow	10.7	5.9–16.9	578	423–681	11	3.81	0.136^{+}	-0.00079^{x}	0.90	−0.12	0.92	3.8
Birch forest	11.9	8.0–15.0	158	56–231	9	4.86	0.053^{+}	-0.00035^{x}	0.80	0.30	0.81	2.0
Snow bed	11.6	4.0–17.2	113	33–184	11	1.59	0.182^{+}	0.0056^{+}	0.87	0.22	0.98	5.2

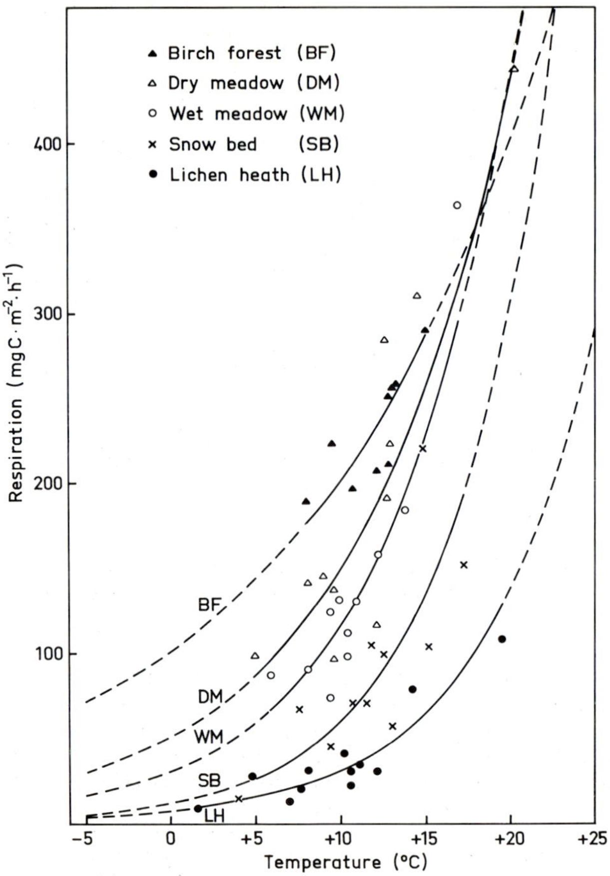

Fig. 1. The mean values of different measurements of respiration over the season in 1972 are plotted against temperature for the Hardangervidda sites, and the graphs predicted from the respiration-temperature regression models are also given

choice of limits for the respiration season, as the soil respiration rate seems to rise immediately after the snow has disappeared.

Since the Swedish data were mainly collected over a 24 hr day, they could be used as mean day values and integrated over the season. The Norwegian data were mainly obtained during daytime when the temperature is high and were thereby too high to be used for integration, since soil respiration seems to follow temperature at daytime (Fig. 2). The figures for the Hardangervidda sites (Table 3) were therefore calculated by using the mean daily temperature in the surface soil layers obtained from the local climatic station, and reading the corresponding respiration values from the graphs in Fig. 1, and summed over the season.

Assuming the systems to be in steady states, gaseous carbon losses from the soil should be equivalent to net primary production plus root respiration. From Table 2 it can be seen that the soil respiration data for the Norwegian lichen heath and wet meadow are somewhat low in relation to the net primary production

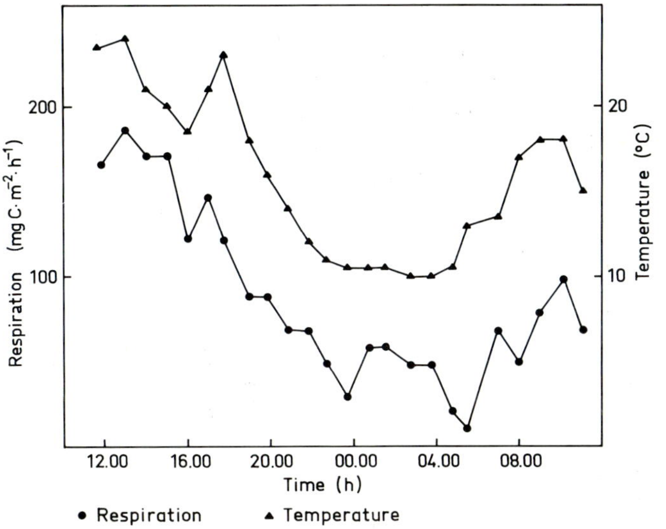

Fig. 2. Variation in respiration and temperature over one 24-hr period in the dry meadow, Hardangervidda, at the beginning of July 1973. The linear correlation coefficient was 0.92

Table 2. Seasonal carbon losses by respiration from the different Norwegian sites are compared with primary production. The season for respiration was taken as being the period without snow at each site (SKARTVEIT *et al.*, 1975). The corresponding lengths of the growing seasons and the primary production values are those given by KJELVIK and KÄRENLAMPI (1975) for the lichen heath and birch forest, and by WIELGOLASKI (1975) for the meadow sites. n.d. indicates not determined

Site	Respiration ($gC\,m^{-2}$)	Length of season (days)	Primary production ($gC\,m^{-2}$)	Growing season (days)
Lichen heath	94	140	114	110
Dry meadow	457	140	225	105
Wet meadow	253	133	351	100
Birch forest	689	154	328	130
Snow bed	177	67	n.d.	60

figures, while those for the dry meadow and birch forest are considerably higher. The low value for the lichen heath may be explained by the fact that the positive moisture effect was not taken into account, only soil respiration against temperature being used for the calculation. Root respiration is probably low at this site in relation to respiration by the decomposer organisms.

The wet meadow is not in a fully steady state, as there is an annual increment of 0.1 mm. With a bulk density of 0.2 $kg\,m^{-3}$ and 80.9% organic matter in the soil (VEUM, 1972), there is an accumulation of about 7 $gC\,m^{-2}$ each year. The amount of organic matter was divided by a factor of 0.648 (SESTAK et al., 1971) to $gCO_2\,m^{-2}$ and then to $gC\,m^{-2}$. Methane evolution was noted at the site (HANS-

Table 3. Estimations of carbon losses (gC m^{-2}) of different habitats at the Stordalen mire during one growing season, 10th May to 20th Sept. 1973, (134 days). The table is adapted from MADSEN and WIDELL (1974) and SVENSSON (1974). The moisture values given are from a single measurement in July 1973. n.d. means not determined

Habitat	Percent cover of the site	Carbon losses (gC m^{-2})			Percent CH_4 of total	Moisture (% d.w.)
		CO_2	CH_4	total		
Elevated ombrotrophic (incl. small ombrotrophic depressions)	63.5	49.9	0.05	50.0	0.1	303–945
Ombrotrophic/minerotrophic	22.0	12.4	1.2	13.6	8.8	1431
Minerotrophic depressions	13.5	8.4	7.3	15.7	46.5	1647
Open water	1.0	n.d.	n.d.	—	—	n.d.
Total	100.0	70.7	8.6	79.3	11.0	—

SEN et al., 1973), but no quantification was made. From the figures for the Stordalen mire (Table 3) it can be seen that considerable amounts of methane can be produced, especially in very wet areas. For the mean moisture (530% d.w.) in the Norwegian wet meadow, a methane production of at least 0.1% of the total carbon losses could be expected on the basis of the Swedish values given for similar soil moisture in Table 3. The figure is probably higher as the wet meadow is a minerotrophic site, and there is a great increase in methane production from the ombrotrophic/minerotrophic to the plain minerotrophic habitat at Stordalen. The annual accumulation of organic matter and the methane production may therefore be due to the difference between the soil respiration and the primary production obtained for the wet meadow.

Primary production in the dry meadow and birch forest was about 50% of soil respiration (Table 2). Since the systems are assumed to be in steady states the most probable explanation for the differences between the input of carbon to the systems, fixed in organic material, and the output, measured as soil respiration, seems to be a high contribution from root respiration. Figures of up to 50% of soil respiration have earlier been reported for root respiration in forests, while lower values are mostly reported for grasslands (MACFADYEN, 1971).

For calculating the carbon losses from the Stordalen site (Table 3), the losses in different plant communities were summed over the season (SVENSSON, 1974), and weighted for 1 m^2 in proportion to the cover of the community in question in the site habitats given by MADSEN and WIDELL (1974).

At Stordalen, there is an increase in methane production with increasing moisture, constituting less than 1% of the total carbon losses on elevated, therefore drier, habitats and up to about 45% for the waterlogged minerotrophic depressions. The increase in methane losses from wetter areas may depend on the anaerobic environments created by stagnant water, as extremely low oxygen tensions are necessary for the growth of methane-forming bacteria (HUNGATE, 1969). The observed contribution by methane to the gaseous carbon losses emphasizes the need for investigations on methane production, when respiration measurements are carried out for obtaining values for carbon flow through an ecosystem. This is especially true for wet areas, which are common at the IBP tundra sites.

A comparison of the Hardangervidda sites with the Stordalen site from the data presented in this paper shows that the subarctic mire with permafrost close to the peat surface has much lower soil activity than alpine heaths, meadows, and forests. The low activity and hence low decomposition results in a low release of nutrients needed for the plants, and consequently in low primary production (about 72 gC m^{-2}, SONESSON et al., 1973). For the minerotrophic Hardangervidda sites more nutrients are available because of higher mineralization rates, which results in higher plant productivity.

References

CLYMO, R. S., REDDAWAY, E. J. F.: Productivity of *Sphagnum* (bog-moss) and peat accumulation. Hidrobiologia **12**, 181–192 (1971).

FLANAGAN, P. W., VEUM, A. K.: The influence of temperature and moisture on decomposition rates in tundra. In: HOLDING, A. J., HEAL, O. W., MACLEAN, S. F., FLANAGAN, P. W. (Eds.): Soil Organisms and Decomposition in Tundra, pp. 249–278. Stockholm: IBP Tundra Biome Steering Committee 1974.

HANSSEN, J. F., LID-TORSVIK, V., TORSVIK, T.: Mikrobiologi. In: IBP in Norway, Annual Report 1972, pp. 75–92. Oslo: Norwegian IBP National Committee 1973. (In Norwegian.)

HUNGATE, R. E.: A role-tube method for cultivation of strict anaerobes. In: NORRIS, J. R., RIBBONS, D. W. (Eds.): Methods in Microbiology, Vol. 3B, pp. 117–132. London-New York: Academic Press 1969.

KJELVIK, S., KÄRENLAMPI, L.: Plant biomass and primary production of Fennoscandian subarctic and subalpine forests and of alpine willow and heath ecosystems. In: WIELGOLASKI, F. E. (Ed.): Fennoscandian tundra ecosystems, Part 1: Plants and microorganisms, pp. 111–120. Berlin-Heidelberg-New York: Springer 1975.

KJELVIK, S., WIELGOLASKI, F. E., JAHREN, A.: Photosynthesis and respiration of plants studied by field technique. In: WIELGOLASKI, F. E. (Ed.): Fennoscandian tundra ecosystems, Part 1: Plants and microorganisms, pp. 184–193. Berlin-Heidelberg-New York: Springer 1975.

MACFADYEN, A.: The soil and its total metabolism. In: PHILLIPSON, J. (Ed.): Methods of study in quantitative soil ecology: population, production and energy flow, IBP Handbook No. **18**, pp. 1–13. Oxford-Edinburgh: Blackwell 1971.

MADSEN, I., WIDELL, S.: A vegetation map of the Stordalen site. Progress Report 1973. Swedish IBP Tundra Biome Project Tech. Rep. **16**, 3–16 (1974).

ROSSWALL, T., VEUM, A. K., KÄRENLAMPI, L.: Plant litter decomposition at Fennoscandian tundra sites. In: WIELGOLASKI, F. E. (Ed.): Fennoscandian tundra ecosystems, Part 1: Plants and microorganisms, pp. 268–278. Berlin-Heidelberg-New York: Springer 1975.

SESTAK, Z., ČATSKY, J., JARVIS, P. G.: Criteria for the selection of suitable methods. In: SESTAK, Z., ČATSKY, J., JARVIS, P. G. (Eds.): Plant photosynthetic production. Manual of methods, pp. 1–49. The Hague: Junk 1971.

SKARTVEIT, A., RYDÉN, B. E., KÄRENLAMPI, L.: Climate and hydrology of some Fennoscandian tundra ecosystems. In: WIELGOLASKI, F. E. (Ed.): Fennoscandian tundra ecosystems, Part 1: Plants and microorganisms, pp. 41–53. Berlin-Heidelberg-New York: Springer 1975.

SONESSON, M., FLOWER-ELLIS, J. G. K., JONSSON, S., ROSSWALL, T., RYDÉN, B. E.: Word model for Abisko (Stordalen) Sweden. Progress report 1972. Swedish IBP Tundra Biome Project Tech. Rep. **14**, 4–13 (1973).

SONESSON, M., WIELGOLASKI, F. E., KALLIO, P.: Description of Fennoscandian tundra ecosystems. In: WIELGOLASKI, F. E. (Ed.): Fennoscandian tundra ecosystems, Part 1: Plants and microorganisms, pp. 3–28. Berlin-Heidelberg-New York: Springer 1975.

SVENSSON, B. H.: Production of methane and carbon dioxide from a subarctic mire. Progress report 1973. Swedish IBP Tundra Biome Project Tech. Rep. **16**, 123–143 (1974).

VEUM, A. K.: Soil description from five intensive studied IBP, PT/UM sites, Hardangervidda, South Norway. In: IBP in Norway. Annual Report 1972, pp. 32–45. Oslo: Norwegian IBP National Committee 1973.

WIELGOLASKI, F. E.: Primary production of alpine meadow communities. In: WIELGOLASKI, F.E. (Ed.): Fennoscandian tundra ecosystems, Part 1: Plants and microorganisms, pp. 121–128. Berlin-Heidelberg-New York: Springer 1975.

WITKAMP, M.: Compatibility of microbial measurements. In: ROSSWALL, T. (Ed.): Modern methods in the study of microbial ecology. Bull. Ecol. Res. Comm. (Stockholm) **17**, 179–188 (1973).

5. Nutrient Cycling

Introduction

P. KALLIO and F. E. WIELGOLASKI

The amount of mineral nutrients originating from bedrock varies in different parts of Fennoscandia, due to the variation in bedrock quality, as described in the site description in the present volume. The Swedish, and particularly the Norwegian IBP sites, have more variable habitats of minerogenic nutrient origin than do the Finnish. There are, however, also habitats with ombrogenic nutrient origin, i.e. dependent on nutrients from precipitation. In this respect, there are further differences in the IBP sites in Fennoscandia as seen in the paper by KALLIO and VEUM (this section). The highest values for most nutrients in precipitation have been measured at the Norwegian sites at Hardangervidda, while the northern sites in Fennoscandia at Abisko and at Kevo show minimum input. This depends partly on the distance of the sites from an ocean, which is one source of nutrients, and also from the big industrial centers of Middle Europe. The latter concerns particularly sulphur, which is both a necessary nutrient and, in higher concentrations, a pollutant.

In many of the Fennoscandian tundra habitats studied, ombrogenic nutrition is very essential, particularly for many cryptogams (mosses and lichens), which are adapted to it. In steady state ecosystems, the ombrogenic input is also an important factor, compensating leakage of minerals from the ecosystems into the oceans.

Nitrogen supply has particular features since the bedrock contains no reserves of it. The atmospheric nitrogen compounds, originating partly from direct solar radiation or electric forces, are not quite homogenously distributed in Fennoscandia. A major part of the nitrogen needed for the function of an ecosystem is of biogenic origin. The nitrogen fixing organisms belong to different groups and are varyingly distributed in the Fennoscandian tundra ecosystems: at some sites, legumes with their rhizobia are relatively important, while free living bluegreen algae and bacteria are more important in other areas, particularly where the soil is wet (GRANHALL and LID-TORSVIK, this section); in drier ecosystems, lichens are the main nitrogen fixers (KALLIO and KALLIO, this section). The dependence of the nitrogenase enzyme activity on temperature, moisture and pollution gives a basis for the study of the differences in the biogenic nitrogen supply at the different IBP sites of Fennoscandia.

Even though the mineral content of the soil may be 100 or 1000 times that incorporated in the phytomass (WIELGOLASKI et al., this section), the latter has a high percentage of the total pool of "easily available" minerals. In the "forest tundra", with birch as its dominant tree, a considerable part of the total "available" mineral resource of the ecosystem is contained in the birch phytomass. In normal cycling, nearly 600 kg leaves fall onto the soil per ha and year in northern Finland, and even more in Norwegian subalpine birch forests (KJELVIK and KÄRENLAMPI, primary producer section of this volume), and the minerals in them are liberated by decomposition. The following year the same mass of minerals in a steady state ecosystem is taken up by the roots of the birches. When an ecocatastrophe takes place, as in the middle of the 1960's, when huge areas of birch forests in northern Fennoscandia were totally killed by an *Oporinia* attack (KALLIO and LEHTONEN, 1975), the main pool of nutrients started to be released. This entailed two developments: the ground vegetation had more nutrients available; particularly some of the gramineous plants (e.g. *Deschampsia flexuosa*) increased their biomass in the area. On the other hand, it also caused increased leaching of minerals for several years, and the soil was impoverished. This happened in northern Fennoscandia between 1969–1974. Normally, however, a steady state prevails in the Fennoscandian tundra ecosystems.

The amount of nutrients varies in different plant organs: typically, roots have a lower nitrogen content compared with tops. The concentrations of N, P, K, Ca, and Mg are generally higher in green than in non-green parts (WIELGOLASKI et al., this section). The seasonal cycling is important; in birch for instance, N-content is high in spring in buds and young leaves, but decreases rapidly in summer, while other minerals show minor seasonal cycling. The nutrient pool is also distributed differently in various plant groups. The highest total mineral content occurs in the forbs and the lowest in woody plants; mushrooms also show a very high mineral content, while lichens represent a group with very low values.

The correlation between the mineral content in soil and mineral uptake by plants is shown e.g. by a comparison of plants from different habitats included in the Norwegian IBP sites. The correlation is manifested particularly in Fe. Fertilization experiments in Norway help us to understand the role of mineral nutrients as a limiting factor in the tundra ecosystems (WIELGOLASKI et al., this section). This strongly emphasizes the necessary interaction between assimilation—primary production studies (as discussed in the previous sections of this volume) and mineral nutrient studies.

Translocation of photosynthates within plants is in the present volume treated as a production problem and is therefore discussed under the primary producer section by BERG et al. The translocation peak of photosynthates from the leaves, as measured by $^{14}CO_2$, is found during the senescence phase in fall (particularly in deciduous woody and herbaceous plants). The same is partly the case for inorganic nutrients. The translocation rate is high; most of the photosynthates labelled are translocated from the leaves within two days. The upward translocation in spring also seems to be rapid in tundra plants, with highest radiocarbon activity recovered in new green leaves only one week after budbreak (^{14}C incorporated in the fall the year before). From this, it is obvious that the content of easily mobilized organic compounds in storage organs increases in fall, as is also found in most Fennoscandian tundra plants (SKRE et al., this section), reaching a maximum in winter.

Part of the organic matter is transported from terrestrial ecosystems to waters where, in some Fennoscandian tundra ecosystems, it is found to have a highly significant influence on the productivity of the freshwater system (LARSSON and TANGEN, this section) both serving as a nutrient for autotrophic organisms by releasing minerals after decomposition, and directly as a nutrient for heterotrophic organisms. The use of organic plant matter for higher trophic levels is discussed in detail in the second Fennoscandian tundra ecosystem volume in the present series. In parts of the Fennoscandian tundra ecosystems, domestic and wild reindeer are important components (GAARE and SKOGLAND, 1975). The biomass of lichens is often critical for the survival of reindeer in winter, being almost the only nutrient available to the animals. The herbivory of other vertebrates may also be important in Fennoscandian tundra areas, as may the carnivory, particularly in years with small mammal peak populations (discussed in some papers in part two). HAUKIOJA and KOPONEN (1975) find a surprisingly high plant intake by invertebrates. This strongly emphasizes the importance of nutrient cycling through various trophic levels in tundra ecosystems. On an average, however, more than 90% of the plant biomass produced in the systems is decomposed directly by microorganisms (GOKSØYR, in the decomposer section) and the mineral nutrients are released again for uptake by the plants.

The complicated problems in nutrient cycling often have to be treated by simulation models. In the second Fennoscandian tundra ecosystem volume, several aspects of cycling of animals, consumption and decomposition are subjected to mathematical systems analysis. Even if the mineral content is not always specifically handled in these models, it is implicit in the analysis as being an integrated part of all ecosystem components. Some energy flow studies may cover just the primary producers, while mineral cycling has to include studies of all trophic levels involved.

References

GAARE, E., SKOGLAND, T.: Wild reindeer food habitats and range use at Hardangervidda, Norway. In: WIELGOLASKI, F. E. (Ed.): Fennoscandian tundra ecosystems, Part 2: Animals and systems analysis. Berlin-Heidelberg-New York: Springer 1975. (In press.)

HAUKIOJA, E., KOPONEN, S.: Birch herbivores and herbivory at Kevo, northern Finland. In: WIELGOLASKI, F. E. (Ed.): Fennoscandian tundra ecosystems, Part 2: Animals and systems analysis. Berlin-Heidelberg-New York: Springer 1975. (in press.)

KALLIO, P., LEHTONEN, J.: Ecocatastrophes of birch forests caused by *Oporinia autumnata* (Bkh.) and the problem of reafforestation. In: WIELGOLASKI, F. E. (Ed.): Fennoscandian tundra ecosystems, Part 2: Animals and systems analysis. Berlin-Heidelberg-New York: Springer 1975. (In press.)

Nitrogen Fixation in Lichens at Kevo, North-Finland

S. KALLIO and P. KALLIO

Lichens represent a very important plant group of primary producers in many arctic and subarctic areas (P. KALLIO and KÄRENLAMPI, 1975). Their importance as nitrogen fixers, particularly in the dry ecosystems in continental climate areas, has been emphasized recently in different proceedings of the IBP Tundra work in Alaska (ALEXANDER and SCHELL, 1973a, b) and in Scandinavia (GRANHALL and SELANDER, 1972; P. KALLIO et al., 1972; LID and HANSEN, 1972; S. KALLIO, 1973; P. KALLIO, 1974). At Kevo (69° 45′ N, 27° 01′ E), which is a typical dry "tundra ecosystem" with very few legumes and alders, the N_2 fixation studies have consequently focused on lichens. In wet ecosystem, free-living blue-green algae are more important (GRANHALL and SELANDER, 1972; HORNE, 1972; STUTZ, 1972; ALEXANDER and SCHELL, 1973a, b; SCHELL and ALEXANDER, 1973).

This paper gives results of the 1972 field experiments with lichens; some supplementary studies were made in summer 1973. The situation, the different ecosystems (pine forest, birch forest, subalpine heath), and the climatic and geological features of the Kevo area as well as human influence are surveyed in the papers by P. KALLIO (1964); P. KALLIO et al. (1969); KÄRENLAMPI (1972); KÄRENLAMPI and KAUHANEN (1972) and SONESSON et al. (1975).

The ecology and physiology of lichens at Kevo, including also the main objects of nitrogen fixation studies, *Nephroma arcticum* (L.) Torss. and *Stereocaulon paschale* (L.) Fr. have been discussed by P. KALLIO and HEINONEN (1971); P. KALLIO et al. (1972); S. KALLIO (1973) and P. KALLIO (1974).

This paper deals with the diurnal and seasonal N_2 fixation in *Nephroma arcticum* and *Stereocaulon paschale*, and attempts to give an estimation of the annual nitrogen input per area in the Kevo ecosystems. An attempt is also made to approximate regression models for ethylene production at three IBP-sites.

Material and Methods

The two most important lichens, which are the objects of this study, are *Nephroma arcticum* (L.) Torss. and *Stereocaulon paschale* (L.) Fr.; *Stereocaulon alpinum* Laur. is rare in the IBP-sites of Kevo (KÄRENLAMPI and KAUHANEN, 1972; KJELVIK and KÄRENLAMPI, 1975). The closely related species to *Stereocaulon paschale*, *S. saxatile* H. Magn. and *S. dactylophyllum* Flk., are absent in the Kevo area (U. LAINE, pers. comm.), so the taxonomical difficulties in the critical genus *Stereocaulon* do not arise here.

Nitrogen fixation was measured by the acetylene reduction method (STEWART et al., 1967; 1968; HARDY et al., 1968). At the birch and pine sites and subalpine heath, one or two series of measurements ($n = 10$ in each series) per week were taken in both lichen species. Sampling and incubations were carried out four times during a period of twenty four hours: between 6–7, 12–13, 18–19, and 23–24 o'clock. The total number of measurements (gas chromatographic analyses) was thus about 3200.

The field experiments began in mid-May, immediately after thaw, and were completed in mid-September, i.e. before the end of the season of activity. In *Nephroma arcticum*, the activity was measured only in separated cephalodia, while top pieces of 1.5–2.0 cm lengths were used in *Stereocaulon paschale*. These tops of *Stereocaulon* grow and photosynthesize actively (KÄRENLAMPI, 1971).

Lichen samples were incubated in situ for an hour in 6 ml serum bottles on the lichen mat, shadowed slightly against direct sunshine. The measurements included the temperature of the lichen layer, moisture content of the samples and light intensity of the growing site. The increase of temperature in the bottles was not more than 1° C compared with temperature on the ground.

The ethylene produced was analyzed using a gas chromatographic technique (P. KALLIO et al., 1972) in the laboratory a few days after incubations. The values of ethylene were calculated to nitrogen using the theoretical coefficient 1.5 (STEWART et al., 1968). On the basis of weekly means of nitrogen values per hour (Figs. 2 and 3) and the biomass (KJELVIK and KÄRENLAMPI, 1975), the total seasonal N_2 input was estimated. The amount, corresponding to the yield of the two last weeks in autumn 1972, was added to these values on the basis of the experiments carried out in autumn 1973.

The nitrogen fixed by the cephalodia of *Nephroma arcticum* was calculated in the results for the total thallus, using a 17/100 experimental ratio of dry weights of cephalodia to the total thallus.

Results

Diurnal Variation in Nitrogen Fixation. The activity of acetylene reduction in *Nephroma arcticum* during June and July, converted to N_2 values, is highest at all sites in the morning. The maximum rate is 5.00 ± 0.56 µg N_2 per g dry lichen per hour in the pine forest, 6.12 ± 0.51 µg in the birch forest and 6.91 ± 0.74 µg at the subalpine heath. The fixation is lowest at all sites at noon: 1.99 ± 0.29 µg at the pine site, 4.08 ± 0.38 µg at the birch site and 0.75 ± 0.12 µg at the subalpine site. In the afternoon, the values in the birch forest are nearly as high (6.03 ± 0.63 µg) at six p.m. as in the morning (6.12 ± 0.51 µg), but lower at the other sites. At the subalpine heath, N_2 fixation still continues during the night.

Light intensities vary from approximately 1000 lux to 30000 lux during the June and July period. Temperatures are highest at noon (about 20° C in the pine and birch forests, but about 17° C at the subalpine heath) and lowest at midnight (about 11° C in the pine and birch forests, but about 8° C at the subalpine heath). During the June–July period, diurnal moisture contents in *Nephroma arcticum* are highest in the morning at the pine site (188% d.w.) and at the birch site

(170% d.w.), but at midnight (130% d.w.) at the subalpine heath. Moisture contents are lowest in lichens (down to about 20% d.w. at the subalpine site) at noon at all the sites.

The daily variation in the N_2 fixation of *Nephroma arcticum* in late summer (August and September) is different from the June–July period. The N_2 fixation is highest at noon in the pine forest (6.62 ± 0.59 μg g^{-1} d.w. h^{-1}) and at the birch site (9.00 ± 0.37 μg), while morning is more favorable to the nitrogenase activity in the subalpine area (8.40 ± 0.62 μg). The mean amount of fixed nitrogen is higher in late summer than during the earlier period.

In the first week of August, the light intensity is still above 100 lux at midnight; "dark periods" of the night begin from the second week of August. The highest light intensities is measured at the subalpine heath. Diurnal temperatures of the *Nephroma* layer are somewhat lower and moisture contents higher in the late summer than in June–July. Moisture contents are relatively uniform throughout day and night.

The diurnal nitrogen fixation in *Stereocaulon paschale* during the June–July period is almost similar to *Nephroma arcticum*. The maximum nitrogenase activity was measured at all the IBP-sites in the morning. The most favorable site for N_2 fixation in *Stereocaulon paschale* is the birch forest; the highest fixation rate is 9.68 ± 0.12 μg N_2 g^{-1} d.w. h^{-1} in the morning. The values at the pine site and subalpine heath are 3.88 ± 0.06 and 2.85 ± 0.06 μg, respectively. Nitrogen fixation in *Stereocaulon paschale* is lowest at noon: 0.65 ± 0.04 μg at the pine site, 3.47 ± 0.07 μg at the birch site and 0.20 ± 0.01 μg at the subalpine site.

The lowest moisture contents in June–July were measured at noon in the birch forest (48% d.w.) and in the low alpine heath (16% d.w.), but at six p.m. in the pine forest (46% d.w.). The highest moisture contents in *Stereocaulon paschale* were at midnight at the pine site (100% d.w.) and at the subalpine heath (110% d.w.), and

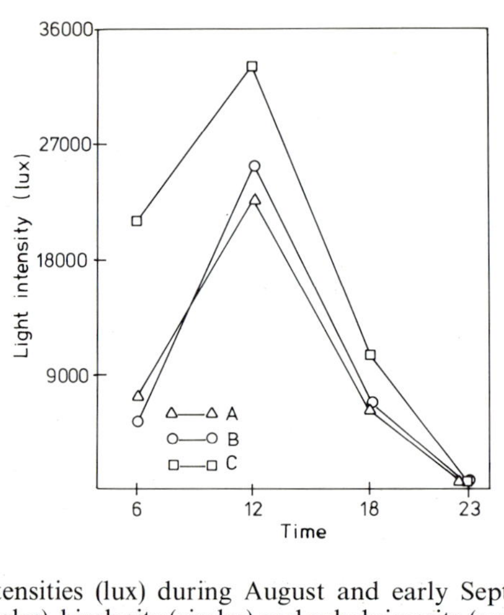

Fig. 1. Diurnal light intensities (lux) during August and early September at the pine site (triangles), birch site (circles) and subalpine site (squares)

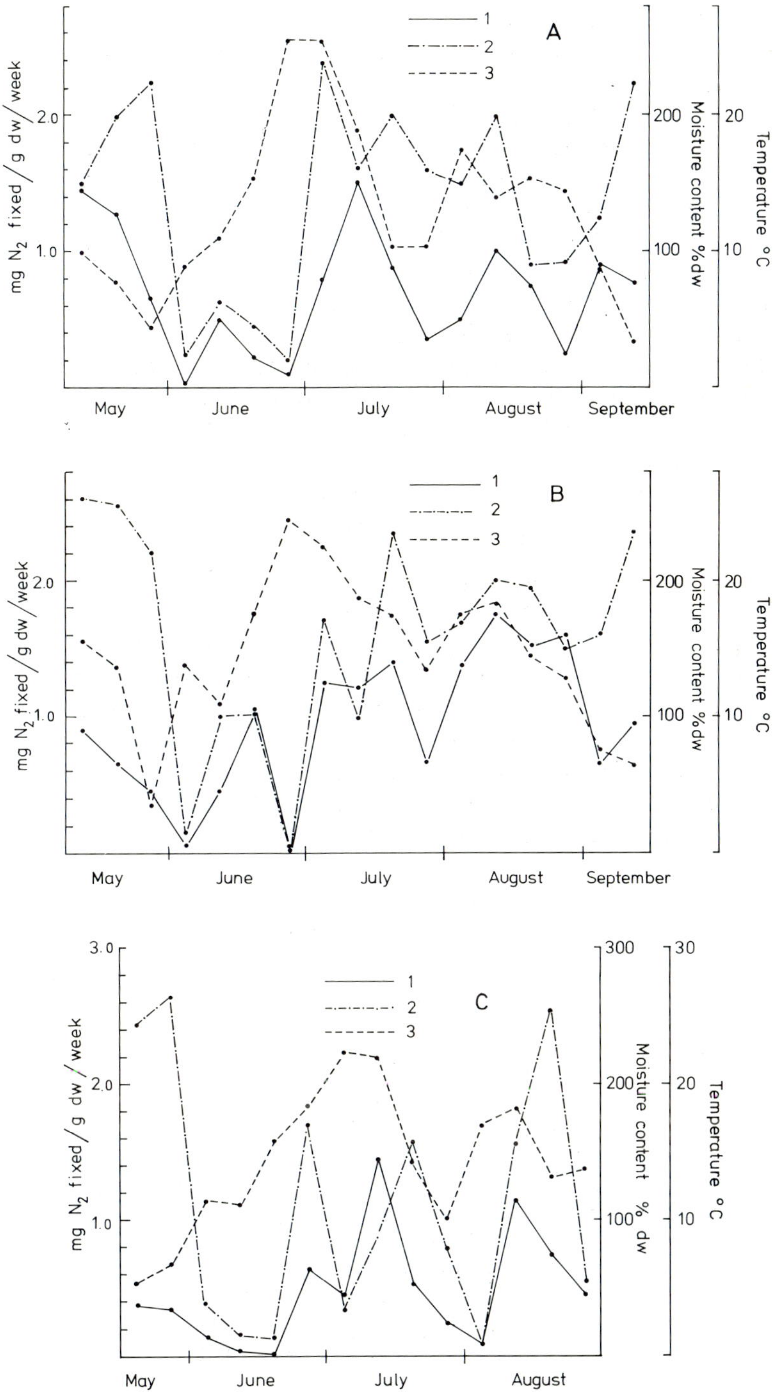

Fig. 2 A–C. Seasonal nitrogen fixation (*1*), moisture contents (*2*) and temperatures (*3*) in *Nephroma arcticum* at the pine site (A), birch site (B) and subalpine site (C)

in the morning at the birch site (137% d.w.). The temperature conditions were nearly the same as in *Nephroma* experiments.

In the late summer (August and September), nitrogen fixation in *Stereocaulon paschale* is highest at noon at the pine site (4.15 ± 0.06 µg N_2 g^{-1} d.w. h^{-1} and at the birch site (6.72 ± 0.07 µg), while the maximum fixation (5.10 ± 0.10 µg) at the subalpine heath is in the morning just as found in *Nephroma.* Diurnal light intensities are seen in Fig. 1. Moisture contents in *Stereocaulon paschale* are rather high throughout the autumn, higher than in the June–July period. Temperatures decrease in August (Fig. 2), especially at night.

Seasonal Variation in Nitrogen Fixation. Seasonal variations in nitrogenase activity and moisture contents in *Nephroma arcticum* and ground temperatures are shown in Figs. 2A, B, C. The values given are the mean rates of the weekly nitrogen fixation. It is clearly seen that the total N_2 fixation in *Nephroma arcticum* is highest in the birch forest and lowest at the subalpine heath (Table 1). In the pine and birch forests, the lowest N_2 fixation during the growing season was measured in June (1.6 and 0.8 mg $month^{-1}$, respectively), and, for the subalpine site, in May (0.7 mg $month^{-1}$). August is the most favorable period for nitrogenase activity in the birch forest (6.2 mg $month^{-1}$) and in the subalpine area (2.5 mg $month^{-1}$). At the pine site, the highest N_2 fixation rate is found in July (3.6 mg $month^{-1}$).

The lowest moisture content in *Nephroma arcticum* is in June. In May and August, moisture conditions are more favorable than in the June-July period. The moisture contents of lichens are lower at the subalpine heath than at the other study sites during the early season.

Seasonal variation in nitrogenase activity and moisture content in *Stereocaulon paschale* and ground temperatures are shown in Figs. 3A, B, C. The nitrogen fixation is highest in the birch forest and lowest at the subalpine heath (Table 1). The nitrogenase activity in *Stereocaulon paschale* is lowest in June in the pine and birch forests (0.3 and 1.3 mg $month^{-1}$, respectively) and also at the subalpine heath (0.5 mg $month^{-1}$). The most favorable periods seem to be July and August. The moisture contents in *Stereocaulon* follow those in *Nephroma.*

As the measurements of the nitrogen input through lichens in the Kevo ecosystems in summer 1972 (Table 1) did not include the production of the latest season, the value for the last two weeks was multiplied by two. This is in accordance with results from field experiments with *Peltigera aphthosa* (L.) Willd. in autumn 1973, and laboratory measurements where all the three lichen species were compared (S. KALLIO and RASKU, unpubl.). The values of nitrogen fixation

Table 1. Nitrogen fixation in *Nephroma arcticum* and *Stereocaulon paschale* at the IBP-sites of Kevo during summer 1972

	Nephroma arcticum	*Stereocaulon paschale*
The pine site	13.8 mg N_2 g^{-1} d.w.	9.9 mg N_2 g^{-1} d.w.
The birch site	18.9 mg N_2 g^{-1} d.w.	17.2 mg N_2 g^{-1} d.w.
The subalpine site	8.9 mg N_2 g^{-1} d.w.	5.2 mg N_2 g^{-1} d.w.

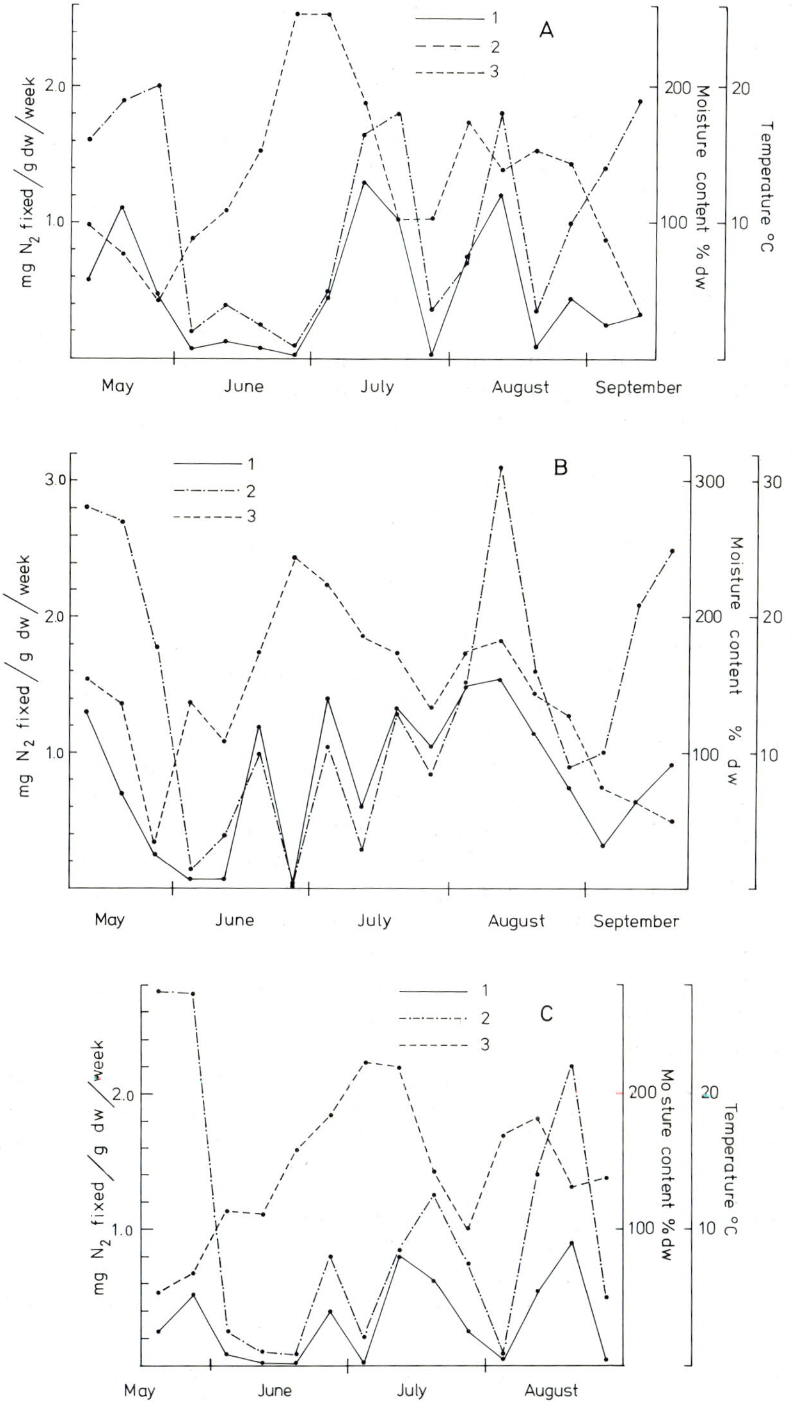

Fig. 3 A–C. Seasonal nitrogen fixation (*1*), moisture contents (*2*) and temperatures (*3*) in *Stereocaulon paschale* at the pine site (A), birch site (B) and subalpine site (C)

Table 2. Annual yield of fixed nitrogen at the IBP-sites of Kevo

	Nephroma arcticum	*Stereocaulon paschale*
The pine site	—	384.1 mg N_2 m^{-2} yr^{-1}
The birch site	104.0 mg N_2 m^{-2} yr^{-1}	46.4 mg N_2 m^{-2} yr^{-1}
The subalpine site	4.5 mg N_2 m^{-2} yr^{-1}	168.0 mg N_2 m^{-2} yr^{-1}

Table 3. Regressions and correlation coefficients *(R)* for ethylene *(E)* formed from acetylene by *Nephroma arcticum* and *Stereocaulon paschale*

	Nephroma arcticum	
The pine site	$E = T\sqrt{M}\,(0.0064 - 0.00017\,T)$,	$R = 0.80$
The birch site	$E = T\sqrt{M}\,(0.0056 - 0.00010\,T)$,	$R = 0.88$
The subalpine site	$E = T\sqrt{M}\,(0.0041 - 0.00007\,T)$,	$R = 0.80$
	Stereocaulon paschale	
The pine site	$E = T\sqrt{M}\,(0.0051 - 0.00012\,T)$,	$R = 0.81$
The birch site	$E = T\sqrt{M}\,(0.0060 - 0.00010\,T)$,	$R = 0.88$
The subalpine site	$E = T\sqrt{M}\,(0.0045 - 0.00014\,T)$.	$R = 0.82$

per square meter per year are given in Table 2, using the average lichen biomasses (KJELVIK and KÄRENLAMPI, 1975).

A Regression Model to Predict Ethylene Production from Acetylene. The dependence of nitrogenase activity on prevailing environmental conditions in *Nephroma arcticum* and *Stereocaulon paschale* was predicted, using a regression model. The most important ecological variables are temperature and moisture content of the lichen layer. The results given above, and also results of the laboratory experiments in lichens, were used to predict ethylene production. The formula used was as follows:

$$E = T\sqrt{M}\,(a - b\,T)\,,$$

where E = amount of ethylene formed, nmol per mg dry weight of lichen per hour, T = temperature of lichen layer (°C), M = moisture content of lichen as per cent of dry weight of lichen, a and b = coefficients, which vary at different study sites.

The regressions and correlation coefficients for ethylene production in *Nephroma arcticum* and *Stereocaulon paschale* at the IBP-sites at Kevo are shown in Table 3.

The ethylene reduced from acetylene by *Nephroma arcticum* and *Stereocaulon paschale* in different environmental conditions was estimated using the regression model; predictions are shown in Figs. 4A, B, C for *Nephroma arcticum* and in Figs. 5A, B, C for *Stereocaulon paschale*.

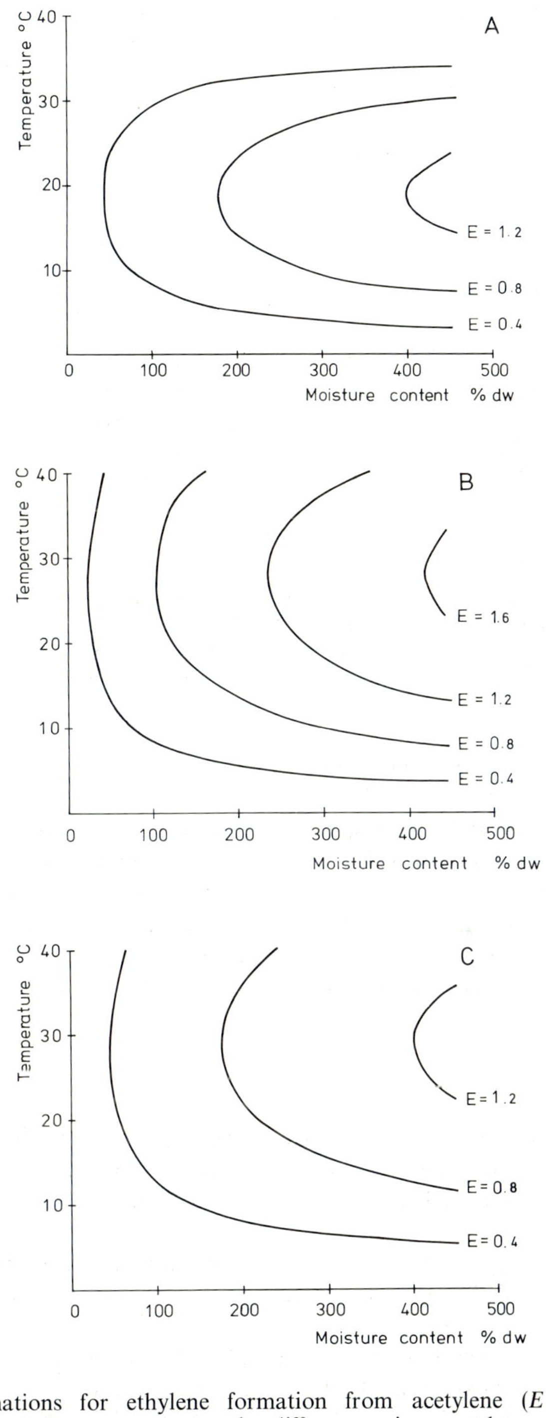

Fig. 4 A–C. Estimations for ethylene formation from acetylene (E = ethylene, nmol mg^{-1} d.w. h^{-1}) in *Nephroma arcticum* under different moisture and temperature conditions at the pine site (A), birch site (B) and subalpine site (C)

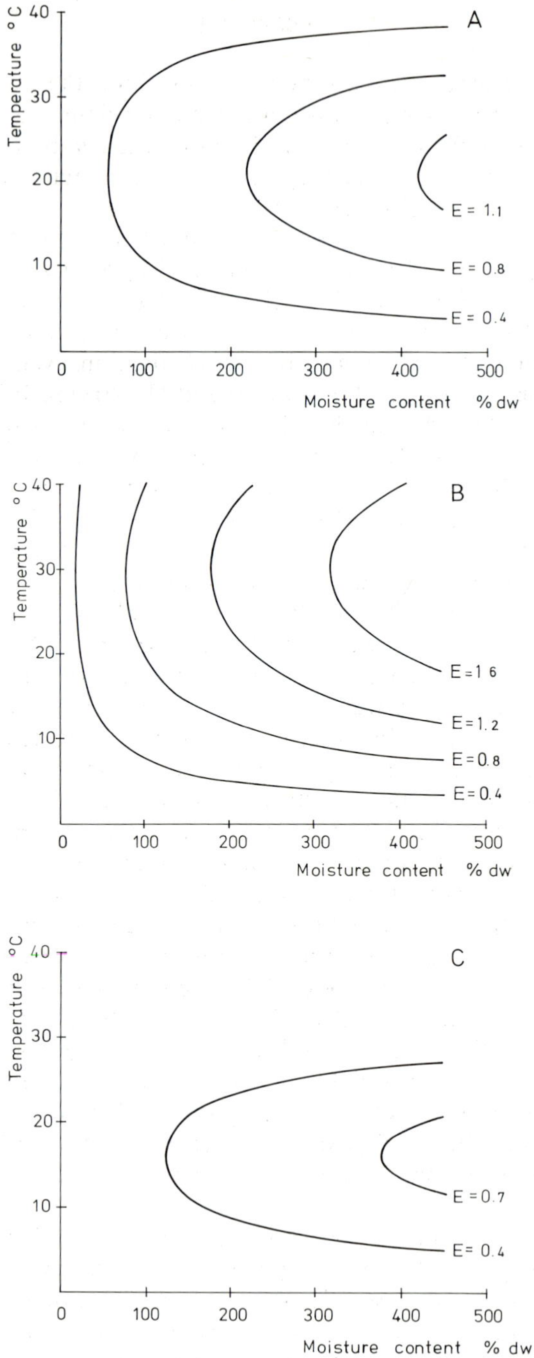

Fig. 5 A–C. Estimations for ethylene formation from acetylene (E = ethylene, nmol mg^{-1} d.w. h^{-1}) in *Stereocaulon paschale* under different moisture and temperature conditions at the pine site (A), birch site (B) and subalpine site (C)

Discussion

The most important ecological factors controlling the nitrogen production through N_2 fixation are temperature, moisture and light. Variations in oxygen content and the pH of the soil are not as important to symbiotic blue-green algae in the present dry ecosystem as they may be to free-living algae under other conditions (ALEXANDER and SCHELL, 1973a).

Is is well known that nitrogenase activity in lichens is inactivated in temperatures near 0° C (FOGG and STEWART, 1968; P. KALLIO et al., 1972; ALEXANDER and SCHELL, 1973b; S. KALLIO, 1973); only one species, *Solorina crocea*, fixes nitrogen below 0° C (P. KALLIO et al., 1972). Sensitivity to low temperatures may be a factor that limits the autumn activity of N_2 fixation, even though conditions for photosynthesis are still good (P. KALLIO and HEINONEN, 1971). Temperatures near 0° C occur at first during the night in autumn, when the lack of light is another limiting factor of N_2 fixation. At this time, nitrogenase activity may continue on the following day. Later on, however, when the frost period begins and temperatures go down to −10° C or −15° C at night, N_2 fixation is reactivated slowly and only in part on the next day (S. KALLIO, 1973), contrary to photosynthesis, which is reactivated very rapidly (P. KALLIO and HEINONEN, 1971). In spring there does not seem to be any lag phase after snow melt for the beginning of the nitrogenase activity in lichens: maximum N_2 fixation has been observed almost immediately after thawing, when temperatures are above zero.

High nitrogenase activity was measured in nature (Figs. 4 and 5) in temperatures above those used in the laboratory experiments (P. KALLIO et al., 1972). When the time of exposure to high temperatures is decreased, the temperature level necessary for maximum activity rises. In the field, high temperatures are combined with desiccation, and therefore at maximum temperature conditions lichens can be active only for very short moist periods.

The temperature relations in the three IBP-sites show some typical differences (KÄRENLAMPI, 1972) important to N_2 fixation of lichens. The growth season at the subalpine site is 15 days shorter than at the pine and birch forest sites, and the mean summer temperature is about 1.5° C lower; the thermal sum is about 410 d.d. (day-degrees above 0° C) at the subalpine heath, in comparison with 577 d.d. in the valley (KÄRENLAMPI and KAUHANEN, 1972). The low values for N_2 fixation measured on the dry weight basis at the subalpine site, i.e. only $^1/_2$–$^1/_3$ of those at the pine and birch sites, are partly due to these differences in temperatures. The differences in the N_2 fixation in the pine and birch forests are mainly dependent on other factors, because temperatures at these two sites are fairly similar.

Low moisture is the most important factor that limits nitrogenase activity in the dry Kevo ecosystem. This is seen particularly in the photosynthetic activity (growth) in lichens (KÄRENLAMPI, 1971). The dependence of N_2 fixation on moisture has been analyzed under laboratory conditions (HENRIKSSON and SIMU, 1971; S. KALLIO, 1973; ALEXANDER and S. KALLIO, unpubl.), and also in some field studies (GRANHALL and SELANDER, 1972; HORNE, 1972). In the Kevo area the dryest month, June, is an inactive period in N_2 fixation, and the peaks of nitrogenase activity tally very well with those of humidity.

The significance of moisture is manifested clearly in the comparison of the three IBP-sites. The birch site is in this respect the best habitat for N_2 fixation. *Stereocaulon paschale* also reaches its highest nitrogenase activity in this ecosystem, although its quantity is very small there. The moss mat normally buffers lichens against desiccation in the birch forests. Birch bushes provide good microclimatic conditions: the evaporation is only half of that of the almost open subalpine heath, where the wind is a strong drying factor (KÄRENLAMPI, 1972). Conditions in the pine forest are between those of the two other sites. It is possible that the water content of lichen thalli is increased in the subalpine heath at night more than at the other IBP-sites, but other factors as well cause the heath to be the most xeric site (KÄRENLAMPI, 1972).

Light intensity has evidently no effect on nitrogenase activity in lichens in the everlasting day period from May to the beginning of August. Light intensity at night is sufficient for N_2 fixation in this period. The fixation continues also in the dark, which lasts only a few hours, e.g. about five hours in August, even though the activity decreases (S. KALLIO, 1973). These results are in accordance with those of HENRIKSSON (1971), who supposes that the relatively short dark period in the summer nights in the North does not stop nitrogen fixation of free-living blue-green algae.

The number of lichen species is higher at the subalpine site than at the other IBP-sites at Kevo. The biomass of *Stereocaulon paschale*, however, is almost equal in the pine forest and at the subalpine heath, which represent a xeric ecosystem. On the other hand, this species has only a very small role at the moister birch site. There *Nephroma arcticum* has its optimum biotope, while this is almost lacking at the two xeric sites.

The representativeness of the three IBP-sites to Northern Lapland has been discussed by KÄRENLAMPI and KAUHANEN (1972). The results given for biomasses of *Nephroma arcticum* may represent below-average values in the birch forests in Lapland. KJELVIK and KÄRENLAMPI (1975) report that *Nephroma arcticum* biomass is 52.9 $g\,m^{-2}$ in a mesic birch forest near the IBP birch site, but only 5.5 $g\,m^{-2}$ in the birch study area. This means that local values of the annual yield of nitrogen fixation may be several times those given for the IBP birch site.

The maximum values for cephalodial nitrogen fixation of *Nephroma arcticum* in Kevo material are about 60 $\mu g\ N_2\,g^{-1}$ d.w. h^{-1} in the optimal laboratory conditions (P. KALLIO et al., 1972). Values of the same order found in the field indicate that the conditions in laboratory may be almost as favorable as in the field, despite the very high sensitivity of nitrogenase to the common air pollution (S. KALLIO and VARHEENMAA, 1974). The same kind of comparison applies to *Stereocaulon paschale* in laboratory (S. KALLIO, 1973) and field conditions.

Only few comparable values of N_2 fixation of *Nephroma arcticum* and *Stereocaulon paschale* are obtainable in literature (GRANHALL and SELANDER, 1972; ALEXANDER and SCHELL, 1973a), but comparisons could be made with *Peltigera aphthosa*. STUTZ (1972) reports 48 $\mu g\ N_2\,g^{-1}\,h^{-1}$ for this species on Devon Island. A lower rate, about 5 $\mu g\,g^{-1}\,h^{-1}$, is known from Sweden (GRANHALL and SELANDER, 1972). Highest values for *Peltigera aphthosa* are known from the arctic tundra at Barrow (Alaska), up to 230 $\mu g\ N_2\,g^{-1}\,h^{-1}$ (in cephalodia), but on the other hand, low activity of 0.4–1.2 μg has been measured in Alaskan alpine tundra

(ALEXANDER and SCHELL, 1973a). Our value for *Peltigera aphthosa* in the field is about 24 μg N_2 g^{-1} d.w. h^{-1} (S. KALLIO and RASKU, unpubl.).

The N_2 input per area is dependent on the lichen species, on ecological conditions, and particularly on the biomasses. For the Finnish IBP birch site (based mainly on *Nephroma arcticum*), this value is found to be about 104 mg N_2 m^{-2} $year^{-1}$. Approximately 10–20% nitrogen should be added to this value in order to obtain a total estimate for the nitrogen input by lichens in the birch forest, including N_2 fixation in *Peltigera aphthosa* and some other species of *Peltigera*. The N_2 fixation in *Nephroma arcticum* and *Stereocaulon paschale* at the subalpine heath gives 173 mg N_2 m^{-2} $year^{-1}$. Nitrogen fixation by these two lichen species covers nearly all the nitrogen fixed at the subalpine site, where the role of other nitrogen fixing lichens is very small. The N_2 fixation in the pine forest (384 mg N_2 m^{-2} $year^{-1}$), based almost entirely on *Stereocaulon paschale*, is nearly twice as much as at the other sites. Maximum potential values can be estimated when the cover of the N_2-fixing lichens is 100%. In the mesic birch forest this maximum nitrogen input would be about 4000 mg m^{-2} $year^{-1}$ for *Nephroma arcticum*, and in the pine forest about 3000 mg m^{-2} $year^{-1}$ for *Stereocaulon paschale*.

Acknowledgements. The authors wish to express their best thanks to Mrs. T. EKLUND for her mathematical modeling work. We also wish to acknowledge the help of the students and many members of the Department of Botany, who assisted in the field work. Financially, the work has been supported by the Finnish State Council for Natural Sciences.

References

ALEXANDER, V., SCHELL, D. M.: Nitrogen fixation in arctic and alpine tundra. Data Report 73–10. U.S. Tundra Biome 1973a.

ALEXANDER, V., SCHELL, D. M.: Seasonal and spatial variation of nitrogen fixation in the Barrow, Alaska, tundra. Arctic and Alpine Res. **5**, 77–88 (1973b).

FOGG, G. E., STEWART, W. D. P.: In situ determinations of biological nitrogen fixation in Antarctica. Brit. Antarctic Surv. Bull. **15**, 39–46 (1968).

GRANHALL, U.: Acetylene reduction by blue-green algae isolated from Swedish soils. Oikos **21**, 330–332 (1970).

GRANHALL, U., HENRIKSSON, E.: Nitrogen fixing blue-green algae in Swedish soils. Oikos **20**, 175–178 (1969).

GRANHALL, U., SELANDER, H.: Nitrogen fixation in a subarctic mire. Swedish IBP Tundra Biome Project Tech. Rep. **11**, 1–24 (1972).

HARDY, R. W. F., HOLSTEN, R. D., JACKSON, E. K., BURNS, R. C.: The acetylene—ethylene assay for N_2 fixation. Laboratory and field evaluation. Plant Physiol. **43**, 1185–1207 (1968).

HENRIKSSON, E.: Algal nitrogen fixation in temperate regions. Plant and Soil, special volume, 415–419 (1971).

HENRIKSSON, E., SIMU, B.: Nitrogen fixation by lichens. Oikos **22**, 119–121 (1971).

HORNE, A. J.: The ecology of nitrogen fixation on Signy Island, South Orkney Islands. Brit. Antarctic Surv. Bull. **27**, 1–18 (1972).

KALLIO, P.: The Kevo Subarctic Research Station of the University of Turku. Reports from the Kevo Subarctic Research Station. Ann. Univ. Turkuensis. II. Biologica-Geographica **32**, 1–9 (1964).

KALLIO, P.: The ecology of the nitrogen fixation in subarctic lichens. Oikos **25**, 194–198 (1974).

KALLIO, P., HEINONEN, S.: Influence of short-term low temperature on net photosynthesis in some subarctic lichens. Rep. Kevo Subarctic Res. Stat. **8**, 63–72 (1971).

KALLIO, P., KÄRENLAMPI, L.: Photosynthesis in mosses and lichens. In: COOPER, J. P. (Ed.): Photosynthesis and productivity in different environments, pp. 393–423. Cambridge: University Press 1975.

KALLIO, P., LAINE, U., MÄKINEN, Y.: Vascular Flora of Inari Lapland. I. Introduction and Lycopodiaceae—Polypodiaceae. Rep. Kevo Subarctic Res. Stat. **5**, 1–108 (1969).

KALLIO, P., SUHONEN, S., KALLIO, H.: The ecology of nitrogen fixation in *Nephroma arcticum* and *Solorina crocea*. Rep. Kevo Subarctic Res. Stat. **9**, 7–14 (1972).

KALLIO, S.: The ecology of nitrogen fixation in *Stereocaulon paschale*. Rep. Kevo Subarctic Res. Stat. **10**, 34–42 (1973).

KALLIO, S., VARHEENMAA, T.: On the effect of air pollution on nitrogen fixation in lichens. Rep. Kevo Subarctic Res. Stat. **11**, 42–46 (1974).

KÄRENLAMPI, L.: Studies on the relative growth rate of some fruticose lichens. Rep. Kevo Subarctic Res. Stat. **7**, 33–39 (1971).

KÄRENLAMPI, L.: Comparisons between the microclimates of the Kevo ecosystem study sites and the Kevo Meteorological Station. Rep. Kevo Subarctic Res. Stat. **9**, 50–65 (1972).

KÄRENLAMPI, L., KAUHANEN, H.: A direct analysis of the vegetation of the surroundings of the Kevo Subarctic Station. Rep. Kevo Subarctic Res. Stat. **9**, 82–98 (1972).

KJELVIK, S., KÄRENLAMPI, L.: Plant biomass and primary production of Fennoscandian subarctic and subalpine forests and of alpine willow and heath ecosystems. In: WIELGOLASKI, F. E. (Ed.): Fennoscandian tundra ecosystems, Part 1: Plants and microorganisms, pp. 111–120. Berlin-Heidelberg-New York: Springer 1975.

LID, V., HANSEN, J. F.: Mikrobiologiske undersøkelser på Hardangervidda. IBP i Norden **8**, 79–83 (1972).

SCHELL, D. M., ALEXANDER, V.: Nitrogen fixation in arctic coastal tundra in relation to vegetation and micro-relief. Arctic **26**, 130–137 (1973).

SONESSON, M., WIELGOLASKI, F. E., KALLIO, P.: Description of Fennoscandian tundra ecosystems. In: WIELGOLASKI, F. E. (Ed.): Fennoscandian tundra ecosystems, Part 1: Plants and microorganisms, pp. 3–28. Berlin-Heidelberg-New York: Springer 1975.

STEWART, W. D. P., FITZGERALD, G. P., BURRIS, R. H.: In situ studies on N_2 fixation using the acetylene reduction technique. Proc. Natl. Acad. Sci. U.S. **58**, 2071–2078 (1967).

STEWART, W. D. P., FITZGERALD, G. P., BURRIS, R. H.: Acetylene reduction by nitrogen fixing blue-green algae. Arch. Mikrobiologie **62**, 336–348 (1968).

STUTZ, R. C.: Nitrogen fixation studies on Devon Island. In: BLISS, L. C. (Ed.): Devon Island IBP-project, high arctic ecosystem. Project Report 1970 and 1971, pp. 252–256. Edmonton: University of Alberta, Dept. of Botany 1972.

Nitrogen Fixation by Bacteria and Free-Living Blue-Green Algae in Tundra Areas

U. GRANHALL and V. LID-TORSVIK

Combined nitrogen is probably the limiting nutrient factor for primary production in many tundra areas. The removal of combined nitrogen from tundra areas by leaching or denitrification seems limited, and even a small nitrogen input from e.g. nitrogen fixation is important for the nitrogen balance of the ecosystem. The main sources of nitrogen input are biological nitrogen fixation, precipitation and dry deposition. Nitrogen-fixing organisms present in tundra areas are free-living, epiphytic and symbiotic blue-green algae, and free-living and symbiotic bacteria.

This paper is concerned with nitrogen-fixing bacteria and nonsymbiotic blue-green algae and their contribution to the nitrogen economy of tundra ecosystems studied at Fennoscandian IBP Tundra Biome sites (SONESSON et al., 1975).

Methods

Nitrogen-fixation measurements *in situ* and in the laboratory were performed by means of the acetylene reduction assay for nitrogenase activity (STEWART et al., 1967) under ambient gas conditions or argon. Ten percent acetylene was injected into flasks of various sizes containing soil, water, moss or root samples.

Incubations were made *in situ* or in the laboratory for a certain period (not exceeding 24 hrs), and the reaction terminated by addition of saturated $(NH_4)_2SO_4$ solution or withdrawal of gas samples from the assay chamber. Ethylene formation was measured by gas chromatography, using flame ionization detectors on columns filled with Porapak. Corrections for impurities in the acetylene and for endogenous ethylene formation were made by including appropiate controls. Estimates of bacterial fixation were made on the basis of dark-incubated samples.

For measurements of global radiation, pH, temperature, CO_2, O_2, alkalinity, total P, total N etc., see methods: KALLIO and KALLIO (1975), LID-TORSVIK and HANSSEN (1971), GRANHALL and SELANDER (1973), GRANHALL and BASILIER (1973) and BASILIER (1974).

The amount of nitrogen fixed was calculated by using the theoretical molar ratio 3:1 for acetylene to nitrogen reduced as a conversion factor (STEWART et al., 1968).

Results

Free-Living Bacteria. Results from investigations on nitrogen fixation by free-living bacteria at Fennoscandian IBP sites are given in Table 1. The annual nitrogen input at these sites was found to vary from 1 to 150 mg N per m^2 per year. The values from Hardangervidda (Norway) and Stordalen (Sweden) were similar, whereas those from Kevo (Finland), were of a smaller magnitude. Bacterial nitrogen fixation thus seems relatively unimportant at the latter site (KALLIO and KÄRENLAMPI, 1972). The surface layers of soils at Kevo are very dry (KÄRENLAMPI, 1972).

Table 1. Nitrogen input by bacteria at Fennoscandian IBP sites in mg N m^{-2} $year^{-1}$

Site	Estimated N fixation	References
Stordalen (Sweden), mire		GRANHALL and SELANDER, 1973
Elevated areas,		
aerobic bacteria	150	
anaerobic bacteria	29	
Wet depression		
aerobic bacteria	32	
Hardangervidda (Norway)		HANSSEN et al., 1973
Wet meadow		
aerobic bacteria	60	
anaerobic bacteria	35	
Dry meadow		
aerobic bacteria	65	
anaerobic bacteria	40	
Lichen heath		
aerobic bacteria	30	
anaerobic bacteria	25	
Birch wood		
aerobic bacteria	50	
anaerobic bacteria	115	
Snow bed		
aerobic bacteria	55	
anaerobic bacteria	85	
Kevo (Finland)		KALLIO and KALLIO, 1975
Pine wood		
aerobic bacteria	1.2	
Birch wood		
aerobic bacteria	1.4	
Subalpine heath		
aerobic bacteria	2.7	

At Hardangervidda free-living bacteria, both aerobic nd anaerobic, contributed similarly at all the sites. No significant difference in fixation rates between the sites was found.

At Stordalen, nitrogen fixation by aerobic bacteria in elevated areas seemed relatively more important than that of anaerobic bacteria. The main contributor was a colorless, acid-tolerant bacterium, occurring in numbers up to 10^4 cells per g dry peat (GRANHALL and SELANDER, 1973).

The appearance of anaerobic photosynthetic bacteria (*Rhodospirillum*) in both ombrotrophic and minerotrophic depressions may be responsible for the anaerobic fixation at Stordalen (ROSSWALL and SVENSSON, 1974), as e.g. *Clostridia* are generally low in numbers (GRANHALL and SELANDER, 1973). *In situ*, fixation rates of peat samples (pH 3.6–3.8) with such spirilli were, in 1970, found to fall within the range 1.0–5.5 ng N g^{-1} h^{-1} (ROSSWALL and MÅRD, 1970). Anaerobic bacteria are considered the main contributors at some tundra sites (STUTZ, 1972).

Symbiotic Bacteria. The dry meadow at Hardangervidda was the only IBP site with a significant distribution of nitrogen-fixing higher plants. At this site, the sole legume species *Astragalus alpinus* occurred. At a riverside near the birch wood, *Lotus corniculatus* also occurred together with *A. alpinus*.

Nitrogen fixation rates up to 270 µg N per dry weight of nodules per hour were found associated with *A. alpinus* root nodules. Highest seasonal activity was found in July and August. The activity was several times lower in September. Although the nitrogenase activity per g dry weight of nodules was high, the numbers of nodules per m^2 were usually low. Their contribution to nitrogen input was therefore estimated to be of the same order of magnitude as nitrogen fixation by free-living bacteria. No significant differences in nitrogen fixation could be found when comparing turfs with and without *A. alpinus*.

Investigations on the influence of temperature on the nitrogenase activity of *A. alpinus* nodules were made in the laboratory. The results (Fig. 1) varied considerably. Laboratory fixation rates were always low compared with field results, probably depending on the assay method used. Nitrogen fixation is found to fall rapidly when turfs with plants are conveyed to the laboratory and roots with nodules removed from the plants. Thus, the maximum values are probably the most representative for the intact nodule-plant system. The results show that nitrogenase activity occurred even at 0° C. Activity increased with a temperature up to 12–15° C, where it reached a maximum. The mean soil temperature (at 2 cm depth) in the dry meadow in July and August was 8–9° C. The relatively low optimum temperature of this nitrogen-fixing system therefore indicated an adaptation to the climatic conditions in agreement with the findings of KALLIO et al. (1972) and BASILIER (1974) for symbiotic and epiphytic blue-green algae respectively. ALEXANDER (1974) also states that nitrogen fixation appears to be maximal for many of the tundra organisms at 15–20° C.

Isolates of clover rhizobia from northern Scandinavia (68–70° N lat.) have also been shown to grow faster, nodulate their host earlier and exhibit higher nitrogenase activities at low temperatures (10° C), in comparison with isolates from southern Scandinavia (EK-JANDÉR and FÅHRAEUS, 1971).

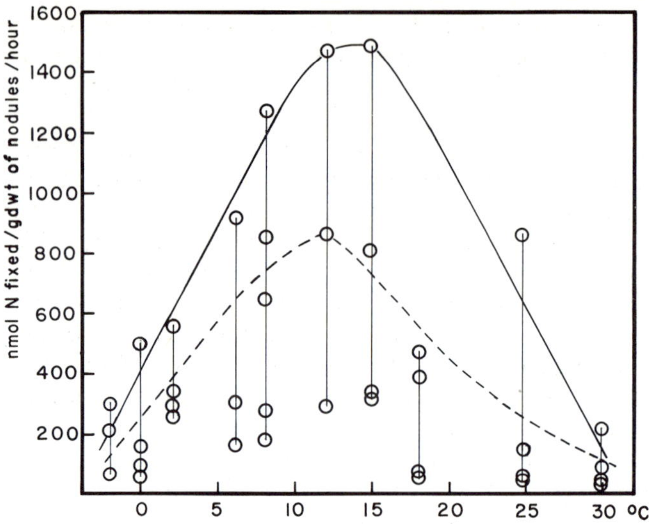

Fig. 1. Laboratory experiment on the effect of temperature on the nitrogenase activity (max and mean) in *Astragalus alpinus* nodules

Two *Astragalus* species occurred at Kevo. Their distribution, however, was so sparse that their contribution to the nitrogen supply was considered relatively unimportant (KALLIO and KÄRENLAMPI, 1972).

Leguminous plants were totally absent at Stordalen.

Nitrogen-fixing vascular plants, especially *Dryas integrifolia* and *Oxytropis*, are important at several other IBP tundra sites, such as Eagle Summit and the Alaska alpine tundra area (ALEXANDER, 1974), but are absent at others. *Dryas octopetala* is widespread in northern Fennoscandia, but no root nodules were found at investigated IBP sites, and nitrogen fixation in soil samples, at the dry meadow at Hardangervidda with *Dryas* roots, was not significantly higher than in samples without *Dryas*. An increase in nitrogen fixation in association with plant roots need not be due to symbiotic relationships, but may just as well be a rhizosphere effect.

Blue-Green Algae. Free-living blue-green algae occur in soil, water and moss communities at several investigated IBP sites in Fennoscandia.

Examined soil samples indicated that free-living heterocystous blue-green algae were rare or absent in soils at the main Kevo and Stordalen sites, possibly due to the low pH (around 4) and ombrotrophic conditions. Such algae occurred, however, quite frequently in soils with higher pH, such as the wet meadow and snow bed at Hardangervidda and in the Abisko (Torneträsk) area surrounding Stordalen (GRANHALL and SELANDER, 1973).

Nitrogen fixation by soil algae has been measured at two sites; a) near a snow bed at Finse, Hardangervidda and b) a *Dryas* heath at Abisko 5 km northwest of Stordalen. Estimated annual fixation rates were 0.9 and 0.1 g N m^{-1}, respectively.

Typical algal crusts were formed at the latter site. These crusts consist of aggregates of fungal hyphae, soil algae and soil particles—which create favorable moisture conditions (CAMERON and FULLER, 1960; CAMERON and DEVANEY, 1970).

Examined water samples from Kevo and Stordalen showed that nitrogen fixation by blue-green algae occurred in tundra-mire pools and brooks, but was generally much lower than in association with mosses. Most of the studies were made at Kevo (KALLIO and KALLIO, 1975). There was a clear difference between the results from Puksalskaidi mire ponds, where the pH of water is 3.8–5.5, and those from the Raessineva mire ponds or Raessijoki brook, where the pH is about 6.1. Nitrogen fixation in water of the latter site (58 ng N 100 ml^{-1} day^{-1}) was nearly 25 times higher than in the more acid water of the Puksalskaidi mire ponds. The algal species in the Raessijoki brook consisted of *Stigonema informe* and *Nostoc sphaericum* (KUPIAS, 1964). The water in these brooks and rivers is cold (5–10° C in summer), clear, oxygenated and oligotrophic. Very much higher fixation rates (26 μg N 100 ml^{-1} day^{-1}) were occasionally found in less acid (5.6–6.4) eutrophic pools at Stordalen (GRANHALL and SELANDER, 1973). The nitrogen-fixing algae consisted of those belonging to the genera *Anabaena*, *Chlorogloea*, *Gloeocapsa*, and *Scytonema*.

The most impressive nitrogen fixation in tundra areas occurs in moss associations (ALEXANDER, 1974). The organisms mainly responsible are free-living, epiphytic and intracellular blue-green algae, associated most commonly with *Sphagnum* and *Drepanocladus* mosses.

A comparison of seasonal variations between mainly *Sphagnum* communities in Kevo mires (Raessineva and Puksalskaidi) and Stordalen is presented in Table 2. Algal fixation and algal species were not studied *per se* at Kevo, but total fixation rates did not seem to vary much with the plant species involved (*S. fuscum*, *girgensohnii*, *compactum*, *russowii*, and *warnstorfii*). Blue-green algae are not predominant at the relatively dry sites at Kevo and the mean monthly fixation rates between the ombrotrophic (Puksalskaidi) and the minerotrophic (Raessineva) sites at Kevo are more or less the same, whereas minerotrophic wet depressions at Stordalen exhibit considerably higher fixation rates. The latter sites have been studied in more detail (GRANHALL and SELANDER, 1973; GRANHALL and BASILIER, 1973; BASILIER, 1974). Nitrogen-fixing blue-green algae have been observed in wet minerotrophic depressions at the Hardangervidda mire site as well. They seem, however, to be relatively more important at Stordalen.

Light microscopic examinations of the predominant mosses *S. riparium* and *Drepanocladus exannulatus* coll. in minerotrophic sites at Stordalen showed that these were both heavily colonized by algae, showing therefore similar fixation rates. The algae consisted mainly of epiphytically growing heterocystous blue-green algae of the genera *Calothrix*, *Chlorogloea*, *Fischerella*, *Hapalosiphon*, *Nostoc*, *Scytonema*, and *Tolypothrix*. *Anabaena* and *Gloeocapsa* predominated among the more loosely associated algae. Intracellular *Nostoc* filaments were sometimes observed within *S. lindbergii* mosses. Their ultrastructure has been the object of a special study (GRANHALL and v. HOFSTEN, unpubl.).

The nitrogenase activity of blue-green algae along moss plants is influenced mainly by two factors: light, and plant effects. Optimal fixation rates on vertically

Table 2. Nitrogen fixation by apical moss parts from mire sites at Kevo and Stordalen

Site	Site classification	Main plant species	pH_{H_2O}	Month	Mean daily acet. red. (nmol $C_2H_4 \cdot$ g d.w.$^{-1} \cdot$day^{-1}) ± S.E.	Mean daily N_2-fix. (ng N $\cdot$ g d.w.$^{-1} \cdot$day^{-1})	No. of rep-licas	References
Raessineva	mire; minerotrophic	*S. girgensohnii*,	6.0–6.3	June	38 ± 3	352	60	KALLIO and KALLIO (1975)
		S. russowii	6.0–6.3	July	38 ± 3	349	120	
			6.0–6.3	Aug	56 ± 9	517	70	
Raessijoki	brook; minerotrophic	*Brachythecium rivulare*	6.1	July	9 ± 2	86	40	KALLIO and KALLIO (1975)
			6.1	Aug	13 ± 3	116	50	
Puksalskaidi	mire; ombrotrophic	*S. fuscum*,	3.7–4.4	June	42 ± 4	392	90	KALLIO and KALLIO (1975)
		S. rubellum	3.7–4.4	July	52 ± 4	485	210	
			3.7–4.4	Aug	4 ± 0.3	33	140	
Stordalen	mire; wet depr; minerotrophic	*S. riparium*	—	June	1040 ± 120	9700[b]	56	BASILIER (1974)
			5.6	July	12240 ± 910	114000[b]	32	
			4.5	Aug	1800 ± 270	16800[b]	32	
			4.0	Sept	176 ± 35	1640[b]	20	
			6.2	July	11760	110000[b]	9	GRANHALL and BASILIER (1973)
			5.2	Aug	3840	35900[b]	18	
		Drepanocladus exannulatus c.	—	June	80 ± 48	746[b]	8	GRANHALL and BASILIER (1973)
			(6.4)	July	8400 ± 1600	78400[b]	8	
			(5.5) 4.9	Aug	352 ± 64	3280[b]	8	
			—	Sept	56 ± 24	523[b]	8	
			6.2	July	3320	31000[b]	6	

Stordalen	minerotrophic	*Drepanocladus exannulatus* c.	5.6	Aug	2880	26900[b]	6	GRANHALL and BASILIER (1973)
Stordalen	mire; wet depr; intermed. trophic	*S. riparium*	—	June	0±0	0	8	BASILIER (1974)
			—	July	880±480	8200[b]	8	
			(6.2) 5.5	Aug	128±32	1190[b]	8	
			—	Sept	0±0	0	8	
Stordalen	mire; wet depr; ombrotrophic	*S. riparium*	6.0	July	55	510	6	GRANHALL and SELANDER (1973)
		S. jensenii	5.7	Aug	336	3140[b]	3	
		S. lindbergii	4.3	Aug	372	3560[b]	6	
Stordalen	mire; elevated area; ombrotrophic	*S. fuscum*,	4.0	July	53±5	492[a]	117	GRANHALL and SELANDER (1973)
		S. balticum	4.0	Aug	184±24	1720	111	

[a] Anaerobic.
[b] Mainly algal fixation (hourly rate ×8 according to GRANHALL and BASILIER (1973)). pH values in brackets refers to investigations the preceding year.

oriented *Sphagnum* plants in wet depressions occurred at 6–8 cm below the water surface (GRANHALL and BASILIER, 1973). At increasing depth, light intensity falls considerably, mainly because of shadowing effects. On the other hand, light inhibition may occur at the water surface. The apex of the moss plant also seems to have an inhibitory effect on algal colonization (BASILIER, 1974), as judged by pigmental analysis (at 625 nm) and light microscopic observations. Fixation was, however, generally higher on living moss plants than on the dead parts of horizontally oriented moss thalli. The plant thus seems to have both positive and negative effects.

The distribution of nitrogen-fixing blue-green algae is influenced by the pH. No blue-green algae were found at any site with a pH of around 4. No correlation could, however, be found *in situ* at Stordalen between pH and nitrogen fixation at moderately higher pH's. This might be explained by laboratory results which indicated a very small increase in fixation rates within the observed field pH interval (4.5–5.6), but a rapid increase above pH 6. The rapid pH decrease in *Sphagnum* communities at Stordalen at the end of the season might, however, have had some effect (Table 2). The total fixation rate (159 mg N m^{-2} yr^{-1}) at Raessineva (pH 6.0–6.3) was significantly higher than at Puksalskaidi mire (pH 3.7–4.4) (KALLIO and KALLIO, 1975).

High oxygen levels may decrease algal fixation (STEWART and LEX, 1970). The oxygen content at Stordalen was generally found to be low and decrease rapidly with depth, especially in *Sphagnum* communities. The oxygen content near the bottom was sometimes less than 3 mg l^{-1}. At these subsites, anaerobic incubations gave slightly higher fixation rates. Local oxygen conditions thus do not seem to be unfavorable, and the methane production observed both at Stordalen (GRANHALL and BASILIER, 1973) and Hardangervidda (HANSSEN, LID-TORSVIK and TORSVIK, 1973) strongly indicates reducing ambient conditions for submerged moss plants.

Carbon dioxide levels, (GRANHALL and BASILIER, 1973; BASILIER, 1974) were generally found to be high, and provided favorable conditions for photosynthesis by algae and mosses.

A special experiment demonstrated that availability of phosphorus may be one of the main factors for intersite variations in nitrogen fixation (BASILIER, 1974).

The influence of light on the nitrogen fixation of moss-associated algae was studied both *in situ* and in the laboratory. A linear relationship was obtained in the laboratory. No light inhibition was observed, even at 14000 lux. The diurnal fixation cycle was found to be correlated with both global radiation and temperature. Maximum fixation occurred around noon, and minimum between 02.40–06.40 h; 32% of total daily fixation occurred between 10.40–12.40 h. Reduction during the night and in dark-incubated samples accounted for about 3% (GRANHALL and BASILIER, 1973). No clear-cut daily fixation cycle was observed at Kevo, and total fixation continued throughout the night (5–8 h dark). This indicates that bacteria were more important than blue-green algae in N-fixation at Kevo.

Temperature is probably the most important environmental factor behind seasonal variations in nitrogen fixation in wet tundra areas (ALEXANDER, 1974). A Q_{10} of about 3 was observed in the laboratory (BASILIER, 1974) in agreement with earlier results (STEWART, 1966). The water temperature curve closely followed that of algal fixation as soon as algal development and colonization on mosses had begun. Negligible fixation values were obtained at Stordalen at the beginning of June and in the middle of September. Highest algal fixation occurred in the middle of July. Highest fixation in *Sphagnum* layers at Puksalskaidi also occurred in July, but the rapid decrease in August seemed to be correlated here with a drought period, which created desiccation of the top layers of *Sphagnum*. The very great temperature variations (sometimes exceeding 30° C) during incubations at Kevo, compared with those at Hardangervidda and Stordalen, might also have had negative effects, at least in August when night temperatures were often below freezing point.

Moisture is limiting the algal nitrogen fixation at many of the Fennoscandian IBP sites, and is probably the main factor for the lower fixation rates obtained in Kevo mires. This is probably also the reason why lichen associations are relatively more important here than at Stordalen.

The lengths of the growing period for free-living blue-green algae at Hardangervidda, Stordalen and Kevo have been estimated at 2.5, 3.5, and 4 months respectively. The annual fixation rates of representative moss communities at Kevo and Stordalen fall within the range 0.2–12 g N m^{-2} yr^{-1}. The higher values apply to epiphytic associations in wet minerotrophic depressions at Stordalen.

Discussion

In most tundra areas, the contribution of symbiotic nitrogen-fixing bacteria is considered to be of less importance than that of free-living bacteria. Blue-green algae, free-living and symbiotic, are the most important nitrogen fixers in the habitats where they occur.

The importance of bacterial fixation *vis-à-vis* precipitation varies with site. At Stigstuv (Hardangervidda), the annual amount of nitrogen input through precipitation was calculated at ca. 200 mg N per m^2, assuming a content of 0.3 mg N per mm and 700 mm per year (LÅG, 1963; SONESSON, WIELGOLASKI, and KALLIO, 1975). This means that the contribution by nitrogen-fixing bacteria may be nearly of the same order of magnitude as the input through precipitation. This applies also to the Kevo region for blue-green algae in lichens in dry areas, and moss-associated algae and bacteria in wet moss layers (input 136 mg N m^{-2} yr^{-1} through precipitation, KALLIO and VEUM, 1975).

At Stordalen the contribution from precipitation is only about 40 mg N m^{-2} yr^{-1}, which is several times lower than the input by aerobic bacteria in elevated areas (GRANHALL and SELANDER, 1973).

Biological fixation at Devon Island and Barrow was considered more important than precipitation (STUTZ, 1972; ALEXANDER, 1974).

Algal fixation in wet, less acid moss communities must be considered highly significant, as its order of magnitude is many times greater than that of the nitrogen input from other sources. Excreted nitrogenous compounds are likely to be transferred directly to the associated plants, and also to other organisms.

Methane production, for instance, closely follows the photosynthetic 24 hrs nitrogen-fixation cycle (SVENSSON, 1973). The main nitrogen source for methane bacteria is NH_4^+-ions.

Mosses covered with epiphytic, nitrogen-fixing, blue-green algae grow remarkably quickly (SONESSON, 1973), and so does *Eriophorum angustifolium*, which is a typical plant for such moss communities. The roots of *E. angustifolium* turfs in wet depressions were often found to be covered with such epiphytes as well.

References

ALEXANDER, V.: A synthesis of the IBP tundra biome circumpolar study of nitrogen fixation. In: HOLDING, A.J., HEAL, O.W., MACLEAN, S.F., FLANAGAN, P.W. (Eds.): Soil organisms and decomposition in tundra, pp. 109–122. Stockholm: IBP Tundra Biome Steering Committee 1974.

BASILIER, K.: Investigations on nitrogen fixation is moss communities. Progress Report 1973. Swedish IBP Tundra Biome Project Tech. Rep. **16**, 83–97 (1974).

CAMERON, R.E., DEVANEY, J.R.: Antarctic soil algae crusts, scanning electron and optical microscope study. Trans. Am. Microscop. Soc. **89**, 264–273 (1970).

CAMERON, R.E., FULLER, N.H.: Nitrogen fixation by some algae in Arizona soils. Proc. Soil Sci. Soc. Am. **24**, 353–356 (1960).

EK-JANDÉR, J., FÅHRAEUS, G.: Adaptation of *Rhizobium* to subarctic environment in Scandinavia. In: LIE, T.A., MULDER, E.D. (Eds.). Biological nitrogen fixation in natural and agricultural habitats. Plant and Soil. Special Volume, pp. 129–137. The Hague: M. Nijhoff 1971.

GRANHALL, U., BASILIER, K.: Nitrogen fixation in tundra moss communities. Progress Report 1972. Swedish IBP Tundra Biome Project Tech. Rep. **14**, 174–190 (1973).

GRANHALL, U., v. HOFSTEN, A.: Intracellular organisms in *Sphagnum* mosses (unpublished).

GRANHALL, U., SELANDER, H.: Nitrogen fixation in a subarctic mire. Oikos **24**, 8–15 (1973).

HANSSEN, J.F., LID-TORSVIK, V., TORSVIK, T.: Mikrobiologi. In: IBP i Norge. Årsrapport 1972, pp. 75–93. Oslo: Norsk IBP 1973 (in Norwegian).

KALLIO, P., KALLIO, S.: Nitrogen fixation by free-living microorganisms in the Kevo district Finnish Lapland. Rep. Kevo Subarctic Res. Stat. **12** (1975) (in press).

KALLIO, P., KÄRENLAMPI, L.: The progress of the Kevo IBP-PT work autumn 1970-autumn 1971. In: WIELGOLASKI, F.E., ROSSWALL, T. (Eds.): Proceedings IBP IV International meeting on the biological productivity of tundra, Leningrad, pp. 276–277. Stockholm: IBP Tundra Biome Steering Committee 1972.

KALLIO, P., SUHONEN, S., KALLIO, H.: The ecology of nitrogen fixation in *Nephroma arcticum* and *Solorina crocea*. Rep. Kevo Subarctic Res. Stat. **9**, 7–14 (1972).

KALLIO, P., VEUM, A.K.: Analysis of precipitation at Fennoscandian tundra sites. In: WIELGOLASKI, F.E. (Ed.): Fennoscandian tundra ecosystems, Part 1: Plants and microorganisms, pp. 333–338. Berlin-Heidelberg-New York: Springer 1975.

KÄRENLAMPI, L.: Comparisons between the microclimates of the Kevo ecosystem study site and the Kevo Meterological Station. Rep. Kevo Subarctic Res. Stat. **9**, 50–65 (1972).

KUPIAS, R.: On filamentous blue-green algae of brooks of Lapland. Rep. Kevo Subarctic Res. Stat. **1**, 236–249 (1964).

LÅG, J.: Plant nutrients in precipitation in Norway. Forskning og Forsøk i Landbruket **14**, 553–563 (1963).

LID-TORSVIK, V., HANSSEN, J.F.: Mikrobiologi. In: IBP i Norge. Årsrapport 1971, pp. 24–44. Oslo: Norsk IBP 1972 (in Norwegian).

ROSSWALL, T., MÅRD, E.: Microbial investigations. Preliminary results 1970. Swedish IBP Tundra Biome Project Tech. Rep. **4**, 1–5 (1970).

ROSSWALL, T., SVENSSON, B.H.: Chemolithotrophic and photosynthetic bacteria at Stordalen. Progress Report 1973. Swedish IBP Tundra Biome Project Tech. Rep. **16**, 73–81 (1974).

SONESSON, M.: Studies in production and turnover of bryophytes at Stordalen 1972. Progress Report 1972. Swedish IBP Tundra Biome Project Tech. Rep. **14**, 66–75 (1973).
SONESSON, M., WIELGOLASKI, F. E., KALLIO, P.: Description of Fennoscandian tundra ecosystems. In: WIELGOLASKI, F. E. (Ed.): Fennoscandian tundra ecosystems, Part 1: Plants and microorganisms, pp. 3–28. Berlin-Heidelberg-New York: Springer 1975.
STEWART, W. D. P.: Nitrogen fixation in plants. London: Athlone Press 1966.
STEWART, W. D. P., FITZGERALD, G. P., BURRIS, R. H.: In situ studies on N_2 fixation using the acetylene reduction technique. Proc. Natl. Acad. Sci. US **58**, 2071–2078 (1967).
STEWART, W. D. P., FITZGERALD, G. P., BURRIS, R. H.: Acetylene reduction by blue-green algae. Arch. Mikrobiologie **62**, 336–348 (1968).
STEWART, W. D. P., LEX, M.: Nitrogenase activity in the blue-green alga *Plectonema boryanum* strain 594. Arch. Mikrobiologie **73**, 250–260 (1970).
STUTZ, R. C.: Nitrogen fixation studies on Devon Island. In: BLISS, L. C. (Ed.): Devon Island IBP project High arctic ecosystem. Project report 1970 and 1971, pp. 252–256. Edmonton: University of Alberta, Dept. of Botany 1972.
SVENSSON, B.: Production of methane and carbon dioxide from a subarctic mire. Progress Report 1972. Swedish IBP Tundra Biome Project Tech. Rep. **14**, 123–143 (1973).

Mineral Content of Tundra and Forest Tundra Plants in Fennoscandia

F. E. WIELGOLASKI, S. KJELVIK, and P. KALLIO

Introduction and Methods

The mineral composition of plant material has been studied in the Finnish and Norwegian IBP tundra projects (SONESSON et al., 1975). Some mineral analyses of tundra plants in USSR were given previously by RODIN and BAZILEVICH (1967), in subarctic Finland by ISOTALO (1971), in Norway by WIELGOLASKI and KJELVIK (1973a) and KJELVIK and WIELGOLASKI (1974), in arctic Canada by MUC (1973) and PAKARINEN and VITT (1973), and in sub-arctic Canada by SCOTTER (1972). Generally, however, less is known of the mineral content of tundra plants than of agricultural plants (LIPMAN and CONYBEARE, 1936; HOMB, 1952, and several more recent publications) and forest plants (DUVIGNEAUD and DENAYER-DE SMET, 1970; COLE et al., 1968; RODIN and BAZILEVICH, 1967).

The present Finnish plant material studies were carried out by generally doing eight analyses throughout the growing season in different plant communities. These studies, carried out in 1971 and 1972, included green parts of some understory vascular plants, mosses, lichens, mushrooms and also various compartments of birch. Analyses at sites in Norway were carried out from 1969 to 1971 and in 1973. Each year two to four analyses were performed of the following compartments: green, non-green above first root, living roots and one compartment of dead material consisting of various dead fractions treated together (WIELGOLASKI and KJELVIK, 1972). This was done for about 30 species of vascular plants, for living parts of lichens and also for green and non-green living bryophytes. The methods used in the Norwegian mineral analyses are described briefly by WIELGOLASKI and KJELVIK (1973a). Similar methods were used in the Finnish subarctic plant analyses.

Results and Discussion

Species Group Variation. The mineral content varies greatly in different plant species (RODIN and BAZILEVICH, 1967; DUVIGNEAUD and DENAEYER-DE SMET, 1970). There is a selective absorbtion of cations by the plants, as observed in physiological studies for several years (COLLANDER, 1941). Some species grouping, however, seems possible on the basis of mineral content from Fennoscandian tundra studies (Table 1). The highest total mineral content (ash) in green parts of vascular plants is found in forbs, and the lowest value is usually found in evergreen dwarf shrubs. High values for forbs and low values in woody plants are also seen from results reported by RODIN and BAZILEVICH (1967) for tundra and forest tundra plants. A similar ash content as given in Table 1 for forbs is found in fruiting bodies of mushrooms in northern Finland, while the values for lichens are mostly lower than even those for green parts of evergreen shrubs (as low as on an average 1.5% of dry weight 2 cm from the plant top in *Cladonia* species in northern Finland). The total mineral content of green parts of monocotyledons and bryophytes is very similar, lower than in forbs and mushrooms.

The variation in the ash content seems to be caused mostly by variation in nitrogen and potassium amounts, which are lowest in evergreen dwarf shrubs and non-nitrogen-fixing lichens (SCOTTER, 1965, 1972) and highest in the mushroom fruiting bodies and in nitrogen-fixing legumes. The nitrogen content found in bryophytes is close to values given by PAKARINEN and VITT (1973) for Canadian high arctic, and reported by RODIN and BAZILEVICH (1973) for USSR and by SMITH (1973) for British montane moorlands. Still lower N-values than found in green material of evergreen shrubs in the Fennoscandian tundra studies are found in *Empetrum nigrum* in the British moorlands (SMITH, 1973). Similarly, the potassium content both in shrubs and monocotyledons seems to be slightly lower in the British studies than in the Fennoscandian tundra studies.

Mushrooms contain on an average less calcium than any other of the plant groups. The calcium content is, as expected, relatively much higher in green material of woody species than in other vascular plants when analyzed at the same site. In green material of woody plants the amount of calcium is normally as high or higher than the amount of potassium at all soil types, while this occurs for herbaceous vascular plants only in calciumrich soil. An especially low calcium content in relation to potassium is found in aboveground live monocotyledons, which is in general agreement with values reported from USSR by RODIN and BAZILEVICH (1967), from Alaska by MILLER (1972) and montane moorlands in Great Britain by SMITH (1973). Calcium and potassium values of the same order are often found in green parts of bryophytes in all soil types. In lichens very low calcium amounts are found both from Fennoscandia (ISOTALO, 1971) and Russian tundra plant studies, even lower than the generally low potassium content.

The amounts of phosphorus and magnesium in the plants are, of course, lower than the N, K, and Ca values, and the variation between the groups is also normally smaller. Very low values for P and Mg are found in lichens, and also for Mg in mushroom fruiting bodies, while the latter have a higher P content than all other plants studied. The amount of magnesium is somewhat lower in green monocotyledons than in other vascular plants, as for calcium. Other small variations between the plant groups seen in Table 1 for P and Mg are not significant, neither are most of the variations in the generally low amounts of Na, Fe, and Mn significant between the plant groups because of species and/or site variation, but the contents are of the same order as reported by RODIN and BAZILEVICH (1967) and SMITH (1973). Possibly there is a tendency to somewhat higher amounts of

Table 1. Mean mineral content (% of dry weight) in some groups of live plants at Hardangervidda, Norway and Kevo, Finland. Order of seasonal variation indicated when observed. Main plants of the groups are: *Betula*: *B. tortuosa* and *B. nana*. *Salix*: Willows and dwarf species. Evergreen Ericaceous: *Empetrum* spp. and *Vaccinium vitis-idaea*. Deciduous Ericaeous: *Vaccinium myrtillus*, *V. uliginosum* and *Arctostaphylos alpina*. Eutrophic monocotyledons: *Anthoxanthum alpinum*, *Deschampsia caespitosa*, *Festuca rubra*, *Phleum commutatum* and *Poa alpina*. Other monocotyledons: *Carex bigelowii* and *C. nigra*, *Deschampsia flexuosa* and *Festuca ovina*. Non-N-fixing forbs: *Alchemilla* spp. *Geranium silvaticum*, *Leontodon automnalis*, *Polygonum viviparum*, *Ranunculus acris*, *Saussurea alpina*, *Silene acaulis* and *Thalictrum alpinum*. N-fixing forbs: *Astragalus alpinus*. *Dryas* forb: *D. octopetala*. Bryophytes: *Barbilophozia lycopodioides*, *Bryum* spp., *Dicranum* spp. *Drepanocladus* spp. *Hylocomium splendens*, *Philonotis fontana*, *Pleurozium schreberi*, *Polytrichum* spp. and *Sphagnum* spp. Non-N-fixing lichens: *Cladonia* (mostly *C. alpestris*, *C. mitis*, and *C. rangiferina*) and *Cetraria* (mostly *C. crispa*, *C. islandica* and *C. nivalis*). N-fixing lichens: *Nephroma arcticum* and *Stereocaulon paschale*. Fruiting bodies of mushrooms: *Boletus* spp., *Lactarius* spp. and *Russula* spp

Plant groups		Plant compart-ments	Ash	N	P	K	Ca[a]	Mg	Na	Fe[a]	Mn[a]
Woody	*Betula*	Green	2.9–4.0	2.2–0.8	0.30–0.10	0.9–0.3	0.4–0.7	0.20–0.30	0.02	0.01	0.10
		Non-green[b]	1.3	0.5–0.3–0.5	0.07	0.2	0.2	0.06	0.01	0.01	0.03
	Salix	Green	4.9–6.3	2.8–1.6	0.30–0.15	1.0–0.6	1.1–1.4	0.15–0.20	0.03	0.15	0.03
		Non-green	4.0	0.9–1.1	0.15	0.3–0.4	1.1–0.9	0.10	0.02	1.00	0.02
Ericaceous	Evergreen	Green	2.7–3.0	0.9–1.1–0.9	0.12	0.4–0.6–0.4	0.3–0.5	0.15	0.02	0.04	0.08
		Non-green	1.6–1.4	0.6–0.8	0.07	0.2	0.2	0.05	0.01	0.06	0.07
	Deciduous	Green	3.0–4.2	1.9–1.0	2.05–0.15	0.9–0.4	0.4–0.6	0.15	0.02	0.04	0.08
		Non-green	2.0–1.3	0.5–0.7	0.07–0.10	0.2	0.2	0.06	0.01	0.06	0.06
Monocot-yledons	Eutrophic	Green	5.5–3.8	2.5–1.4	0.25–0.15	2.1–1.0	0.4	0.13	0.03	0.05	0.04
		Roots	5.0–3.7	1.1	0.12	0.4	0.6–0.3	0.05	0.04	0.50	0.05
	Other	Green	5.6–4.9	2.3–1.7	0.20–0.13	1.6–0.9	0.3	0.12	0.03	0.04	0.03
		Roots	3.7	0.8	0.10	0.4–0.2	0.3	0.05	0.02	0.60	0.02
Forbs	Non-N-fixing	Green	7.6–10.0	2.4–1.8	0.26–0.17	2.2–1.4	1.0–1.6	0.30	0.04	0.08	0.04
		Roots	6.5–5.3	1.1–1.2	0.15–0.16	0.7–0.6	1.1–0.9	0.20	0.05	0.30	0.06
		Green	8.6–8.0	4.5–3.7	0.28–0.16	2.4–1.1	1.4–1.5	0.20	0.05	0.11	0.02
	N-fixing	Roots	6.2–3.0	1.2–2.6	0.11	0.3	1.8–1.2	0.11	0.04	0.07	0.04
		Green	5.0–3.9	2.4–1.4	0.25–0.15	0.7–0.4	1.3	0.10	0.03	0.03	0.01
	Dryas	Non-green	3.8–3.3	0.9–1.1	0.08	0.2	0.9–0.7	0.05	0.02	0.04	0.02
Bryophytes		Green	5.5[a]	1.2	0.15–0.12	0.6	0.06	0.12	0.07	0.35	0.07
		Non-green	9.0[a]	1.1	0.14	0.3	1.0–0.9	0.10	0.05	0.95	0.14
Lichens	Non-N-fix.	Tops	2.5[a]	0.6	0.08	0.2	0.2	0.05	0.03	0.08	0.02
	N-fixing	Tops	1.9	1.5	0.11	0.4	0.1	0.05	—	—	—
Mushrooms		Fruit bodies	8.2	3.5	0.45	2.9	0.3	0.08	—	—	—

[a] Great variations between sites (see Tables 3–5).

[b] Non-green = Live aboveground non-green (woody and non-woody) and live belowground (rhizomes and roots).

iron in green parts of bryophytes than in the green material of other plant groups, and a tendency to a particularly strong variation of both Fe and Mn in green parts of monocotyledons.

Variations in plant mineral content are, of course, found on both a plant group level, as seen in Table 1, and on a single species level (WIELGOLASKI and KJELVIK, 1973a). It is obvious that Ericaceous dwarf shrub species, and especially the evergreen species studied within this family, generally have a lower content of most nutrients (except Mn) than the other woody tundra plants analyzed in Fennoscandia, particularly lower than the most eutrophic *Salix* species, e.g. willows (mostly *S. lapponum*) and the dwarf shrub *S. reticulata*. The nutrients in *Dryas octopetala* are very much of the same order as in *S. reticulata*, and thus lower than in the other forbs studied. In many ways *Dryas* functions ecologically more as a shrub than as a forb, as also seen for the mineral content. The monocotyledons in Table 1 are subdivided into two groups. The tendency is to highest content of most nutrients in the more eutrophic grass species groups, although not in the total ash content of green material. For nitrogen the tendency would have been clearer if the *Carex* species were not combined with the more oligotrophic grass species. The amount of nitrogen in green material of *Carex* is in the order of 2.8–2.1%, which is comparable to N-values in *Carex* from Alaska (MILLER, 1972) and Canadian high arctic (MUC, 1973), while in oligotrophic grasses (*Festuca ovina, Deschampsia flexuosa*) it is only 1.7–1.3% throughout the growing season, which is of the same order as is found for oligotrophic monocotyledons in British montane moorlands (SMITH, 1973), and for *Deschampsia flexuosa* in other mountain meadow studies in Norway (SELSJORD, 1968). *Ranunculus acris* shows generally higher K and Na content than most of the other species studied. The content of Mg and Na is found to be highest in *Equisetum palustre*. This species also has a very high total ash content, as generally found by ISOTALO (1971) for all species of the genus, as well as a high N content. In the bryophytes studied a somewhat higher phosphorus content is found in the green material of *Pleurozium schreberi* than in other species, at all sites where the species is found both in Finland and Norway. A tendency towards higher calcium content is seen in *Cetraria nivalis* than in the *Cladonia* reindeer lichens studied at the same sites. This also causes a higher ash content in *Cetraria* than in *Cladonia* although smaller differences are found between these non-nitrogen fixing species for most minerals than in N-fixing lichens. Naturally they all have a higher N-content than the non-N-fixing species, but while the value is less than twice that for non-N-fixing lichens in *Stereocaulon paschale*, it is more than three times as high for *Nephroma arcticum* as for the non-N-fixing species of *Cladonia* and *Cetraria* studied. Still larger differences in N content are found between N-fixing and non-N-fixing lichens by SCOTTER (1965) in Canada. The present results indicate that *Nephroma* is a good N-fixer, which is seen from the Finnish studies (KALLIO and KALLIO, 1975). Higher amounts of the other minerals are also registered in *Nephroma* than in *Stereocaulon*, as well as higher contents of P and K in *Nephroma* than in the non-N-fixing species. The fruiting bodies of 12 species of mushrooms were analyzed. The strongest variation in minerals was found for calcium, (from 0.03% in *Boletus testaceoscaber* to 0.80% in *Russula vinosa*), possibly caused as much by local soil variation as by selective plant absorbtion.

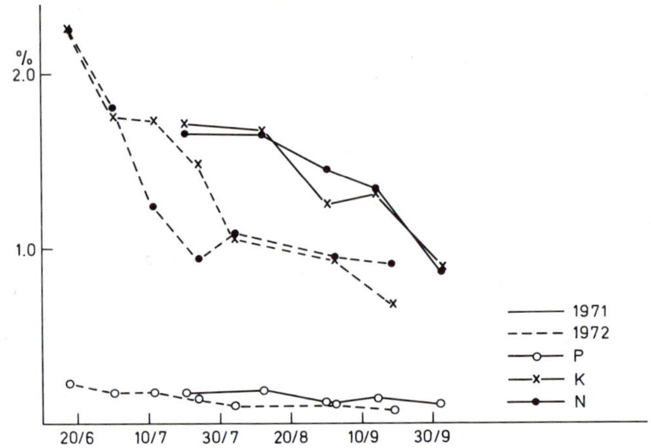

Fig. 1. Seasonal variation in mineral content in green parts of *Deschampsia flexuosa* at the *Oporinia* damaged birch forest site at Kevo, northern Finland, in 1971 and 1972 as percentage of dry weight

Annual and Seasonal Variation. A relatively strong variation is often found in the mineral content at the same plant phenophase in different years. The highest values in aboveground plant components, particularly in green material, were very often found in 1971 for nitrogen and potassium, and partly for phosphorus, especially for herbaceous vascular plants both in southern Norway and northern Finland (Fig. 1), while the amount of calcium for this year was often lower than in the other years. The mineral content in *Nephroma arcticum*, however, varies in the opposite direction to the herbaceous plants in the different years. The summer of 1971 had very low temperatures, both in southern Norway and northern Finland. Under these climatic conditions the herbaceous plants are poorly developed, even though flowering may occur. Therefore the mineral content of N, K, and P is not reduced to the same low levels during the season as in warmer summers. The nitrogen fixation is reduced in low temperatures, and therefore low N-content is found in *Nephroma* in 1971 compared to 1972. See also SKRE et al. (1975).

It is well known from both agriculture and forestry that the content of, for instance N, P, and K, is highest in young green plant material, while the content of Ca in green material increases during the growing season (RIPPEL, 1927; MITCHELL, 1936; OLSEN, 1948; TAMM, 1951; SELSJORD, 1968; DUVIGNEAUD and DENAEYER-DE SMET, 1970 and their references). This is also obviously the general trend for most of the plant groups studied in the present investigations (Table 1, Figs. 1–3), but there are great variations caused by, for instance, environmental factors. As found, for example by TAMM (1951), the most rapid decrease in N, P, and K in green plant parts normally takes place in the first part of the growing season, which means before the time of peak aboveground biomass (usually at the end of July — beginning of August). A particularly rapid decrease in spring is found in the N and K content in leaves of eutrophic grass species in the alpine meadows in southern Norway, species which generally end their vegetative growth early. The evergreen dwarf shrubs start their new leaf growth later in the spring than decid-

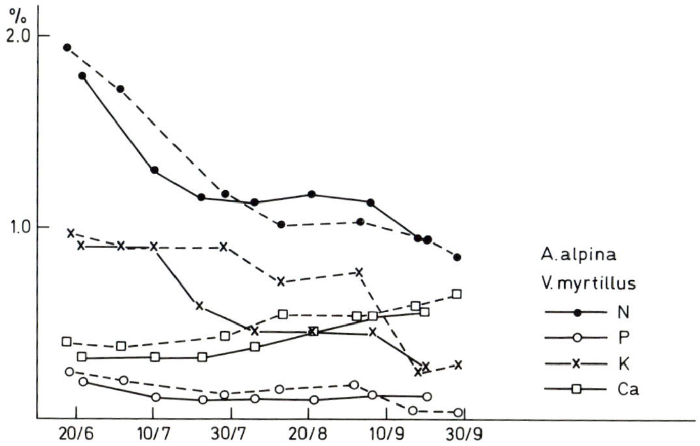

Fig. 2. Seasonal variation in mineral content in *Arctostaphylos alpina* (broken line) at the low alpine heath and *Vaccinium myrtillus* (continuous line) at the birch forest site at Kevo, northern Finland, in 1971 as percentage of dry weight

uous plants (WIELGOLASKI and KÄRENLAMPI, 1975). This may be the reason for the increase in N and K found in the green parts of this plant group early in the summer (Table 1).

In some cases, a slight increase in N in green material is also found early in the season for other plant species, both in the present and other studies (COUPLAND, 1973a). This may be caused by translocation from other plant parts. Calcium often shows an increase in green material particularly late in the growing season (Fig. 2), but the increase is usually less drastic than the decrease found in N and K, and may even be absent (Table 1). Because of the soil variation, the amount of Ca in plants also varies strongly from site to site. Because of small amounts of the other minerals analyzed in the plants, and the site and species variation, seasonal trends are usually not found for these minerals. The total ash content decreased somewhat during the season in green material of monocotyledons, as also found by SELSJORD (1968), but increased, for instance in green material of woody plants, partly because of the increased Ca content.

The seasonal variation in green and live non-green material (branches, stems and roots of woody material and roots of herbaceous plants) may be expected to exhibit nearly the opposite trends because of the translocation taking place between organs, as shown by, for instance, OLSEN (1948), for several minerals. As seen in Table 1 this is also often found for the macronutrients, although usually less significantly for non-green than for green material.

An example of seasonal cycling of minerals is given for nitrogen in birch in northern Finland (Fig. 3). In early spring the content decreases in branches and also in stems (trunk) and roots, while it strongly increases in buds. Just after bud break the amount of nitrogen in leaves is high, but then decreases. An increase in the branches during summer indicates a translocation of nitrogen back to woody compartments. A similar seasonal trend as for branches is also found for stems (trunks) and thick roots, but the N-content is generally lower than in branches

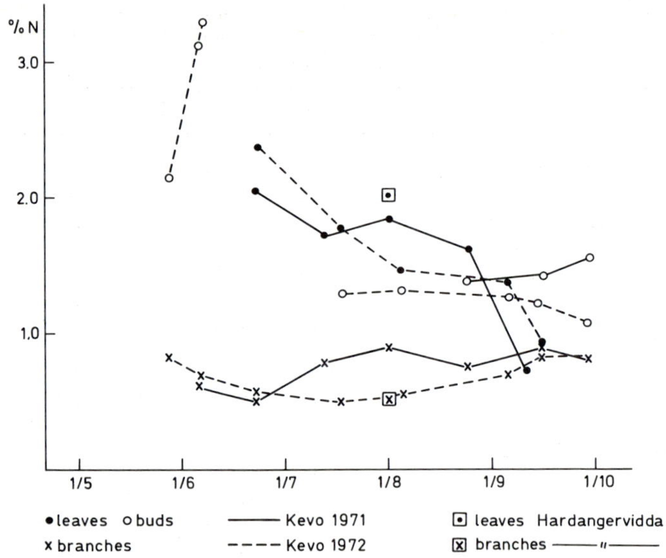

Fig. 3. Seasonal variation in nitrogen content in leaves, buds and branches of birch at Kevo, northern Finland, in 1971 and 1972 in percentage of dry weight. Values from Hardangervidda, Norway, in 1970 are shown in the squares

(Table 2). The amount of nitrogen in thin roots is higher than in branches until early fall (beginning of September). The seasonal variation in the thin roots is unclear, and possibly depends on both the uptake rate of nitrogen and the production rate of new roots. Part of the spring nitrogen content in the leaves is also translocated to the new buds formed in the middle of July, and also to catkins. Even if the loss of nitrogen by leaf fall is considerable, a relatively high percentage is accumulated through the winter in various non-green live parts of the tree. Other authors have found that 35–45% of the nitrogen in tree leaves is reabsorbed by the tree prior to their abscission (OLSEN, 1948).

Compartment Variation. In living plants the concentrations of both N, P, K, Ca, and Mg are usually higher in green than in non-green parts (aboveground non-green and roots), as seen in Tables 1–4. This is in accordance with other studies in similar plant communities (RODIN and BAZILEVICH, 1967; SMITH, 1973). MUC (1973) found the roots to contain less than half the amount of N found in green parts of sedge in the Canadian high arctic. The difference is still greater in the present study. The nitrogen content of live *Carex* roots is only 0.8–0.5% of the dry weight, which is three to four times lower than the amount in the tops. Similar differences are found between N in green material and roots in *Betula tortuosa* and *Betula nana* in subalpine and low alpine regions of southern Norway, while smaller differences are found in northern Finland (Table 2). For most of the other species studied in the Norwegian tundra project, the N-content of the roots is also about half the content of the tops. As seen in Tables 1–4, the differences between the amounts in green parts and roots of the other elements mentioned are of similar order, with the greatest difference generally occurring for potassium.

Table 2. Comparisons between macronutrients (% of dry weight) in various plant parts of *Betula tortuosa* in the beginning of August at a subalpine site in Southern Norway and a subarctic site in Northern Finland

		Ash	N	P	K	Ca	Mg
Hardangervidda, Southern Norway							
Roots		1.73	0.40	0.08	0.29	0.17	0.11
Lower part stem (trunk)	wood	0.38	0.16	0.03	0.08	0.07	0.05
	bark	1.51	0.58	0.05	0.13	0.42	0.08
Stem (trunk)	wood	0.48	0.22	0.05	0.15	0.07	0.04
	bark	1.41	0.54	0.07	0.14	0.41	0.07
Branches		1.10	0.49	0.07	0.18	0.22	0.07
Current year twigs		2.10	1.40	0.19	0.48	0.34	0.17
Leaves		3.78	2.00	0.24	0.71	0.66	0.39
Catkins		2.84	1.81	0.23	0.79	0.32	0.20
Kevo, Northern Finland							
Thin roots		2.10	0.83	0.07	0.29	0.28	0.11
Thick roots		1.04	0.30	0.05	0.19	0.16	0.05
Lower part stem (trunk)		0.95	0.24	0.04	0.11	0.14	0.04
Stem (trunk)		0.89	0.21	0.03	0.19	0.15	0.05
Branches		1.48	0.72	0.08	0.27	0.22	0.08
Buds		2.65	1.37	0.20	0.36	0.41	0.18
Leaves		3.20	1.74	0.14	0.53	0.55	0.29
Catkins		2.93	1.46	0.16	0.39	0.32	0.17

The detailed compartmental separation carried out for *Betula* shows high mineral content variation between wood and bark of stems (trunk), as also reported by DUVIGNEAUD and DENAEYER-DE SMET (1970). For the macronutrients analyzed with the exception of potassium, higher values are found in the bark than in the wood (Table 2). This is also true for the micronutrients Na, Mn, and Zn, while the content of Fe is highest in the wood. An about six times higher amount of Mn and Ca is found in the bark than in the wood, while the differences are smaller for other minerals. There is a generally decreasing tendency in all macronutrients from leaves to current year twigs, younger branches and older branches. The thin roots (Table 2) have a higher content of all nutrients than thicker roots (DUVIGNEAUD and DENAEYER-DE SMET, 1970), and except for Na the roots are found to have a higher mineral content than the stem (trunk) of *Betula*. The nutrient amount in catkins is for most elements somewhat lower than in the leaves, but the differences are small for P, K, and Na, indicating a relative increase in the catkins of these minerals.

The elements Na and Mn, as the macronutrients, existed in generally higher amounts in green than in non-green parts of woody plants, as did Na in some other plant groups (Table 1). In some plant groups, however, the content of Na and Mn is higher in non-green compartments. For Fe this is the rule, and particularly high amounts are found in roots of vascular plants and in non-green bryophytes, but there is great variation between sites (Tables 3 and 4).

From Table 3 it is also clear that the content of K, P, Mg, and Na is usually lower in mixed dead vascular plant material than on an average for different

Table 3. Average mineral content of all species for various plant compartments in percentage of dry weight at different sites

Compartments		Ash	N	P	K	Ca	Mg	Na	Fe	Mn
Green Vascular plants	Low alpine heath North. Finl.	2.5	0.8	0.10	0.4	0.4	0.12	—	—	—
	Pine forest North. Finl.	3.0	0.8	0.09	0.4	0.4	0.11	—	—	—
	Birch forest North. Finl.	3.2	1.3	0.13	0.5	0.5	0.22	—	—	—
	Subalp. birch for. South Norw.	3.5	1.6	0.20	0.7	0.6	0.26	0.02	0.03	0.15
	Alp. lichen heath South. Norw.	2.8	1.2	0.10	0.4	0.5	0.13	0.02	0.02	0.06
	Alp. dry meadow South. Norw.	6.0	1.9	0.18	1.0	1.3	0.17	0.04	0.03	0.02
	Alp. wet meadow South. Norw.	5.0	2.8	0.26	0.9	0.7	0.20	0.03	0.11	0.03
	Alp. will. thicket South. Norw.	6.8	2.2	0.26	1.1	1.0	0.22	0.04	0.20	0.17
Green Bryophytes	Pine forest North. Finl.	3.0	0.7	0.11	0.4	0.3	0.09	—	—	—
	Birch forest North. Finl.	3.0	0.8	0.12	0.5	0.3	0.12	—	—	—
	Subalp. birch for. South. Norw.	6.7	1.2	0.23	0.9	0.4	0.20	0.06	0.15	0.11
	Alp. lichen heath South. Norw.	14.4[a]	1.0	0.12	0.4	0.2	0.07	0.04	0.15	0.01
	Alp. dry meadow South. Norw.	6.5	1.4	0.13	0.5	1.3	0.11	0.07	0.12	0.03
	Alp. wet meadow South. Norw.	6.3	1.2	0.14	0.8	1.2	0.11	0.08	1.10	0.05
	Alp. will. thicket South. Norw.	5.2	1.2	0.23	0.8	0.7	0.13	0.06	0.19	0.14
Live non-N-fixing lichens	Low alpine heath North. Finl.	2.1	0.4	0.06	0.2	0.2	0.04	—	—	—
	Subalp. birch for. South. Norw.	2.7	0.7	0.16	0.3	0.2	0.07	0.03	0.06	0.03
	Alp. lichen heath South. Norw.	2.6	0.5	0.05	0.1	0.2	0.04	0.02	0.08	0.01
	Alp. dry meadow South. Norw.	11.3[a]	1.3	0.11	0.3	3.9	0.06	0.04	0.13	0.02
Non-green above ground Vascular plants	Subalp. birch for. South. Norw.	0.7	0.4	0.05	0.2	0.2	0.06	0.02	0.02	0.03
	Alp. lichen heath South. Norw.	1.7	0.7	0.06	0.2	0.2	0.06	0.02	0.04	0.02
	Alp. dry meadow South. Norw.	4.6	1.1	0.12	0.3	0.3	0.09	0.03	0.03	0.02
	Alp. wet meadow South. Norw.	3.7	1.3	0.18	0.6	0.6	0.10	0.05	0.47	0.03
	Alp. will. thicket South. Norw.	2.5	0.9	0.12	0.2	0.8	0.07	0.02	0.10	0.14
Non-green Bryophytes	Subalp. birch for. South. Norw.	7.4	1.2	0.14	0.3	0.6	0.18	0.03	0.16	0.18
	Alp. lichen heath South. Norw.	11.0[a]	0.9	0.08	0.2	0.2	0.04	0.02	0.16	0.01
	Alp. dry meadow South. Norw.	9.8[a]	1.9	0.14	0.3	1.7	0.09	0.04	0.17	0.09
	Alp. wet meadow South. Norw.	11.1	1.5	0.15	0.3	1.5	0.06	0.07	3.51	0.13
	Alp. will. thicket South. Norw.	6.2	0.9	0.17	0.5	1.2	0.11	0.06	0.31	0.27
Live below ground Vascular plants	Subalp. birch for. South. Norw.	2.3	0.5	0.10	0.2	0.2	0.09	0.02	0.07	0.04
	Alp. lichen heath South. Norw.	10.5[a]	0.7	0.06	0.1	0.1	0.04	0.02	0.20	0.02
	Alp. dry meadow South. Norw.	4.1	1.1	0.08	0.3	0.8	0.06	0.04	0.11	0.02
	Alp. wet meadow South. Norw.	6.3	1.3	0.12	0.2	0.6	0.06	0.06	2.54	0.01
	Alp. will. thicket South. Norw.	6.2	0.9	0.17	0.4	0.7	0.12	0.06	1.63	0.15
Dead Vascular plants	Subalp. birch for. South. Norw.	5.6	1.1	0.08	0.1	0.5	0.10	0.02	0.13	0.11
	Alp. lichen heath. South. Norw.	11.8[a]	1.0	0.06	0.1	0.2	0.04	0.01	0.18	0.01
	Alp. dry meadow South. Norw.	7.9	1.5	0.10	0.2	1.7	0.07	0.03	0.16	0.05
	Alp. wet meadow South. Norw.	7.5	1.8	0.11	0.1	1.0	0.04	0.04	2.76	0.02
	Alp. will. thicket South Norw.	6.3	1.2	0.16	0.2	1.1	0.14	0.02	0.48	0.29

[a] Uncertain, sand in some samples

Subalp. birch for. South. Norw. = Subalpine birch forest Southern Norway

Alp. lichen heath South. Norw. = Alpine lichen heath Southern Norway

Alp. will. thicket South. Norw. = Alpine willow thicket Southern Norway

living material, as also seen for many plant species in RODIN and BAZILEVICH (1967), SMITH (1973), and MUC (1973). The differences in mineral content between living and dead are found to be greatest for potassium. Often there are 4–8 times lower amounts in dead than in green parts, which is of the same order as differences found by MUC (1973) in sedges in the Canadian high arctic. He also reports half as much N in dead as in green sedges, as found in the present studies, although smaller differences are found for some vascular plant species. The content of Fe, and to some extent Ca and Mn, increased somewhat in dead parts, indicating a slower recycling than for the other elements mentioned. This is possibly caused by more rapid movement of carbon compounds out of leaves after death (COUPLAND, 1973b). The increase of Ca in dead material is, however, smaller in many species than found by MUC (1973). At the Norwegian alpine oligotrophic lichen heath site, even lower values are found for Ca and Mn in dead parts of evergreen dwarf shrubs than in green leaves.

Site Variation. There are often great differences in the content of mineral elements in plants at the various sites studied (Tables 3 and 4). They generally reflect variations in soil nutrients (HINNERI et al., 1975, and Table 5). The lowest content of most of the macronutrients (except P) is found at the oligotrophic heath sites both in Finland and Norway, while the highest amounts are found at eutrophic meadow sites and at the willow thicket site. This variation is most pronounced for calcium; often values three times higher are found in a plant compartment analyzed from one of the eutrophic sites than in the same compartment of the same plant group taken from one of the oligotrophic sites. Extremely high Fe content (% of soil weight) in the waterlogged wet meadow and partly also at the wet willow thicket (20–70 times higher Fe percentages in the humus layer at these sites than in drier soil), caused a luxury uptake of this element at these sites, up to 10- to 20 times more in belowground and dead material of vascular plants and in non-green bryophytes than at the other sites. The highest amount of Mn, both in the soil and in all plant compartments, is found in the willow thicket and in the Norwegian birch forest. The phosphorus content is higher, expecially in the mineral soil, at the Norwegian *Salix herbacea* snow bed site than at the other sites studied, and also 2–3 times higher in the plants at this site than in the same species at other sites (Table 4).

There seems to be a similar level for the macronutrients in green parts of Ericaceous shrubs in the birch forests studied in Finland and Norway, respectively on the alpine heaths in the two countries (Table 4). The level in green material is slightly higher in the Norwegian birch forest (which also has somewhat more nutrients in the soil, Table 5) than at the two heaths. The mineral content of *Cladonia* reindeer lichens is very similar at the two heath sites studied, but generally somewhat higher in the Norwegian birch forest. Highest values are also found at this site for the moss *Pleurozium schreberi* and for P and K in *Deschampsia*, analyzed from the different forest sites in the two countries. The N content of green parts of this species and in *Vaccinium myrtillus*, however, was, in contrast to other species, higher at the Finnish forest sites than in the Norwegian subalpine birch forest, showing the selective uptake by various plant species (COLLANDER, 1941) in relation to environmental factors (possibly both soil and climate). Similar variations between species, regarding the amount of N in green material, are seen

Table 4. Examples on variation in mineral content (yearly mean % of dry weight) of the same plant species at different sites in Northern Finland and Southern Norway

Species, compartments and sites		Ash	N	P	K	Ca	Mg	Na	Fe	Mn
	Vaccinium vitis-idaea									
Green	Low alpine heath, North. Finland	2.7	0.9	0.11	0.4	0.4	0.13	—	—	—
	Birch forest, North. Finland	2.8	0.8	0.09	0.3	0.4	0.13	—	—	—
	Subalpine birch forest, South. Norway	2.8	1.1	0.12	0.4	0.5	0.13	0.03	0.05	0.08
	Alpine lichen heath, South. Norway	2.9	1.0	0.10	0.3	0.5	0.14	0.02	0.03	0.09
	Alpine dry meadow, South. Norway	3.3	1.5	0.13	0.4	0.8	0.13	0.03	0.05	0.07
Roots	Subalpine birch forest, South. Norway	1.6	0.5	0.09	0.2	0.2	0.06	0.02	0.07	0.06
	Alpine lichen heath, South. Norway	4.1	0.6	0.06	0.1	0.2	0.06	0.01	0.11	0.04
	Alpine dry meadow, South Norway	2.2	0.8	0.08	0.2	0.7	0.05	0.02	0.04	0.06
	Vaccinium myrtillus									
Green	Birch forest, North. Finland	3.8	1.6	0.12	0.5	0.4	0.15	—	—	—
	Subalpine birch forest, South. Norway	2.9	1.4	0.15	0.6	0.4	0.12	0.02	0.03	0.20
	Alpine dry meadow, South. Norway	3.4	1.2	0.17	0.6	0.7	0.14	0.02	0.05	0.10
Roots	Subalpine birch forest, South. Norway	1.5	0.6	0.07	0.2	0.1	0.05	0.01	0.04	0.11
	Alpine dry meadow, South. Norway	1.1	0.5	0.09	0.2	0.1	0.03	0.01	0.04	0.07
	Salix herbacea									
Green	Alpine snow bed, South. Norway	5.9	2.1	0.33	1.7	0.4	0.23	0.02	0.13	0.03
	Alpine lichen heath, South. Norway	6.9	2.0	0.19	1.0	0.4	0.17	0.03	0.07	0.04
	Alpine dry meadow, South. Norway	5.3	1.9	0.18	1.1	1.2	0.15	0.03	0.03	0.03
	Alpine wet meadow, South. Norway	5.1	2.6	0.24	0.7	1.3	0.23	0.03	0.17	0.04
	Alpine willow thicket, South. Norway	5.5	1.3	0.15	1.5	0.9	0.21	0.03	0.06	0.17
Roots	Alpine snow bed, South. Norway	4.5	—	0.24	0.4	0.2	0.09	0.01	0.51	0.01
	Alpine lichen heath, South. Norway	5.5	1.0	0.11	0.3	0.2	0.08	0.02	0.10	0.01
	Alpine dry meadow, South. Norway	3.2	1.2	0.09	0.3	1.1	0.08	0.02	0.03	0.02
	Alpine wet meadow, South. Norway	5.2	0.9	0.16	0.3	1.0	0.12	0.05	1.50	0.02
	Alpine willow thicket, South. Norway	3.3	1.0	0.16	0.5	0.5	0.09	0.01	0.05	0.13
	Deschampsia flexuosa									
Green	Low alpine heath, North. Finland	4.8	1.5	0.11	1.1	0.2	0.11	—	—	—
	Birch forest, North. Finland	5.3	1.4	0.14	1.4	0.2	0.09	—	—	—
	Subalpine birch forest, South. Norway	5.5	1.0	0.25	1.6	0.2	0.11	0.04	0.03	0.02
	Alpine dry meadow, South. Norway	4.6	1.6	0.13	1.0	0.7	0.09	0.06	0.02	0.03
Roots	Subalpine birch forest, South. Norway	3.3	0.7	0.12	0.4	0.1	0.03	0.02	0.02	0.02
	Alpine dry meadow, South. Norway	2.7	0.8	0.07	0.2	0.4	0.03	0.03	0.04	0.02
	Ranunculus acris									
Green	Subalpine birch forest, South. Norway	13.9	1.9	0.23	3.2	1.2	0.25	0.06	0.17	0.12
	Alpine dry meadow, South. Norway	9.4	2.2	0.19	2.3	1.3	0.27	0.06	0.06	0.02
	Alpine wet meadow, South. Norway	9.7	2.0	0.25	1.2	1.8	0.29	0.06	0.13	0.05
	Alpine willow thicket, South. Norway	11.2	2.1	0.23	3.3	1.2	0.30	0.05	0.17	0.05
Roots	Subalpine birch forest, South. Norway	8.3	1.0	0.19	1.9	0.6	0.18	0.10	0.28	0.13
	Alpine dry meadow, South. Norway	5.1	1.4	0.15	0.9	1.0	0.18	0.04	0.06	0.03
	Alpine wet meadow, South. Norway	5.8	1.0	0.19	0.6	0.8	0.21	0.08	0.79	0.03
	Alpine willow thicket, South. Norway	5.1	1.1	0.27	0.7	0.6	0.24	0.05	0.24	0.17
	Pleurozium schreberi									
Green	Pine forest, North. Finland	3.0	0.7	0.11	0.4	0.3	0.09	—	—	—
	Birch forest, North. Finland	3.1	0.8	0.11	0.5	0.3	0.12	—	—	—
	Subalpine birch forest, South. Norway	6.0	1.3	0.29	0.8	0.4	0.20	0.05	0.15	0.10
	Alpine willow thicket, South. Norway	6.1	1.2	0.15	0.7	0.7	0.11	0.06	0.30	0.16
Non green	Subalpine birch forest, South. Norway	7.0	1.3	0.18	0.3	0.6	0.18	0.03	0.16	0.18
	Alpine willow thicket, South. Norway	6.9	1.0	0.13	0.5	0.9	0.11	0.07	0.30	0.36
	Cladonia rangiferina									
Live tops	Low alpine heath, North. Finland	1.3	0.5	0.05	0.2	0.1	0.03	—	—	—
	Subalpine birch forest, South. Norway	2.7	0.7	0.16	0.3	0.2	0.07	0.03	0.06	0.03
	Alpine lichen heath, South. Norway	2.4	0.5	0.05	0.1	0.1	0.03	0.01	0.09	0.01

Table 5. Amount of minerals per ground area in soil down to 35 cm ($kg\,m^{-2}$) in live perennial biomass and dead plants ($g\,m^{-2}$) and annual mineral uptake ($g\,m^{-2}$). Total soil weight ($kg\,m^{-2}$), average dry weight total live perennial biomass and annual production ($g\,m^{-2}$). Five plant communitites Hardangervidda, Norway

Site	Element	Minerals accumulated in perennial live plants			in dead plants	Annual uptake	Soil minerals (total)
		above ground	below ground	total			
Subalpine birch forest	Ash	23.7	42.1	65.8	59.5	22.3	—
	N	9.4	9.4	18.8	11.2	8.1	1.10
Live per. biom.: 4595 $g\,m^{-2}$	P	1.4	1.8	3.2	0.9	1.1	0.70
Ann. prod.: 780 $g\,m^{-2}$	K	4.1	4.6	8.7	1.2	3.5	1.10
Soil weight: 379 $kg\,m^{-2}$	Ca	5.7	3.4	9.1	4.8	2.7	1.27
	Mg	1.7	1.6	3.3	1.1	1.3	1.60
	Na	0.5	0.4	0.9	0.2	0.2	0.05
	Fe	0.6	1.3	1.9	1.4	0.3	11.70
	Mn	0.9	0.7	1.6	1.1	0.7	0.12
	Zn	0.3	0.2	0.5	0.1	0.1	—
Alpine lichen heath	Ash	12.0	20.0	32.0	31.9	15.7	—
	N	2.4	1.4	3.8	2.7	2.1	0.24
Live per. biom.: 625 $g\,m^{-2}$	P	0.2	0.1	0.3	0.2	0.2	0.33
Ann. prod.: 270 $g\,m^{-2}$	K	0.6	0.2	0.8	0.2	0.5	0.47
Soil weight: 520 $kg\,m^{-2}$	Ca	1.0	0.2	1.2	0.5	0.7	0.75
	Mg	0.2	0.1	0.3	0.1	0.2	0.51
	Na	0.1	0.04	0.1	0.03	0.1	0.04
	Fe	0.3	0.4	0.7	0.5	0.3	4.30
	Mn	0.04	0.03	0.1	0.03	0.1	0.08
	Zn	0.02	0.01	0.03	0.02	0.01	—
Alpine dry meadow	Ash	7.6	22.1	29.7	36.4	27.6	—
	N	1.5	5.8	7.3	7.1	7.4	1.16
Live per. biom.: 725 $g\,m^{-2}$	P	0.1	0.4	0.5	0.5	0.7	0.35
Ann. prod.: 534 $g\,m^{-2}$	K	0.5	1.3	1.8	0.9	3.0	0.66
Soil weight: 411 $kg\,m^{-2}$	Ca	1.8	4.4	6.2	8.0	5.8	1.44
	Mg	0.1	0.3	0.4	0.3	0.6	1.20
	Na	0.1	0.2	0.3	0.2	0.2	0.04
	Fe	0.1	0.6	0.7	0.7	0.4	5.50
	Mn	0.04	0.1	0.1	0.2	0.1	0.11
	Zn	0.02	0.1	0.1	0.1	0.1	—
Wet meadow	Ash	20.6	82.4	103.3	119.5	51.4	—
	N	3.3	16.6	19.9	28.1	13.1	1.69
Live per. biom.: 1605 $g\,m^{-2}$	P	0.4	1.7	2.1	1.8	1.3	0.15
Ann. prod.: 833 $g\,m^{-2}$	K	1.0	2.0	3.0	2.1	3.6	0.05
Soil weight: 70 $kg\,m^{-2}$	Ca	2.9	8.5	11.4	15.5	6.5	0.62
	Mg	0.2	0.8	1.0	0.7	0.8	0.11
	Na	1.5	0.7	2.2	0.6	0.5	0.06
	Fe	5.9	33.4	39.3	44.3	14.7	7.74
	Mn	0.2	0.2	0.4	0.4	0.3	0.01
	Zn	0.03	0.1	0.1	0.1	0.1	—
Willow thicket	Ash	33.3	76.0	109.0	54.2	51.1	—
	N	8.5	10.9	19.4	10.1	11.2	1.00
Live per. biom.: 2390 $g\,m^{-2}$	P	1.3	2.1	3.4	1.4	1.7	0.64
Ann. prod.: 884 $g\,m^{-2}$	K	2.7	4.3	7.0	1.7	5.2	0.39
Soil weight: 280 $kg\,m^{-2}$	Ca	8.4	8.4	16.8	9.3	7.7	1.68
	Mg	0.8	1.5	2.3	1.2	1.3	1.24
	Na	0.3	0.8	1.1	0.2	0.4	0.04
	Fe	1.4	20.0	21.4	4.1	6.3	5.11
	Mn	1.6	1.8	3.4	2.5	1.5	0.36
	Zn	0.1	0.7	0.8	0.1	0.3	—

at the various eutrophic sites. The content is considerably higher in the wet meadow than in the dry meadow and willow thicket for, for instance *Salix herbacea* (Table 4), but only small differences are seen in the amount of N at these sites for *Ranunculus acris*. For both species, however, the contents of Ca and P are highest, and the content of K lowest in the wet meadow. The relatively low K content in green parts of many species studied from the peaty wet meadow site (Table 4) may reflect the low total potassium amount in this soil (Table 5), although relatively high amounts of readily soluble potassium are present in the soil. On the other hand, relatively more of the heavier soluble soil minerals, such as P, seem to be available to the plants in the normally waterlogged wet meadow, even if analyses show relatively small amounts of both readily soluble and total P in the wet meadow soil.

It has been found that the mineral level in green material generally increases with decreasing light intensity, partly because of physiologically younger plants (MYHR and SÆBØ, 1969). This may be one reason why the nutrient amount of *Festuca rubra* is higher below the shrubs at the willow thicket site than in the more exposed wet meadow. On the whole, there may be several explanations for variations in mineral content found in plants at different sites.

Fertilization Effect. In a separate study some soil plots of dry alpine meadow vegetation is fertilized with, respectively, nitrate and mixed fertilizer (50 and 100 g m^{-2}). Plant analyses of green material (Table 6) show that the P and K content in the plants has increased with the use of mixed fertilizer, and the N content has increased with all fertilizer treatments. The Ca content has decreased. Three levels of P fertilization tagged with ^{32}P up to 5 gP m^{-2} (as in the strongest mixed fertilization) have all given strong increases in P uptake (the uptake was 2.34 mgP g^{-1} d.w. with an amount of P as in 50 g m^{-2} mixed fertilizer and 5.85 mgP g^{-1} d.w. as in 100 g m^{-2}).

From the plant analyses and dry weights of the meadow plants (WIELGOLASKI, 1975), the amount of nutrients taken up by the plants may be roughly calculated. Approximately 38%, 54%, and 33% of P, K, and Mg, respectively, of the mixed fertilizer added in spring 1973 at the 100 g m^{-2} plots with mixed fertilizer is taken up, in addition to what is found in the standing crop at the control plots. The calculation also shows that the amount of P and K in the standing crop has increased 10 times on a ground area basis, while the standing crop of green material has increased 4.5 times per m^2. This is in accordance with results of RIPPEL (1927), showing that N, P, and K are taken up faster by growing plants than dry matter is produced. Ca has increased only 2.4 times, Mn 2.1 times, Mg and Na 3.5 times and Fe 4.5 times in the standing crop. As the standing crop has increased strongly with the amount of mixed fertilizer given, there should be a possibility of further increases with additional fertilizing. The uptake indicates that the amounts given of Ca, Mg, Na, and Mn possibly have the least effect, while N, P. K are, as expected, more important. The N-fertilization results and the tagged P study indicate, in addition to the mixed fertilization results, a combined effect of the elements.

Mineral Uptake and Accumulation. Calculations of the minerals accumulated per unit ground area in perennial compartments of living plants, and in dead plants at the non-fertilized Norwegian sites (Table 5), are based on the weighed

Table 6. Mineral content in percent of dry weight of green material of some species at control and fertilized plots

Treatment and species	Ash	N	P	K	Ca	Mg	Na	Fe	Mn
Control									
Anthoxanthum alpinum	4.6		0.18	0.86	0.46	0.15	0.06	0.03	0.07
Carex bigelowii	5.2	1.64	0.20	1.21	0.48	0.08	0.04	0.01	0.05
Deschampsia flexuosa	3.8	1.24	0.16	0.49	0.37	0.13	0.07	0.02	0.07
Poa alpina	2.8		0.16	0.89	0.24	0.06	0.04	0.02	0.01
Festuca rubra	5.4	1.32	0.16	0.88	0.38	0.11	0.07	0.01	0.04
Other monocotyledons	4.4		0.15	0.59	0.36	0.10	0.08	0.02	0.04
Polygonum viviparum	4.4		0.24	0.69	1.02	0.35	0.05	0.03	0.10
100 g m^{-2} mixed fertilizer									
Anthoxanthum alpinum	4.8	1.88	0.49	1.89	0.22	0.08	0.04	0.02	0.04
Carex bigelowii	4.4	1.48	0.30	1.79	0.31	0.10	0.04	0.02	0.04
Deschampsia flexuosa	5.2	1.52	0.46	1.89	0.23	0.10	0.05	0.02	0.03
Poa alpina	4.5	1.44	0.53	1.81	0.17	0.09	0.02	0.02	0.01
Festuca rubra	5.0	1.36	0.52	1.91	0.27	0.10	0.06	0.03	0.02
Other monocotyledons	5.8		0.40	1.88	0.27	0.12	0.08	0.02	0.03
Polygonum viviparum	3.5		0.31	1.16	0.36	0.17	0.02	0.02	0.02
100 g m^{-2} N fertilizer									
Anthoxanthum alpinum	3.2	2.24	0.18	1.10	0.35	0.08	0.05	0.02	0.02
Carex bigelowii	4.3	2.28	0.22	1.07	0.60	0.10	0.05	0.02	0.02
Deschampsia flexuosa	3.9	1.92	0.17	0.81	0.36	0.10	0.07	0.03	0.02
Poa alpina	2.0		0.12	0.69	0.26	0.04	0.05	0.02	0.01
Festuca rubra	4.6	1.96	0.17	1.06	0.41	0.09	0.06	0.02	0.02
Other monocotyledons	3.1		0.15	0.67	0.38	0.07	0.06	0.02	0.01
Polygonum viviparum	2.9		0.17	0.66	0.20	0.18	0.06	0.02	0.01

average biomass of the various plant species (KJELVIK, 1973; WIELGOLASKI and KJELVIK, 1973b; KJELVIK and WIELGOLASKI, 1974; KJELVIK and KÄRENLAMPI, 1975; WIELGOLASKI, 1975).

The similarity in accumulation of most minerals in the living and dead material, respectively, is striking, although generally slightly lower in dead material for most minerals except calcium in the two eutrophic meadows. The mineral reservoir in the organic material is, in most cases, small compared to the readily soluble soil minerals (HINNERI et al., 1975), but in the oligotrophic birch forest a higher amount of phosphorus, in particular, is accumulated in the plant biomass than is readily soluble in the soil.

The amounts of total minerals (ash) accumulated in the perennial plant compartments are, for all sites, higher than found for arctic tundra in USSR according to data given by RODIN and BAZILEVICH (1967, p. 22). Their data for shrub tundra, however, indicate a higher amount of ash in plants per m^2 than found in the Norwegian alpine dry meadow and lichen heath sites, and in the subalpine birch forest. Because of the high mineral amount accumulated in the live roots in the Norwegian alpine wet meadow and willow thicket, the total in perennial compartments at these sites is closer to values reported by RODIN and BAZILEVICH (1967) for forest tundra. The total mineral content per m^2 in perennial aboveground compartments seems to be high at all the Norwegian sites, compared with the Russian data for both arctic and shrub tundras.

Nitrogen is accumulated to a higher degree than the other minerals in the Norwegian tundra plants, as is also found in tundra areas in USSR (RODIN and BAZILEVICH, 1967), except at sites where Fe is a locally dominating element (Table 5). A high amount of calcium is also found in the plants per m^2, particularly at the eutrophic sites, but the content of potassium is nearly as high per m^2 in the subalpine birch forest as is the amount of phosphorus. This order of magnitude is somewhat different from the values found for Douglas fir forests (COLE et al., 1968), spruce forests (RODIN and BAZILEVICH, 1967) and deciduous forests (DUVIGNEAUD and DENAEYER-DE SMET, 1970). In these forest types the amount of calcium in plants per m^2 is higher than nitrogen, and the amount of phosphorus in relation to the other macronutrients is lower than found in the tundra plant communities.

The annual uptake of minerals compared to their accumulation in plants is, in contrast to results from forest areas (RODIN and BAZILEVICH, 1967; COLE et al., 1968; DUVIGNEAUD and DENAEYER-DE SMET, 1970), high at the Norwegian tundra sites (Table 5). The same tendency is shown by RODIN and BAZILEVICH (1967) for the low-productive, compared to higher productive areas of the world, and also for relatively dry grasslands (COUPLAND, 1973b) compared to forests. The uptake of minerals on an area basis is in the present tundra studies mostly found to be lower than in the forest studies mentioned, except for N at the wet eutrophic areas, and to some degree also for P. The uptake per m^2 of N, P, and Ca is higher for aboveground parts at all the eutrophic sites than found in Canadian grasslands (COUPLAND, 1973b), but the total uptake of minerals (ash) is lower.

The annual uptake per m^2 of some elements, particularly potassium, is found to be higher at some of the Norwegian sites than the reserves in dead organic material plus the amount contributed by precipitation. This indicates that the release of potassium from plants and soil is higher than for the other elements. The amount of readily soluble minerals in the soil (HINNERI et al., 1975) is always higher than the annual uptake, and the total reservoir of soil minerals (Table 5) is of the order 100–1000 times greater than the annual uptake of plants.

As nearly all plant species in tundra areas are perennials, it is difficult to compare the annual mineral uptake with values for removal by crops in agricultural studies. The total removal of nitrogen in agricultural studies in USA has been estimated at 40–45 kg ha^{-1} (BLACK, 1968), of which about 30 kg ha^{-1} is by the crop itself. If only the green material is harvested at the peak of the growing season at the Norwegian alpine wet meadow site, an almost comparable amount of nitrogen is found to be removed, while 20 kg ha^{-1} is estimated to be removed from the dry meadow. Only 1/10 of these amounts is found to be removed for phosphorus in green material at the same sites, which is lower than the above 4 kg ha^{-1} estimated for agricultural crops in USA by LIPMAN and CONYBEARE (1936). For potassium about 9–10 kg ha^{-1} may be estimated as removed by green material in the two Norwegian meadows, which is about half the amount estimated by LIPMAN and CONYBEARE (1936) as being removed from crop plants in USA.

The Norwegian sites seem to be in an approximately steady state. This means that the amount of minerals returned to the soil per year and unit area are of the same order as the uptake. Very little is retained in plants, as also reported by RODIN and BAZILEVICH (1967) for other tundra communities. In the forest studies of DUVIGNEAUD and DENAEYER-DE SMET (1970) more of the nutrients are retained in the plants by the growth increment. In the present study death of old material nearly compensates for new growth, even of woody material, and therefore very little of the mineral uptake can be said to be retained in the plants.

Acknowledgement. Most of the chemical analyses from the Norwegian sites were carried out by the technical assistants Miss A. POLDER and Mrs. E. M. LUNDE, and we wish to thank them both.

References

BLACK, C. A.: Soil-plant relationship. New York-London-Sidney: John Wiley and Sons 1968.

COLE, D. W., GESSEL, S. P., DICE, S. F.: Distribution and cycling of nitrogen, phosphorus, potassium and calcium in a second-growth Douglas-fir ecosystem. In: YOUNG, H. E. (Ed.): Symposium on primary productivity and mineral cycling in natural ecosystems, pp. 197–232. Orono: Univ. Maine Press 1968.

COLLANDER, R.: Selective absorbtion of cations by higher plants. Plant Physiol. **16**, 691–720 (1941).

COUPLAND, R. T.: Producers: II. Nutrient and energy contents of aboveground parts. Matador Project Tech. Rep. **32**. Saskatoon: CCIBP 1973a.

COUPLAND, R. T.: Producers: III. Rates of dry matter production and of nutrient and energy flow through shoots. Matador Project Tech. Rep. **33**. Saskatoon: CCIBP 1973b.

DUVIGNEAUD, P., DENAEYER-DE SMET, S.: Biological cycling of minerals in temperate deciduous forests. In: REICHLE, D. E. (Ed.): Analysis of temperate forest ecosystems, pp. 199–225. Berlin-Heidelberg: Springer 1970.

HINNERI, S., SONESSON, M., VEUM, A. K.: Soils of Fennoscandian IBP tundra ecosystems. In: WIELGOLASKI, F. E. (Ed.): Fennoscandian tundra ecosystems. Part 1: Plants and microorganisms, pp. 31–40. Berlin-Heidelberg-New York: Springer 1975.

HOMB, T.: Kjemisk sammensetning og fordøyelighet av engvekster. Ås: NLH, Foringsforsøkene 71. Beretning 1952.

ISOTALO, A.: The value of the natural fodder plants on the reindeer feeding. The Research Society of Lapland, Year book **11**, 36–45 (1971).

KALLIO, S., KALLIO, P.: Nitrogen fixation in lichens at Kevo, North Finland. In: WIELGOLASKI, F. E. (Ed.): Fennoscandian tundra ecosystems. Part 1: Plants and microorganisms, pp. 292–304. Berlin-Heidelberg-New York: Springer 1975.

KJELVIK, S.: Biomass and production in a willow thicket and a subalpine birch forest, Hardangervidda, Norway. In: BLISS, L. C., WIELGOLASKI, F. E. (Eds.): Primary production and production processes, Tundra Biome, pp. 115–122. Edmonton-Oslo: IBP Tundra Steering Committee 1973.

KJELVIK, S., KÄRENLAMPI, L.: Plant biomass and primary production of Fennoscandian subarctic and subalpine forests, and alpine willow and heath ecosystems. In: WIELGOLASKI, F. E. (Ed.): Fennoscandian tundra ecosystems. Part 1: Plants and microorganisms, pp. 111–120. Berlin-Heidelberg-New York: Springer 1975.

KJELVIK, S., WIELGOLASKI, F. E.: Biomass, nutrient content and energy of some dwarf shrubs in a Norwegian subalpine birch forest. Rep. Kevo Subarctic Res. Stat. **11**, 47–51 (1974).

LIPMAN, J. G., CONYBEARE, A. B.: Preliminary note on the inventory and balance sheet of plant nutrients in the United States. New Jersey Agr. Exp. Sta. Bul. **607** (1936).

MILLER, P. C.: A model to incorporate minerals into tundra plant production. In: BOWEN, S. (Ed.): Proceedings 1972 US Tundra Biome Symposium, pp. 51–54. Hannover, New Hampshire: USA CRREL 1972.

MITCHELL, H. L.: Trends in the nitrogen, phosphorus, potassium and calcium content of the leaves of some forest trees during the growing season. Black Rock Forest Paper **1**, 30 (1936).

MUC, M.: Primary production of plant communities of the Truelove Lowland, Devon Island, Canada. Sedge meadows. In: BLISS, L. C., WIELGOLASKI, F. E. (Eds.): Primary production and production processes, Tundra Biome, pp. 3–14. Edmonton-Oslo: IBP Tundra Steering Committee 1973.

MYHR, K., SÆBØ, S.: Verknaden av skygging på vekst, utvikling og kjemisk samansetning hos nokre grasarter. Forskn. Fors. Landbr. **20**, 297–315 (1969).

OLSEN, C.: The mineral, nitrogen and sugar content of beach leaves and beach sap at various times. Compt. Rend. Trav. Lab. Carlsberg, Ser. Chim. **26**, 197–230 (1948).

PAKARINEN, P., VITT, D. H.: Primary production of plant communities of the Truelove Lowland, Devon Island, Canada. Moss communities. In: BLISS, L. C., WIELGOLASKI, F. E. (Eds.): Primary production and production processes, Tundra Biome, pp. 37–46. Edmonton-Oslo: IBP Tundra Biome Steering Committee 1973.

RIPPEL, A.: Über den Zusammenhang zwischen dem Aufnahmeverlauf der Bodennährstoffe bei den höheren Pflanzen und der Beweglichkeit dieser Stoffe in der Pflanze. Biochem. Z. **187**, 272–282 (1927).

RODIN, L. E., BAZILEVICH, N. I.: Production and mineral cycling in terrestrial vegetation. Edinburgh-London: Oliver and Boyd 1967.

SCOTTER, G. W.: Chemical composition of forage lichens from Northwestern Saskatchewan as related to use by barren-ground caribou. Canad. Journ. Plant Sci. **45**, 246–250 (1965).

SCOTTER, G. W.: Chemical composition of forage plants from the reindeer preserve, Northwest Territories. Arctic **25**, 21–27 (1972).

SELSJORD, I.: Kjemiske analyser av beiteplanter. Forskn. Fors. Landbr. **19**, 1–7 (1968).

SKRE, O., KJELVIK, S., BERG, A., WIELGOLASKI, F. E.: Nitrogen content in Norwegian alpine plants. In: VIK, R. (Ed.): IBP in Norway. Methods and results. Sections PT-UM Grazing project Hardangervidda. Botanical investigations, pp. 159–194. Oslo: Norwegian National IBP Committee 1975.

SMITH, R. A. H.: The environmental parameters of IBP experimental sites at Moor-House. Aspects of the ecology of the northern Pennines. Moor-House occasional papers **4**, 1–63 (1973).

SONESSON, M., WIELGOLASKI, F. E., KALLIO, P.: Description of Fennoscandian tundra ecosystems. In: WIELGOLASKI, F. E. (Ed.): Fennoscandian tundra ecosystems. Part 1: Plants and microorganisms, pp. 3–28. Berlin-Heidelberg-New York: Springer 1975.

TAMM, C. O.: Seasonal variation in composition of birch leaves Physiol. Plantarum **4**, 461–469 (1951).

WIELGOLASKI, F. E.: Primary productivity of alpine tundra meadow communities. In: WIELGOLASKI, F. E. (Ed.): Fennoscandian tundra ecosystems. Part 1. Plants and microorganisms, pp. 121–128. Berlin-Heidelberg-New York: Springer 1975.

WIELGOLASKI, F. E., KÄRENLAMPI, L.: Plant phenology of Fennoscandian tundra areas. In: WIELGOLASKI, F. E. (Ed.): Fennoscandian tundra ecosystems. Part 1: Plants and microorganisms, pp. 94–102. Berlin-Heidelberg-New York: Springer 1975.

WIELGOLASKI, F. E., KJELVIK, S.: The methodology of net primary production investigations in Norwegian IBP tundra studies. Rep. Kevo Subarctic Res. Stat. **9**, 24–27 (1972).

WIELGOLASKI, F. E., KJELVIK, S.: Mineral elements and energy of plants at Hardangervidda, Norway. In: BLISS, L. C., WIELGOLASKI, F. E. (Eds.): Primary production and production processes, Tundra Biome, pp. 231–238. Edmonton-Oslo: IBP Tundra Biome Steering Committee 1973a.

WIELGOLASKI, F. E., KJELVIK, S.: Production of plants (vascular plants and cryptogams) in alpine tundra, Hardangervidda, Norway. In: BLISS, L. C., WIELGOLASKI, F. E. (Eds.): Primary production and production processes, Tundra Biome, pp. 75–86. Edmonton-Oslo: IBP Tundra Biome Steering Committee 1973b.

Analysis of Precipitation at Fennoscandian Tundra Sites

P. Kallio and A. K. Veum

Nutrient input through precipitation is noticeable particularly in ombrotrophic habitats such as the oligotrophic IBP sites dominated by lichens at Hardangervidda, Norway and at Kevo in Finland; these sites are scarcely affected by minerals in ground water. The ombrotrophic nutrition is also typical of the Swedish IBP mire area at Stordalen (Sonesson et al., 1975). There are, however, also minerotrophic patches in the latter area, which are characterized by some indicator plants (Sonesson and Bergman, 1972). The micro- and macronutrients in precipitation restore a balance to the ecosystem, and affect the production and composition of vegetation. The macronutrients N, K, Ca, and Mg in precipitation have been analyzed for at all the IBP sites mentioned above. Further, phosphorus levels have been measured at Kevo and Stordalen.

In addition to nutrients, the aerosol contains different kinds of pollutants of which sulfur compounds are most important. These are transported especially from emission centres like towns or industrial areas (long distance emission), and only a small part of the sulfur as even Na, Mg, and Cl originates from the sea and other natural sources. Southern Fennoscandia, particularly, is affected by the East and West European pollution centres (Nordø, 1973; Ottar, 1973), and this fact must have an ecological importance now and in the future.

The sulfuric acid apparently has an effect particularly on lichen heaths, and owing to the low buffering capacity and low pH the leaching of nutrient minerals increases; on the other hand, this also has an immediate effect on lichens as an inhibitor. Therefore it is important that the sulfur content and the pH of precipitation have been measured at all the IBP tundra sites. The short term observations, however, do not provide good material for detailed comparison of the real differences, because there are large yearly as well as seasonal variations between the sites. The results received during the IBP period at the IBP tundra sites give only some rough trends. The national networks for observation and analysis of precipitation (e.g. meteorological stations and the activity coordinated by OECD), however, give much general information also concerning the large scale distribution of the solid components of the atmosphere.

Material

The precipitation was collected and analyzed at Hardangervidda at the IBP tundra lichen heath site during 1970–1972. The detailed technique has been reported by VEUM (1973).

The precipitation analyses at Kevo have been organized by the National Board of Waters. The rainwater was collected at the Kevo meteorological station, which is situated close to all the IBP sites. Results are available from the years 1971 and 1972, but only partly from the year 1973. A review of the methods and results from 1971 is given by HAAPALA (1972).

In Sweden, precipitation analyses were carried out from 1972, but in this paper only preliminary results based on extrapolated values from Abisko meteorological station from 1971 were used (RYDÉN, 1973).

Although the nutrient and pollutant content of snow may be high (VIRO, 1955), these have been omitted because comparable data do not exist from all stations.

Results and Discussion

The annual material from the three years at Hardangervidda is presented in Table 1. Values based on the precipitation which falls during the growth period is shown in Table 3. In these 3.5 months (WIELGOLASKI and KÄRENLAMPI, 1975) the amount of precipitation is about 250 mm, which is close to one third of the yearly value (SONESSON et al., 1975). In Table 2, monthly values from Kevo are given for the years 1971 and 1972 and the growth period of 1973.

In Table 3 also, the values from Kevo are given for the growth period June 1 st—September 15th (KÄRENLAMPI, 1972), and in this Table the three areas are roughly comparable.

It can be seen that all the values are highest at Hardangervidda. This agrees well with the fact that the content of solids in the air is dependent on the prevailing winds, the human activity and, particularly as regards sulfur, on the effect of the pollution centers further south (HAAPALA, 1973; NORDØ, 1973). Thus all the contents except calcium diminish from south to the north in Finland (BUCH,

Table 1. The results from precipitation analyses in the years 1970, 1971 and 1972 at Hardangervidda, Norway (annual values)

Year	SO_4–S	Cl	Na	K	Ca	Mg	NO_3–N	NH_3–N	pH
				$mg\ m^{-2}$					
1970			336	658	176	62			5.5
1971	562		616	686	532	140			5.3
1972	700		146	420	78	42			4.9
$\bar{x}$	630	623[a]	366	588	260	84	70[a]	56[a]	5.2

[a] KJELVIK and WIELGOLASKI, pers. comm.

Table 2. Results from the analyses of precipitation at Kevo. Monthly results given for the year 1971 (upper) and 1972 (lower). In the months 6–9 the third value is from the year 1973

Month	Precipitation mm	SO_4	Cl	Org. C	Na	K	Ca	Mg	Tot.-N	NO_3–N	NH_4–N	Tot.-P	pH
		mg m^{-2}											
1	26.5	—	—	—	—	—	—	—	—	—	—	—	—
	6.2	—	—	670	—	—	—	—	—	—	—	—	6.7
2	12.7	—	—	584	10	20	27	3.8	—	2.3	—	—	5.6
	4.8	—	—	209	14	—	—	—	3	2.2	—	—	6.8
3	12.4	97	—	198	—	—	—	—	15	—	1.6	0.4	4.0
	31.7	124	222	951	95	32	143	19.0	10	7.0	1.3	0.9	5.9
4	6.2	17	56	174	8	9	14	6.2	2	0.3	0.1	0.1	5.3
	20.3	81	32	244	16	8	8	2.6	8	4.7	0.2	0.3	5.5
5	2.8	—	—	—	13	—	—	—	3	1.2	—	—	5.4
	11.6	—	43	244	19	6	23	2.9	6	2.0	0.3	0.2	6.1
6	9.2	74	48	—	26	—	17	3.7	6	2.3	—	—	6.4
	25.3	94	86	405	10	23	13	3.3	6	0.3	0.3	2.0	4.1
	56.9	80	17	57	6	17	23	5.7	16	4.3	3.1	1.0	4.6
7	70.7	141	42	537	28	57	28	4.2	34	20.0	0.8	—	—
	118.9	309	83	238	6	12	12	7.1	33	10.7	4.8	2.4	5.5
	30.9	56	28	93	6	9	3	3.0	6	3.9	4.6	0.7	4.5
8	120.9	157	48	242	25	12	12	7.3	30	10.0	6.0	1.5	6.0
	47.3	114	132	—	5	5	24	2.8	19	5.2	5.7	0.2	4.5
	39.1	90	51	117	23	12	12	5.1	11	3.9	3.7	0.3	4.8
9	45.7	64	55	137	27	5	5	2.7	218	2.3	0.9	0.7	6.4
	41.9	88	46	—	8	13	13	2.5	23	1.7	0.4	0.4	5.7
	26.6	40	19	186	11	16	3	2.7	5	0.1	0.3	0.4	5.0
10	26.6	40	40	452	16	11	11	1.6	17	1.3	0.3	0.3	6.3
	37.9	57	—	76	19	11	30	4.9	16	4.2	1.1	2.0	5.8
11	21.5	—	9	—	—	—	—	—	12	3.7	—	0.2	6.6
	34.2	41	51	68	14	2	3	0.0	8	3.4	1.0	0.1	4.7
12	19.0	89	72	66	65	—	—	—	9	3.6	2.7	3.6	6.5
	24.2	44	—	48	29	2	17	4.8	8	6.1	0.6	0.4	—

Table 2 (continued). Annual averages for the years 1971 and 1972 n = sample numbers

Month	Precipitation mm	SO_4	Cl	Org. C	Na	K	Ca	Mg	Tot.-N	NO_3–N	NH_4–N	Tot.-P	pH
						mg m^{-2}							
$\bar{x}$ (1971)		85	46	299	24	19	16	4.2	15	4.7	1.9	1.0	5.8
$\bar{x}$ (1972)		106	67	315	21	11	29	5.0	13	4.3	1.5	0.9	5.6
n (1971)	12	8	8	8	9	6	7	7	10	10	7	8	10
n (1972)	12	9	8	10	11	10	10	10	11	11	10	10	11

Table 3. Chemical analyses of precipitation from Hardangervidda (Norway), Kevo (Finland) and Abisko (Sweden) in mg m^{-2} of the growing season

	SO_4–S	Cl	Org.-C	Na	K	Ca	Mg	Tot.-N	NO_3–N	NH_4–N	Tot.-P
					mg m^{-2}						
Hardangervidda											
1970–1972 growth period	225	223[a]		131	210	93	30		25[a]	20[a]	
Kevo											
1971–1973 growth period	145	218	671	60	60	55	17	136	22	10	3
Abisko											
Approx. value of the growth period (annual values)	80 (160)			20 (40)	25 (50)	80 (160)	15 (30)	20 (40)			

[a] KJELVIK and WIELGOLASKI, (pers. comm.)

1960). The same direction of gradient is clearly seen also in pH in Fennoscandia (ODÉN, 1968). The isolines of minerals and pH of air in Scandinavia follow the direction of the seashore and the gradient from west to east. The differences seen in Table 3 between Stordalen and Kevo may be based on the limited amount of material.

Annal total input of N through precipitation does not vary much at KEVO. These values are 3–4 times lower than the average in Finland (HAAPALA, 1972). When the biological input of N by nitrogen fixing bacteria at Kevo is compared with the input through precipitation, the latter is much higher. At Stordalen, on the other hand, with higher activity of nitrogen fixing microorganisms and lower total N values in the air, the role of these two N sources is similar (GRANHALL and LID-TORSVIK, 1975). When, however, the total "biogenic" nitrogen input is compared with that added through precipitation, the latter is lower (KALLIO and KALLIO, 1975). Very roughly, however, the two sources of nitrogen input in the Fennoscandian tundra ecosystems are comparable.

The amount of P input through precipitation (3 mg/m^2/growth period) at Kevo is approximately one third of that in S. Finland (HAAPALA, 1972).

A typical difference is seen in the sulfur amounts. Although there are variations between years and seasons, the differences agree well with the data report e.g. by BUCH (1960), HAAPALA (1972) and OTTAR (1973). The isopleths indicating the limits of areas of the same degree of pollution go largely from SW to NE in Scandinavia, and Hardangervidda thus has a higher pollution than Stordalen and Kevo. Hardangervidda is at present clearly above the isopleths of 500 mg SO^4 m^{-2} $year^{-1}$, while the other two stations are north of the 400th isopleth. On the question of whether there are any biological differences between the two areas which could be explained by the differences in sulfur content of the air, the best indicator might be the nitrogenase (nitrogen content) activity in lichens, which is very sensitive to air pollution (apparently sulfur) (KALLIO and VARHEENMAA, 1974). It seems that a real difference exists between the nitrogen fixation results from Kevo and Hardangervidda. Only one lichen, however, *Stereocaulon paschale*, is comparable. This lichen shows a much lower activity at Hardangervidda than at Kevo (HANSSEN et al., 1973; KALLIO and KALLIO, 1975). This needs more detailed study. The pH values of precipitation at Hardangervidda and Kevo, 5.3 and ca. 5.7 respectively, seem to show a real difference. Stordalen and Kevo represent a more "undisturbed" area than Hardangervidda, which has a stronger influence on the large scale transport of pollutants.

References

BUCH, K.: Zusammensetzung des atmosphärischen Niederschlages in Finnland. Soc. Sci. Fenn. Comment. Physico-Mathematiae **24**, 1–16 (1960).

GRANHALL, LID-TORSVIK, V.: Nitrogen fixation by bacteria and blue-green algae in tundra areas. In: WIELGOLASKI, F. E. (Ed.): Fennoscandian tundra ecosystems, Part 1: Plants and microorganisms, pp. 305–315. Berlin-Heidelberg-New York: Springer 1975.

HAAPALA, K.: Sadeveden laatu Suomessa vuonna 1971 (english summary: The quality of rainwater in Finland according to observations made during 1971). Report 26. Helsinki: National Board of Waters 1972.

HANSSEN, J. F., LID-TORSVIK, V., TORSVIK, T.: Microbiology. In: IBP i Norge. Årsrapport 1972. pp. 75–93. Norsk IBP 1973.

KALLIO, S., KALLIO, P.: Nitrogen fixation in lichens at KEVO, North Finland. In: WIELGOLASKI, F. E. (Ed.): Fennoscandian tundra ecosystems, Part 1: Plants and microorganisms, pp. 292–304. Berlin-Heidelberg-New York: Springer 1975.

KALLIO, S., VARHEENMAA, T.: On the effect of air pollution on nitrogen fixation in lichens. Rep. Kevo Subarctic Res. Stat. **11**, 42–46 (1974).

KÄRENLAMPI, L.: Comparison between the microclimates of the Kevo ecosystem study sites and the Kevo meteorological station. Rep. Kevo Subarctic Res. Stat. **9**, 50–65 (1972).

NORDØ, F.: Meso-scale and large-scale transport of air pollutants. In: Proc. of the third international clean air congress. **B**, pp. 105–108. Düsseldorf 1973.

ODÉN, S.: Nederbördens och luftens försurning, des orsaker, förlopp och värkan i olika miljöer. Bull. **1**. Stockholm: Ekologikommitten 1968.

OTTAR, B.: The long range transport of air pollutants. In: Proc. of the third international clean air congress. **B**, pp. 102–104. Düsseldorf 1973.

RYDÉN, B. E.: Abiotic studies, Stordalen 1972. Progress Report 1972. Swedish IBP Tundra Biome Project Tech. Rep. **14**, 14–25 (1973).

SONESSON, M., BERGMAN, H.: Phytomass changes between two samplings at Stordalen 1970. Swedish IBP Tundra Biome Project Tech. Rep. **2** (1972).

SONESSON, M., WIELGOLASKI, F., KALLIO, P.: Introduction to the Fennoscandian IBP tundra sites. In: WIELGOLASKI, F. E. (Ed.): Fennoscandian tundra ecosystems, Part 1: Plants and microorganisms, pp. 3–28. Berlin-Heidelberg-New York: Springer 1975.

VEUM, A. K.: Jordundersøkelser og nedbrytning av organisk materiale. In: IBP i Norge. Årsrapport 1972, pp. 29–74. Oslo: Norsk IBP 1973.

VIRO, P.: Loss of nutrient and the nutrient balance of the soil in Finland. Comm. Inst. Forest Fenn. **41**, 128 (1953).

VIRO, P.: Loss of nutrient and the natural nutrient balance of the soil in Finland. Comm. Inst. Forest. Fenn. **42**, 1–45 (1955).

WIELGOLASKI, F. E., KÄRENLAMPI, L.: Plant phenology of Fennoscandian tundra areas. In: WIELGOLASKI, F. E. (Ed.): Fennoscandian tundra ecosystems, Part 1: Plants and microorganisms, pp. 94–102. Berlin-Heidelberg-New York: Springer 1975.

Organic Compounds in Alpine Plants

O. Skre, A. Berg, and F. E. Wielgolaski

Analyses of e.g. carbohydrate content of plants taken at various times throughout the season are important for the study of photosynthesis and primary production. Studies of the amounts of reserve carbon and protein in various plant compartments may increase the knowledge of translocation and growth, while the cellulose and lignin content, especially of cell wall tissues, gives information about the decomposition process, as well as plant growth. The present study was done on various plant species and groups grown under natural conditions at one subalpine and five alpine sites at Hardangervidda, southern Norway. The sites have been described by Sonesson et al. (1975), the plant community by Lye (1975), the geology and soil by Hinneri et al. (1975) and the weather conditions by Skartveit et al. (1975).

Material and Methods

Sampling of the field layer was carried out according to Wielgolaski and Kjelvik (1972), while samples of the shrub and tree layers were taken by cutting twigs at random throughout the summer. At four three weekly intervals throughout the growing season 1972, which lasted 100–130 days, material (10 replicates) was collected. For phenology see Wielgolaski and Kärenlampi (1975). The first sampling was done at the beginning of July. A fifth collection was made shortly after budbreak the following year, 1973. Green tissue (leaves and young stems) and non-green tissue (woody stems, rhizomes, and roots) were separated. The dominant species on each site were studied separately (Skre, 1975), otherwise the plants were divided into woody, forbs, monocotyledons, mosses and lichens in proportion to their field occurrence, (Wielgolaski, 1975; Kjelvik and Kärenlampi, 1975). Subsamples of blended fresh plant material (two parallels) were extracted with 70% ethanol (Thomas et al., 1949) to determine the monosaccharides. Oligosaccharides were determined according to Bell (1955) after mild hydrolysis (0.15 N H_2SO_4, 15 min, 100° C). The storage polysaccharides (starch, fructosans, xylans, glucosides etc.) were found in the residue after strong hydrolysis (Webster et al., 1963; Pirt and Whelan, 1951) (2 N H_2SO_4, 1 hr, 100° C). The total amounts of sugar in the three solutions were found (as glucose) by copper reduction test (Nelson, 1944). The relative amounts of different sugars were found by paper chromatography (Block, Durrum and Zweig, 1955) in n-butanol:pyridin:water (5:3:2) and the spots exposed by p-anisidine · HCL (HOUGH et al., 1950). Some samples were also chromatographed by electrophoresis in borate buffer solutions to determine mannose (RIBEIRO et al., 1961). For more details see SKRE (1975).

The method used in cellulose and lignin determination is modified after JERMYN (1955). Dried and ground plant material were extracted for about 20 hrs in 90% ethanol (Soxhlet appatatus) until no sugar could be detected in the anthrone test (DREYWOOD, 1946) by further extraction. The residue was then air-dried and extracted for another 20 hrs in water. The compounds removed by ethanol and water are mainly sugar, starch, pectins, lipids and some proteins. The lignins were removed by heating in water (4 hrs, 75° C; 1.5 g $NaClO_2$ and 10 drops of acetic acid added per hour). The remaining holocellulose in the filter was dried at 85° C after careful washing and then weighed. In addition to the lignins, the water-insoluble proteins were removed by the last extraction and the nitrogen content was therefore determined in the residue before delignification by the Kjeldahl method, giving the protein content by multiplying with the factor 6.25. The values of all compounds studied are given as percentage of plant dry weight.

Results and Discussion

Chromatography. The chromatograms indicated glucose, fructose and sucrose in the ethanol soluble fraction of all samples. Fructose seemed to be present in greater amounts than glucose (Table 1). In addition glyceraldehyde was found in roots of *Salix lapponum*, galacturonic acid in leaves of *Silene acaulis*, galactose in leaves of *Salix reticulata* and arabinose and rhamnose in *Melampyrum pratense*. Mannose was not found (CLEMENTS, 1932). Extracts from the latest harvesting contained a little more glucose relative to fructose than in the mid growing season. In *Empetrum*-leaves an oligosaccharide was found which by hydrolysis yielded equal amounts of glucose, arabinose and xylose. Sucrose was the dominating oligosaccharide in all samples.

In the storage polysaccharide fraction varying amounts of glucose, arabinose and xylose were found in the hydrolysate of nearly all samples (Table 1). Less common were fructose and galactose, the latter only in extracts from the latest harvesting of forb roots, monocotyledons, and in lichens. Xylose was the dominating sugar in the fraction of non-green tissue from woody plants (GERHARDT, 1929) and monocotyledons, arabinose dominated often in leaves from woody plants, and glucose in forb roots. The glucose seemed to come from a water-soluble polysaccharide (starch), while the arabinose and xylose came from water-insoluble components (probably xylans). Bryophytes and lichens contained more glucose and fructose and less arabinose in the hydrolysate of the storage polysaccharide fraction than vascular plants. WAKSMAN and STEVENS (1928) report sucrose and fructosans as the most common ethanol-soluble sugars in *Sphagnum*, while galactose, glucose and fructose occurred after hydrolysis. The analysis seemed to point to xylans as more common storage polysaccharides than starch in woody plants and monocotyledons (MURNEEK, 1929).

Table 1. Percentage sugar as monosaccarides in the hexose fraction of ethanol-soluble carbohydrates (A) and in the hydrolysate from the storage polysaccaride fraction (B). Approximate values. LH = Lichen heath, WT = Willow thicket, BF = Birch forest, WM = Wet meadow, Glc = glucose, Frc = fructose, Gal = galactose, Arab = arabinose, Xyl = xylose

	Site	A		B				
		Glc	Frc	Gal	Glc	Frc	Arab	Xyl
Woody green	LH	28	72		7		13	79
	WT	37	63		18		72	10
	BF	28	72		27		22	51
	WM	36	64		18		72	10
Woody non-green	LH	16	84		21		13	66
	WT	68	32		21		39	40
	BF	34	66		42		12	46
	WM	69	31		25		37	38
Forbs green	DM	34	66		48	4	29	19
	WT	48	52		20		43	37
	WM	35	65		31		44	25
Forbs non-green	DM	60	40		70	4	18	8
	WT	45	55		63		12	25
	WM	46	54		48	1	26	25
Monocot. green	DM	13	87		15	4	30	51
	WT	9	91		29		31	40
	BF	10	90				30	70
	WM	55	45		6		34	60
Monocot. non-green	DM	28	72		26	1	13	60
	WT	11	89		7		21	72
	BF	16	84		36		17	47
	WM	54	46		23		20	57
Bryophytes	WT	40	60		—	—	—	—
	BF	15	85		50	28	22	
	WM	40	60		50	50		
Lichens	LH	20	80	28	38	15	15	10

Seasonal Variation. Average values and seasonal range for the different groups of organic compounds studied are given in Table 2. Seasonal variations show quite different patterns from one plant group to another, as shown in Fig. 1, where some of the results are plotted on a time scale. For more detailed information see BERG et al. (1973) and SKRE (1975). The variation pattern within each plant group was quite similar from one site to another, the only differences being the seasonal amplitude (Table 2) and small time deviations in spring. Smallest variations through the season were found in lichens from the lichen heath and bryophytes from the wet meadow.

A general increase in the lignin content towards the end of the growing season was found in cryptogams and all groups of vascular plants in green as well as in non-green material. In green parts of monocotyledons a peak was also found in early August. The lignification of plants in fall is in agreement with e.g. PHILIPS

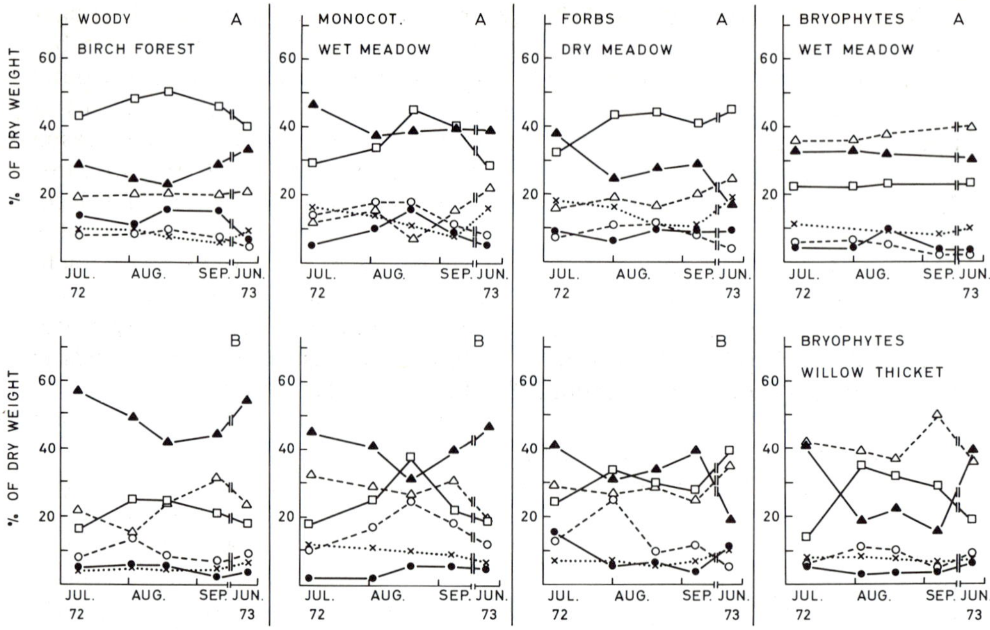

Fig. 1A and B. Seasonal variations in the percentage of different organic compounds in green (A) and non-green (B) tissue. Open squares = ethanol-soluble compounds. Filled triangles = holocellulose. Open triangles = lignin. Filled circles = ethanol-soluble sugars. Open circles = starch and hemicellulose. Crosses = protein

and GOSS (1935) and KOSTRUBIN (1956). Particularly strong lignin increase throughout the season was found in forbs at the willow thicket site.

The holocellulose content generally decreased during the growing season in all plant groups and compartments. Strongest reduction was found in July, while the content increased somewhat again in the fall in some cases (Fig. 1). Usually the amount of holocellulose varied in inverse proportion to the amount of hydrolyzable storage polysaccharides. Minimum amounts of holocellulose were often found in late July—beginning of August. This was commonly the time of maximum storage polysaccharide content in green as well as in non-green tissue. This high storage polysaccharide content may be seen as a result of high temperatures (SKARTVEIT et al., 1975) and high net assimilation rates during the flowering stage in July, as suggested by RUSSEL (1948).

During the first part of August the amount of storage polysaccharides in non-green tissue frequently decreased, and relatively more of the carbohydrates were found as ethanol-soluble sugars. This is perhaps again a temperature reaction. The first half of August was relatively wet and cold (SKARTVEIT, pers. comm.), causing decreased rates of photosynthesis. Energy requirements are, however, high in plants during the period of flowering and fruit setting (WIELGOLASKI and KÄRENLAMPI, 1975) as reported by RUSSEL (1940) and SVOBODA (1972), and part of the reserves might be transformed for use in metabolism.

MUC (1972) and SVOBODA (1972) generally found the highest soluble sugar contents of green tissue in late July—mid August in the Canadian high arctic, but the time of the peak varied from year to year. MCCOWN and TIESZEN (1972) found a relatively low polysaccharide content in leaves of arctic monocotyledons at peak season and a relatively constant amount of sugar throughout the season which was, however, somewhat lower than polysaccharides. In addition to transformation from e.g. starch, caused by low temperatures, the soluble sugar accumulation in leaves frequently found in the middle of August in the present studies (Fig. 1) may be caused by reduced night transport to other parts of the plant (LASCOMBES, 1952). The maximum values for the total ethanol-water-soluble compounds (including pectic and lipic compounds) were also found normally in August, while MCCOWN and TIESZEN (1972) found the highest content of lipids in arctic monocotyledons (up to about 20% of dry weight) in spring.

Ageing generally causes sucrose formation and transport out of the leaves, and therefore the easy soluble sugar content in green tissue often showed a decreasing tendency in fall. The increase in storage polysaccharides was, however, generally smaller than the decrease in the soluble sugar, indicating losses by respiration and leaching or consumption for e.g. root growth. The root growth hypothesis is supported by the fact that the general decrease in storage polysaccharide content in late season seemed to be most striking at the wet meadow, where root growth was most pronounced (WIELGOLASKI and KJELVIK, 1973). The relative increase in the holocellulose content, which takes place simultaneously, may also indicate root growth. In some forb species, particularly at the dry meadow and the willow thicket, the storage polysaccharide content increased strongly in roots at the last harvest in fall, indicating transport from leaves (Fig. 1), as reported for grass species by e.g. KOSTRUBIN (1956) and BLAKE and RICHARDS (1970). The sugar content in birch roots decreased in fall, while it increased in young stems. This may mean that stems are more active storage organs than roots (SVOBODA, 1972; GIBBS, 1940).

During the winter season a change to mono- and oligosaccharides generally takes place in the roots (CHEYMOL, 1931), which has the ecological significance of protection against frost damage (e.g. SMYTH, 1938). In e.g. forb roots (Fig. 1) and non-green compartments of *Salix lapponum*, relatively high amounts of ethanol-soluble sugars are thus found in the early June harvesting, probably also indicating mobilization for translocation and growth. A remarkable feature is a relative increase in the content of both ethanol-soluble sugars and storage polysaccharides of evergreen leaves (*Empetrum hermaphroditum* and *Vaccinium vitis-idaea*) during the winter months at the lichen heath (SKRE, 1975), which is also reported in *Hedera* and *Ilex* leaves (PREISING, 1930). The evergreen shrubs in the present investigations were found to have a late budbreak (WIELGOLASKI and KÄRENLAMPI, 1975), and the depletion in storage polysaccharides may then take place somewhat later in the season. For other plant groups (e.g. monocotyledons and forbs from the dry meadow) growth initiation took place earlier, and minimum amounts of storage polysaccharides, particularly in non-green compartments, and also often soluble sugars in green material were found in June. This normal carbohydrate depletion in spring is reported by several authors, in tundra areas e.g. by MOONEY and BILLINGS (1960), FONDA and BLISS (1966) and MUC (1972). In

Table 2. Mean values $\bar{x}$ and seasonal amplitude Δx of some organic compounds measured as percentages of dry weight for different groups of species and sites. The fraction "soluble in ethanol and water" constitutes e.g. of "ethanol soluble sugars", partly also of "polysaccharides" and "proteins" (see under methods)

	Plant group	Green material											
		Lignin		Holocellulose		Solub. in ethanol & water		Ethanol soluble sugar		Storage polysaccharide		Proteins	
		$\bar{x}$	Δx	$\bar{x}$	Δx	$\bar{x}$	Δx	$\bar{x}$	Δx	$\bar{x}$	Δx	$\bar{x}$	Δx
Birch forest	Trees and shrubs	14	7–18	32	26–35	44	39–52	13	8–15	6	5–7	11	9–14
	Dwarf shrubs	20	19–20	28	25–33	45	40–50	11	4–15	8	7–10	8	7–10
	Monocotyledons	11	5–18	45	41–55	38	30–45	12	6–17	13	8–17	8	6–16
	Bryophytes	35	23–46	33	24–43	28	19–36	4	1–8	8	3–15	7	6–8
Willow thicket	Shrubs	17	8–23	28	20–39	40	32–47	12	4–22	8	6–13	18	13–24
	Monocotyledons	13	6–20	38	30–48	41	30–55	13	6–18	14	10–20	11	7–14
	Forbs	20	14–26	22	17–29	48	43–51	14	11–17	10	5–16	13	8–20
	Bryophytes	41	36–51	28	16–41	26	14–35	5	3–7	7	5–12	8	7–8
Lichen heath	Dwarf shrubs	21	16–26	26	23–31	49	38–54	13	10–17	10	5–19	7	6–7
	Monocotyledons[a]	19	19–19	44	37–51	30	23–37	10	7–16	17	11–20	10	9–12
	Lichen[e]	15	10–21	68	60–73	14	12–17	0.6	0.4–0.8	13	10–15	4	3–4
Dry meadow	Dwarf shrubs	24	18–29	27	23–32	38	34–41	14	11–18	7	4–9	13	11–16
	Monocotyledons	13	7–27	39	32–46	38	18–47	11	6–15	12	7–18	14	8–28
	Forbs	19	17–25	27	18–38	41	32–47	8	5–9	8	4–12	15	11–18
	Bryophytes	35	30–37	37	32–42	20	19–22	2	1–4	11	6–18[b]	12	12–12
Wet meadow	Dwarf shrubs	26	18–32	21	15–25	39	32–45	13	10–17	8	6–15	17	12–25
	Monocotyledons	15	7–23	41	38–48	36	28–46	9	5–15	14	8–18	13	8–16
	Forbs	19	17–22	23	17–32	44	33–50	13	9–18	8	4–12	18	12–22
	Bryophytes	38	36–40	32	31–33	23	22–23	5	3–10	5	3–7	10	8–11
Snow bed	Dwarf shrubs[d]	27	25–32	25	23–27	36	31–41	12	8–17	8	6–12	11	10–14
		Non-green material											
Birch forest	Trees and shrubs	19	10–26	57	46–71	20	16–25	7	5–10	9	7–14	4	3–6
	Dwarf shrubs	24	15–31	49	42–57	22	17–25	4	3–5	9	6–15	4	4–5
	Monocotyledons	26	23–32	47	34–56	24	17–32	5	3–8	12	5–21	6	5–9
Willow thicket	Shrubs[c]	24	22–26	41	36–46	31	28–34	8	5–12	8	6.0–10.0	5	5–5
	Monocotyledons	29	27–30	41	32–51	26	17–33	8	5–15	16	12–23.5	5	5–6
	Forbs	37	28–49	21	17–31	40	29–50	18	11–23	15.5	7.0–26.0	5	4–6
Lichen heath	Dwarf shrubs	28	21–34	49	44–59	19	11–27	4	2–8	6	1–9	4	3–5
	Monocotyledons	25	21–29	50	41–63	23	13–30	5	3–6	14	11–20	4	3–5
Dry meadow	Dwarf shrubs	26	21–34	40	28–57	30	19–34	4	3–6	8	6–10	5	4–6
	Monocotyledons	24	21–29	45	43–56	25	15–30	5	2–8	13	10–20	7	5–9
	Forbs	29	25–36	33	20–41	25	11–34	7	5–10	12	6–26	7	6–11
Wet meadow	Dwarf shrubs	32	26–38	34	27–41	29	23–33	8	3–13	13	4–22	5	5–7
	Monocotyledons	28	19–32	41	32–48	25	18–39	4	2–6	16	10–25	6	5–7
	Forbs	34	28–41	22	14–30	39	32–44	10	6–12	12	6–15	8	6–11
Snow bed	Dwarf shrubs[d]	29	25–31	28	21–32	40	36–46	9	7–10	11	6–16	5	4–6

[a] Only two samples of green tissue (July 30th 1972 and June 23rd 1973).
[b] Only two samples (July 5th and August 1st 1972).
[c] Only two samples of non-green tissue (August 18th 1972 and June 23rd 1973).
[d] No June sampling was taken in 1973.
[e] Lichenin and chitin are part of the holocellulose fraction.

the early birch forest (bud break at an average on May 30th according to WIELGOLASKI and KÄRENLAMPI, 1975), much of the depletion had possibly taken place before the June harvest and may be one reason why the reserve carbon values in non-green parts at this site was not lower than observed in the fall either in monocotyledons or woody plants.

The protein content showed a reduction tendency throughout the season. Shortly after growth initiation remarkably high protein values were obtained in green material, especially in monocotyledons and forbs. The seasonal fluctuation in the protein content was, however, small in non-green material of vascular plants and in cryptogams, particularly small in lichens (Table 2).

Site Variation. The site-variation in plant lignin content was relatively small, but there seemed to be a weak tendency to lower lignin content in the subalpine birch forest than in plants at the alpine sites. This tendency seemed to hold true both for green and non-green compartments, although the lowest lignin content was always found in green material. The highest lignin contents were measured at the wet eutrophic sites, and particularly in non-green material at the willow thicket (Table 2). The amount of holocellulose in plants seemed to be more dependent on soil moisture conditions than on climate. Highest values, particularly in the non-green vascular plant parts, were generally found at the driest sites, lichen heath, birch forest and dry meadow, and usually the values were higher in non-green than in green material. In bryophytes, greater seasonal variations in holocellulose took place at the willow thicket *(Polytrichum)* and the birch forest sites than in the wet meadow (Table 2). In forbs the holocellulose values were lower at the willow thicket than in the dry meadow. On the other hand, the highest relative amounts of ethanol-water-soluble compounds were found in the willow thicket plants, and as expected were higher in green than in non-green tissue.

The relative amounts of storage polysaccharides and total sugar content (storage polysaccharides and ethanol-soluble sugars) were highest at the lowest elevated site (the birch forest), e.g. in *Deschampsia flexuosa* and ericaceous plants (BERG et al., 1973), probably due to higher temperatures and net assimilation. This is in accordance with results found by MOONEY and BILLINGS (1965). SVOBODA (1972) suggests a similar trend toward lower reserve carbon in plants with increase in latitude. The water and nutrient level in soil also seemed to play an important role, giving higher total sugar content in plants from the eutrophic wet willow thicket than from the other sites at 1200–1250 m elevation, e.g. *Salix lapponum*, *Anthoxanthum alpinum*, *Alchemilla vulgaris*, and *Ranunculus acris* (BERG et al., 1973). The values for e.g. leaves of *Carex bigelowii* were found to be higher at the mesotrophic dry meadow site than at the very dry oligotrophic lichen heath site (maximum 37% and 31%, respectively). Also in bryophytes a high storage polysaccharide content was found at the subalpine forest site and at the willow thicket site compared to the alpine wet meadow (Table 2). The level of mono- and oligosaccharides in forb roots from the eutrophic wet willow thicket was considerably higher than from the mesotrophic dry meadow, again indicating a nutrient and water influence. When various species from different sites were compared in the same phenophase (SKRE, 1975), higher amounts of mono- and oligosaccharides were usually found (especially in green tissue) at the higher elevated sites with lowest temperature, in accordance with EAGLES (1967). This was found e.g. for green parts of *Vaccinium uliginosum* at the alpine lichen heath compared to the subalpine birch forest, and for both leaves and roots of *Deschampsia flexuosa* in the alpine dry meadow compared to the birch forest. In the dry meadow the sum of mono- and oligosaccharides of green parts of the species was much higher than the amount of polysaccharides. Generally, however, the amount of ethanol-soluble sugars is of the same order as storage polysaccharides at all the sites in the present study (Table 2). This is in contrast to measurements by MUC (1972) and SVOBODA (1972) in monocotyledons and cushion plants in the Canadian arctic. They found highest oligosaccharide, smaller polysaccharide and nearly no monosaccharide content in the plants, probably because of the low temperatures.

Plant Group and Compartment Variation. The lignin content was generally much lower in green tissue of monocotyledons than for the other plant groups studied (Table 2). The values for grasses and sedges in the present studies were close to those found for Montana grass species by PATTON and GIESEKER (1942). In all groups of vascular plants, and particularly the herbaceous groups, lower lignin content was found in green than in non-green material. Particularly high lignin content was observed in green bryophytes, considerably higher than the 2–14% lignin found by BLACK et al. (1955) in *Sphagnum* species.

The holocellulose values were considerably higher for green monocotyledons than other vascular plants (Table 2), and also higher than the 20–30% reported for forage grass species (BUSTON, 1934; STODDARD, 1945). The difference in holocellulose content between green and non-green tissue seemed to be most pronounced in woody plants with highest content in the non-green compartments. The amount of holocellulose in non-green tissue of woody plants was of the same order as the 40–50% reported by ROSTOVTSEV (1964) e.g. in *Salix*. The amount of holocellulose in bryophytes was higher than the 3–19% reported from *Sphagnum* species by BLACK et al. (1955). Still higher was the holocellulose fraction in lichens from the lichen heath. Most of this is probably chitin, a very resistant compound occuring in the cell wall of fungi, and lichenin, a resistant polysaccharide composed mainly of glucose units. Lichenins comprise 30–40% of the biomass in *Cetraria islandica* (DRAKE, 1943).

In the perennial green leaves of woody plants particularly from the lichen heath (*Empetrum hermaphroditum* and *Vaccinium vitis-idaea*), relatively high amounts of excess ethanol-water-soluble compounds in addition to mono- and disaccharides were found, indicating high content of lipids and pectins. The relative amounts of ethanol-soluble sugars in leaves of woody plants were found to be somewhat higher than the storage polysaccharide content, often higher than in other plant groups studied (Table 2). The mono- and oligosaccharides in monocotyledons are of the same order as the 3–14% found for forage grass species by BUSTON (1934), and MYHR and SÆBØ (1969). Somewhat higher values were in some cases measured in the monocotyledons in the present study, particularly in the willow thicket and the birch forest (Table 2). This may be caused by the dominance of *Deschampsia caespitosa* and *Deschampsia flexuosa* at these sites, the latter was found by SELSJORD (1968) to have a high content of ethanol-soluble sugars (up to 25%). The amount of mono- and oligosaccharides in green tissue of forbs is of the same order as e.g. found by RUSSEL (1940) in arctic *Oxyria digyna*. The maximum content of ethanol-soluble sugars in forb roots was somewhat higher than in non-green tissue from other vascular plants. In cryptogams, very low values were found for ethanol-soluble sugars, lowest of all in lichens (Table 2). Low content of reducing sugars was also found in *Sphagnum* by BLACK et al. (1955).

The storage polysaccharide content of green tissue was slightly higher in monocotyledons than in other vascular plants, and close to values given for starch and hemicellulose (15–20%) in forage grasses by BUSTON (1934) and by WEBSTER et al. (1963). The averages found in the forbs seemed to be lower than the starch content reported by WARREN WILSON (1954) for arctic *Oxyria*. The amount of storage polysaccharides found in forb roots in the present study was generally higher than in the tops (BERG et al., 1973), particularly in species with storage roots. This varied strongly, however, throughout the season (Table 2, Fig. 1). It was also generally higher than in roots of other vascular plants. Lichens seem to have a higher content of storage polysaccharides than bryophytes. SMITH (1962) reported 21–27% total sugar in the lichen *Peltigera*. PUEYO (1960) found that sugar alcohols may occur in the ethanol-soluble fraction in lichens. The seasonal average protein content in perennial green leaves of woody plants (ericaceous species dominating at the lichen heath and birch forest) was 6–8%, in good agreement with nitrogen data of ericaceous plants from subarctic Finland (WIELGOLASKI et al., 1975) and for arctic and alpine USSR populations of *Empetrum* and *Vaccinium* (MARCHENKO and KARLOV, 1962 in RODIN and BAZILEWICH, 1967; MANAKOV, 1972). Higher protein content was found in leaves of *Salix* species (dominating woody plants e.g. in the wet meadow), and the values were of the same order as those given by RODIN and BAZILEVICH (1967) for *Salix* leaves in the USSR tundra. The protein values calculated for green tissue of monocotyledons (Table 2) were similar to values reported by MYHR and SÆBØ (1969) and several other authors for forage grasses, and by SELSJORD (1968) in subalpine *Deschampsia flexuosa* and *Deschampsia caespitosa*. The protein content of non-green material was lower than of green tissue. In woody plants, somewhat higher nitrogen values were found in stems than in roots (WIELGOLASKI et al., 1975). Slightly higher protein values were found in the roots of monocotyledons than in other plants, and are in good agreement with nitrogen data given by RODIN and BAZILEVICH (1967) and MANAKOV (1972). The protein content of cryptogams was lower than in vascular plants, and a particularly low amount was found in the lichens (Table 2), which were all non-nitrogen fixing (WIELGOLASKI et al., 1975).

Acknowledgement. We wish to thank the technical assistant H. TØNNESEN for valuable help with the chemical analysis.

References

BELL, D. J.: Mono- and Oligosaccharide Derivatives. In: PAECH, K., TRACEY, M. V. (Eds.): Modern methods of plant analysis, Vol. II, pp. 1–63. Berlin-Göttingen-Heidelberg: Springer 1955.

BERG, A., SKRE, O., WIELGOLASKI, F. E., KJELVIK, S.: Leaf areas and angles, chlorophyll and reserve carbon in alpine and subalpine plant communities, Hardangervidda, Norway. In: BLISS, L. C., WIELGOLASKI, F. E. (Eds.): Primary production and production processes,

Tundra Biome, pp. 239–254. Edmonton-Oslo: IBP Tundra Biome Steering Committee 1973.

BLACK, W. A. P., CORNHILL, W. J., WOODWARD, F. N.: A preliminary investigation on the chemical composition of *Sphagnum* moss and peat. J. Appl. Chem. **5**, 484–492 (1955).

BLAKE, J. D., RICHARDS, G. N.: Polysaccharides of tropical pasture herbage I. Studies on the distribution of the major polysaccharide components of speargrass (*Heteropogon contortus*) during growth. Australian J. Chem. **23**, 2353–2360 (1970).

BLOCK, P. J., DURRUM, E. L., ZWEIG, G.: Paper chromatography and paper electrophoresis. New York: Academic Press 1955.

BUSTON, H. W.: The polyuronide constituents of forage grasses. Biochem. J. **28**, 1028–1037 (1934).

CHEYMOL, J.: Variation dans la composition glucidique de la racine de benoite (*Geum urbanum* L.), au cour de la végétation d'une année. Bull. Soc. Chim. Biol. **13**, 470–476 (1931).

CLEMENTS, H. F.: Mannose and the first sugar of photosynthesis. Plant Physiol. **7**, 547–550 (1932).

DRAKE, B.: Some polyglucosides of lichens, particularly lichenin and the newly discovered pustulin. Biochem. Z. **313**, 388–399 (1943).

DREYWOOD, R.: Qualitative test for carbohydrate material. Ind. Eng. Chem. Anal. Ed. **18**, 499 (1946).

EAGLES, C. F.: Variation in the soluble carbohydrate content of climatic races of *Dactylis glomerata* at different temperatures. Ann. Botany **31**, 645–651 (1967).

FONDA, R. W., BLISS, L. C.: Annual carbohydrate cycle of alpine plants on Mt. Washington, N. H. Bull. Torrey Botan. Club **93**, 268–277 (1966).

GERHARDT, F.: Effect of acid and alkaline hydrolysis on the estimation of hemicellulose and associated groups in young apple wood. Plant Physiol. **4**, 373–383 (1929).

GIBBS, R. D.: Studies in tree physiology II. Seasonal changes in the food reserves of field birch (*Betula populifolia* Marsh.) Can. J. Res. Sect. C. Bot. Sci. **18**, 1–9 (1940).

HINNERI, S., SONESSON, M., VEUM, A. K.: Soils of Fennoscandian IBP tundra ecosystems. In: WIELGOLASKI, F. E. (Ed.): Fennoscandian tundra ecosystems, Part 1: Plants and microorganisms, pp. 31–40. Berlin-Heidelberg-New York: Springer 1975.

HOUGH, L., JONES, J. K. N., WADMAN, W. H.: Quantitative analysis of mixtures of sugars by the method of partition chromatography V. Improved methods for the separation and detection of their methylated derivatives on the paper chromatogram. J. Chem. Soc. 1702–1706 (1950).

JERMYN, M. A.: Cellulose and hemicellulose. In: PAECH, K., TRACEY, M. V. (Eds.): Modern methods of plant analysis, Vol. II, pp. 197–225. Berlin-Göttingen-Heidelberg: Springer 1955.

KJELVIK, S., KÄRENLAMPI, L.: Plant biomass and primary production of Fennoscandian subartic and subalpine forests and of alpine willow and heath ecosystems. In: WIELGOLASKI, F. E. (Ed.): Fennoscandian tundra ecosystems, Part 1: Plants and microorganisms, pp. 111–120. Berlin-Heidelberg-New York: Springer 1975.

KOSTRUBIN, M. V.: The production and transformation of hemicellulose in wheat stalks. Biokhimia **20**, 360–367 (1956). In Russian.

LASCOMBES, G.: Action des variations quotidiennes de température sur le métabolisme de la betterave cultivée a haute altitude. Compt. Rend. **235**, 894–896 (1952).

LYE, K. A.: Survey of the main plant communities on Hardangervidda. In: WIELGOLASKI, F. E. (Ed.): Fennoscandian tundra ecosystems, Part 1: Plants and microorganisms, pp. 68–73. Berlin-Heidelberg-New York: Springer 1975.

MANAKOV, K. N.: The productivity and biological turnover in the tundra biogeocoenoses in the Kola peninsula. Leningrad: Nauka 1972. In Russian.

MCCOWN, B. H., TIESZEN, L. L.: A comparative investigation of periodic trends in carbohydrate and lipid levels in arctic and alpine plants. In: BOWEN, S. (Ed.): Proceedings 1972 US Tundra Biome Symposium, pp. 40–45. Hannover, New Hampshire: USA CRREL 1972.

MOONEY, H. A., BILLINGS, W. D.: The annual carbohydrate cycle of alpine plants as related to growth. Am. J. Botany **47**, 594–598 (1960).

MOONEY, H. A., BILLINGS, W. D.: Effects of altitude on carbohydrate content of mountain plants. Ecology **46**, 750–751 (1965).

MUC, M.: Vascular plant production in the sedge meadows of the Truelove Lowland. In: BLISS, L.C. (Ed.): Devon Island IBP project high arctic ecosystem. Project Report 1970 and 1971, pp. 113–145. Edmonton: University of Alberta, Dept. of Botany 1972.

MURNEEK, A.E.: Hemicellulose as a storage carbohydrate in woody plants with special reference to the apple. Plant Physiol. **4**, 251–264 (1929).

MYHR, K., SÆBØ, S.: Verknaden av skygging på vekst, utvikling og kjemisk samansetning hos nokre grasartar. Forskn. Fors. Landbr. **20**, 297–315 (1969).

NELSON, N.: A photometric adaption of the Somogyi method for the determination of glucose. J. Biol. Chem. **153**, 375–380 (1944).

PATTON, A.R., GIESEKER, L.: Seasonal changes in the lignin and cellulose content of some Montana grasses. J. Anim. Sci. **1**, 22–36 (1942).

PHILIPS, M., GOSS, M.J.: Composition of the leaves and stalks of barley at successive stages of growth, with special reference to the formation of lignin. J. Agr. Res. **51**, 301–319 (1935).

PIRT, S.J., WHELAN, W.J.: The determination of starch by acid hydrolysis. J. Sci. Food Agr. **2**, 224–228 (1951).

PREISING, F.A.: Untersuchungen über den Kohlenhydratstoffwechsel des immergrünen Laubblattes im Laufe eines Jahres. Botan. Arch. **30**, 241–306 (1930).

PUEYO, G.: Recherches sur la nature et l'évolution des glucides solubles chez quelques lichens du bassin Parisien. Année Biol. **36**, 117–169 (1960).

RIBEIRO, L.P., MITIDIARI, E., AFFONSO, O.R.: Paper electrophoresis. A review of methods and results. Amsterdam: Elsevier Publ. Comp. 1961.

RODIN, L.E., BAZILEVICH, N.I.: Production and mineral cycling in terrestrial vegetation. Edinburgh-London: Oliver and Boyd 1967.

ROSTOVTSEV, S.A.: Cellulose content and the length of the wood fiber in willow. Les Khoz **17**, 1–22 (1964). In Russian.

RUSSEL, R.S.: Physiological and ecological studies on an arctic vegetation II. Observations on carbon assimilation, carbohydrate storage and stomatal movement in relation to the growth of plants on Jan Mayen Island. J. Ecol. **28**, 289–309 (1940).

RUSSEL, R.S.: The effect of arctic and high mountain climates on the carbohydrate content of *Oxyria digyna* J. Ecol. **36**, 91–95 (1948).

SELSJORD, I.: Kjemiske analyser av beiteplanter. Forskn. Fors. Landbr. **19**, 1–7 (1968).

SKARTVEIT, A., RYDÉN, B.E., KÄRENLAMPI, L.: Climate and hydrology of some Fennoscandian tundra ecosystems. In: WIELGOLASKI, F.E. (Ed.): Fennoscandian tundra ecosystems, Part 1: Plants and microorganisms, pp. 41–53. Berlin-Heidelberg-New York: Springer 1975.

SKRE, O.: Methods and results of carbohydrate measurements on Norwegian alpine plants. In: VIK, R. (Ed.): IBP Norway. Methods and results. Sections PT-UM Grazing project, Hardangervidda. Botanical investigations, pp. 195–235. Oslo: Norwegian National IBP Committee 1975.

SMITH, D.G.: The biology of lichen thalli. Biol. Rev. **37**, 537–570 (1962).

SMYTH, E.: The seasonal cycles of ash, carbohydrate and nitrogenous constituents in the terminal shoots of apple trees and the effect of five vegetatively propagated rootstocks on theme II. Carbohydrate fractions and lignin. J. Pomol. **16**, 185–200 (1938).

SONESSON, M., WIELGOLASKI, F.E., KALLIO, P.: Description of Fennoscandian tundra ecosystems. In: WIELGOLASKI, F.E. (Ed.): Fennoscandian tundra ecosystems, Part 1: Plants and microorganisms, pp. 3–28. Berlin-Heidelberg-New York: Springer 1975.

STODDARD, D.A.: Chemical composition of wheatgrass and its response to the season. Farm and Home Sci. **6**, 5–15 (1945).

SVOBODA, J.: Vascular plant productivity studies of raised beach ridges (semi-polar desert) in the Truelove Lowland. In: BLISS, L.C. (Ed.): Devon Island IBP project, high arctic ecosystem, Project report 1970 and 1971, pp. 146–184. Edmonton: University of Alberta, Dept. of Botany 1972.

THOMAS, J.W., MELIN, C.G., MOORE, L.A.: Determination of sugar in forage plants. Anal. Chem. **21**, 1363–1365 (1949).

WAKSMAN, S.A., STEVENS, K.R.: The chemical composition of peat. I. Chemical nature of organic complexes in peat and methods of analysis. Soil Sci. **26**, 113–137 (1928).

WARREN WILSON, J.: The influence of "midnight sun" conditions on certain diurnal rhytms in *Oxyria digyna*. J. Ecol. **42**, 81–94 (1954).
WEBSTER, J. E., SCHRYOCK, G., COX, P.: The carbohydrate composition of two species of grama grasses. Oklahoma Agr. Exp. Sta. Techn. Bull. **T-104**, 1–16 (1963).
WELLS, S. D.: Cellulose from cereal straws. Industr. & Engineer. Chem. **21**, 275–278 (1929).
WIELGOLASKI, F. E.: Primary productivity of alpine meadow communities. In: WIELGOLASKI, F. E. (Ed.): Fennoscandian tundra ecosystems, Part 1: Plants and microorganisms, pp. 121–128. Berlin-Heidelberg-New York: Springer 1975.
WIELGOLASKI, F. E., KÄRENLAMPI, L.: Plant phenology of Fennoscandian tundra areas. In: WIELGOLASKI, F. E. (Ed.): Fennoscandian tundra ecosystems, Part 1: Plants and microorganisms, pp. 94–102. Berlin-Heidelberg-New York: Springer 1975.
WIELGOLASKI, F. E., KJELVIK, S.: The methodology of net primary production investigations in Norwegian IBP tundra studies. Rep. Kevo Subarctic Res. Sta. **9**, 24–27 (1972).
WIELGOLASKI, F. E., KJELVIK, S.: Production of plants (vascular plants and cryptogams) in alpine tundra, Hardangervidda. In: BLISS, L. C., WIELGOLASKI, F. E. (Eds.): Primary production and production processes, Tundra Biome, pp. 75–86. Edmonton-Oslo: IBP Tundra Biome Steering Committee 1973.
WIELGOLASKI, F. E., KJELVIK, S., KALLIO, P.: Mineral content of tundra and forest tundra plants in Fennoscandia. In: WIELGOLASKI, F. E. (Ed.): Fennoscandian tundra ecosystems, Part 1: Plants and microorganisms, pp. 316–332. Berlin-Heidelberg-New York: Springer 1975.

The Input and Significance of Particulate Terrestrial Organic Carbon in a Subalpine Freshwater Ecosystem

P. LARSSON and K. TANGEN

In a freshwater ecosystem there are four contributors to the pool of primary produced plant material available for animals, fungi and bacteria: phytoplankton, macrophytes, benthic and epiphytic algae, and allochthonous plant material. Both the relative and absolute amounts of these four components are different in various water bodies.

In deep lakes with slow renewal of the water masses the phytoplankton is the predominant primary producer. With decreasing depth of the water body, macrophytes and benthic and epiphytic algae represent an increasing proportion of the total primary production (RICH et al., 1971). Increased water-flow through a freshwater system may reduce the relative contribution from the autochthonous plants compared with allochthonous material.

In alpine and arctic freshwater ecosystems, which are characterized by low autochthonous production, long periods with ice cover and an intense spring spate, allochthonous material may play a relatively more important role than it does in many temperate and lowland water bodies.

In this paper, results from the Norwegian IBP/PF project will be used to suggest the importance of the input of particulate organic material from the surrounding environment, for a subalpine-alpine freshwater ecosystem.

The lake, Øvre Heimdalsvatn, is situated 1090 m a.s.l. in the eastern part of the Jotunheim Mountains in southern Norway. The catchment area of the lake is 24 km^2, and its highest point 1840 m a.s.l. Approximately $^3/_4$ of the catchment area lies in the mid alpine zone, and the upper border of the low alpine zone occurs at about 1250 m a.s.l. In some places along the streams *Salix* spp. (willow) are found above 1250 m.

On the slopes facing south the low alpine zone is very narrow, and the subalpine zone with birch trees extends to about 1200 m a.s.l. On the slopes facing north, however, the low alpine zone goes down almost to the lake, to about 1100 m a.s.l.

The bedrock in the catchment area north of the lake consists mainly of gabbro, south of the lake of gneiss and to the west there is a smaller area of late precambrian-eocambrian sediments. The valley floor is covered by moraine material. The lake has a surface of 0.78 km^2, is 3 km long and has a maximum depth of 13 m.

The theoretical renewal time is relatively short, the yearly average being about 8 weeks. Marked seasonal variations do, however, occur, from about 2 days renewal time in the spring spate to the theoretical value of 400 days during the late winter.

Material and Methods

The data used in this paper have all been collected during 1972, with the exception of the macrophytic data, which are mainly from 1969 (BRETTUM, 1971). Input and output of particulate organic carbon, POC, were measured daily from the end of April to the end of October, and during the rest of the year approximately monthly. Primary production was measured, and POC, phytoplankton, bacterioplankton and zooplankton samples were taken simultanously at intervals of from 5 to 30 days. Sampling was most frequent in May and June and less frequent during the winter months.

Water-flow through the lake has been calculated from continuous water level readings and measurements using a current meter. The amount of POC was determined by water filtering (pre-ignited glass fibre filters). Ash-free dry weight was found by loss on ignition (500° C). The fraction of particles, >1 mm, were collected by nets (mesh-size 0.9 mm), suspended in the outlet and the main inlet. The volume of water filtered was estimated from flow measurements carried out in the net opening.

The input of POC was measured in the main inlet stream of the lake. The POC concentrations in the other inlet streams did not differ significantly from that in the main inlet. Input directly onto the lake surface was collected by traps on the lake, but seems to be of minor importance except for the zone near land. Input and output of POC have been converted to yearly amount per m^2 lake surface in the following way: The daily concentrations in the water were multiplied by the mean water flow that day; all the daily totals in the year were summed up, and the resulting figure divided by the surface area in m^2.

Phytoplankton production was measured by ^{14}C technique (NIELSEN, 1952). Phytoplankton biomass was determined by the inverted microscope method and cell volume calulations (BRETTUM, 1972). The bacterioplankton biomass was determined in stained and filtered water samples (acridine orange and dark membrane filters, pore size 0.45 μm). The bacteria were counted by using an epifluorescence microscope (FRANCISCO, 1970).

Zooplankton numbers have been estimated from the counting of samples filtered through nylon filters with mesh size of 20 μm (LARSSON, 1972). Biomass estimates were based on the numbers and the actual dry weights of single individuals. Production estimates were based on detailed studies of growth and recruitment (LARSSON, in prep.). Phytoplankton and zooplankton biomass and production and bacterioplankton biomass were converted to amounts per surface area according to LARSSON (1972). Particulate organic carbon (POC) ash-free dry weights, phytoplankton volumes, bacterioplankton numbers, zooplankton dry weights and macrophyte dry weights were converted to carbon weights according to WINBERG (1971).

Results and Discussion

The density of POC in the inlet water, the outlet water, the net input of POC per unit volume of water and the water flow through the lake, are shown in Fig. 1. At the beginning of the melting period, in early May, even small increases in water-flow produced marked increases in POC densities in the inlet. Subsequently, more and more water was needed to increase the POC density. When the spring spate culminated, the amount of POC per unit volume of water was reduced to about 0.1–0.5 g ash-free dry weight per m^3, where it seemed to stabilize during the rest of the summer and autumn.

The spring spate is caused by melting of ice and snow in the catchment area, and the melting period ends in the middle of June. The first heavy rain after disappearance of the snow cover from the majority of the catchment also brings large amounts of material into the lake. This is demonstrated during the period 28.6–3.7 (Fig. 1). Subsequently, precipitation does not generally increase input sufficiently to produce a net input to the lake. Thus the main input of POC to the lake comes in the spring spate during May and June.

During the summer there is a net output probably due to autochthonous production and resuspension of the sediment by wind action. The input in the autumn associated with leaf fall seems to be limited, and might be restricted to that falling directly on the lake surface.

On a yearly basis there is a large net input of allochthonous POC into the lake due to the extremely large volume of water flowing through the lake during May and June.

The size distribution of the particles in the inlet water varies during the year, and in the spring spate more large particles were found than in the rest of the year (Fig. 2). In the outlet almost no larger particles (>1 mm) were recorded, even in the spring spate.

The particles flowing into the lake are mainly fragments from terrestrial plants. Aquatic algae and animals form only a minor fraction. The material collected in the drift nets has been analysed. Leaves and stems of vascular plants occurred in approximately equal amounts and constituted more than 90% of the total ash-free dry weight (Table 1). It was found to consist largely of fragments of *Betula nana*, *Betula tortuosa* and *Salix* spp. (willow). Other plants such as mosses and lichens seem to be of minor quantitative significance (Table 1).

The yearly input of POC to the lake, expressed as grams of carbon per square metre lake surface, was in 1972 22 $gC\,m^{-2}\,yr^{-1}$ and 8 $gC\,m^{-2}\,yr^{-1}$ left the lake in the outflowing water. Output includes allochthonous material just going through the lake, material resuspended from the lake sediments and autochthonous material produced in the lake itself. Net input into the lake system during the year is thus more than 14 $gC\,m^{-2}\,yr^{-1}$.

The chemical composition of the sediments has been analysed and the sedimentation rate estimated, and it has been found that about 3 $gC\,m^{-2}\,y^{-1}$ is stored on the bottom of the lake (KLOSTER, pers. comm.). This permanent sedimentation also includes autochthonous-produced particles, and thus more than 11 $gC\,m^{-2}\,yr^{-1}$ of allochthonous POC must be utilized in the lake.

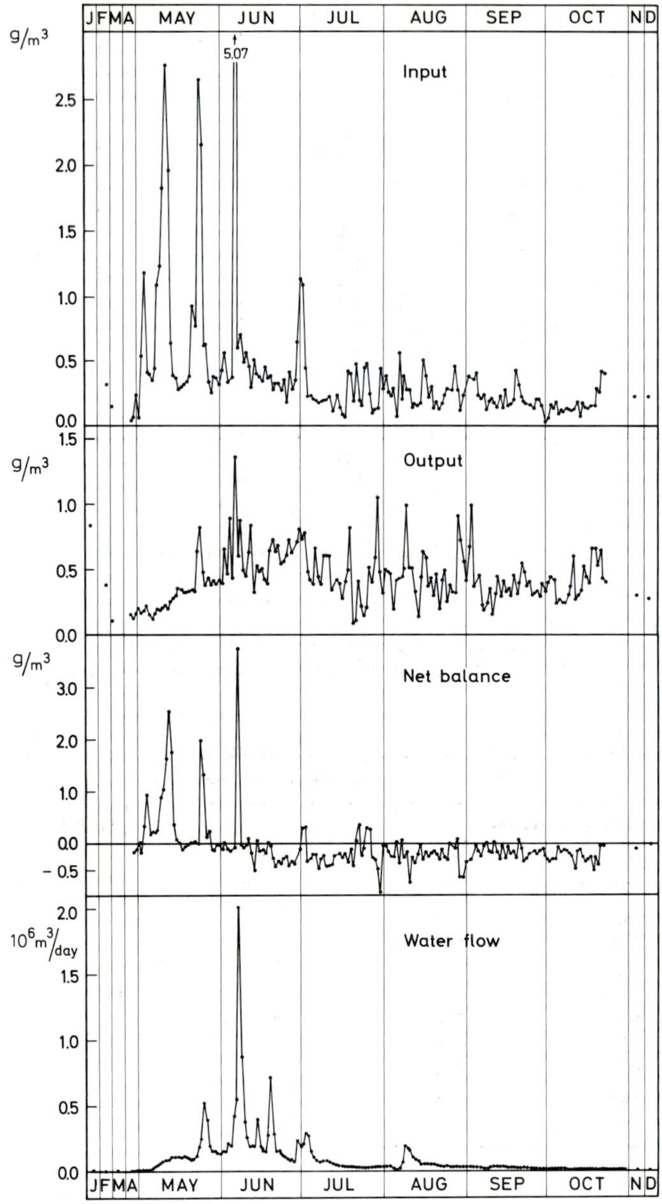

Fig. 1. Input, output and resulting net balance of the Particulate Organic Carbon (POC) concentration in the inlet and the outlet water measured as ash-free dry weight per m^3. The water flow refers to the outflowing water. The peaks in the spring spate until about 10th June are mainly caused by snow melting. The peaks on 15th and 19th June are caused both by melting and precipitation, while the subsequent peaks are caused solely by precipitation

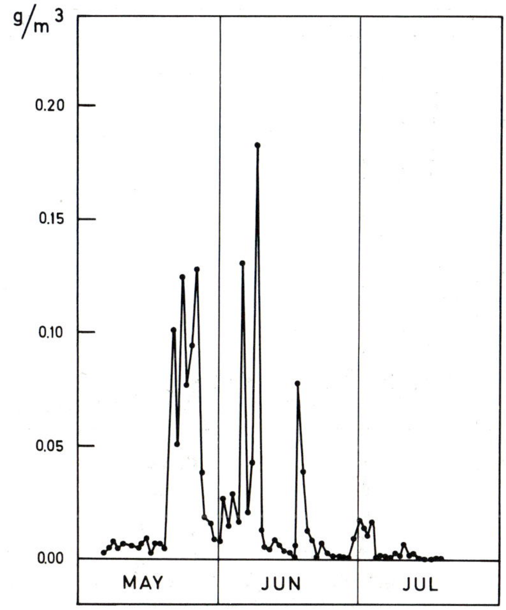

Fig. 2. Input of particles collected in the drift net in the main inlet. The amounts found per day are converted to concentration in terms of ash-free dry weight per m^3 water filtered by the net

Table 1. Percentage in terms of ash-free dry weight of different plant fragments collected in the drift net in the main inlet stream

Date	Vascular plants			Mosses	Lichens
	leaves	stems	roots		
18/5	38.8	61.2	0.0	0.0	0.0
23/5	39.8	56.8	1.5	0.8	1.2
2/5	47.4	47.3	2.7	1.5	1.2
25/5	55.9	40.1	1.4	0.9	1.7
26/5	48.6	48.4	1.0	1.3	0.6
5/6	40.5	55.8	1.4	1.4	0.8
6/6	51.7	41.4	2.1	3.2	1.7
7/6	32.3	65.4	0.7	0.9	0.8
8/6	29.9	67.6	1.1	0.9	0.5

The measurements of phytoplankton production in the lake gave an estimate of 11 gC m^{-2} yr^{-1} for 1972. Photosynthetic material secreted as dissolved organic carbon is not included in this figure, but this fraction is obviously of minor importance (WETZEL et al., 1972).

The main macrophyte in the lake is *Isoëtes lacustris*. The biomass of this perennial species is on average 20 gC m^{-2} in September (BRETTUM, 1971). Probably less than 25% of the biomass is produced per year, i.e. less than 5 gC m^{-2} yr^{-1}. The aquatic moss *Scorpidium scorpioides* found in the western part of the lake

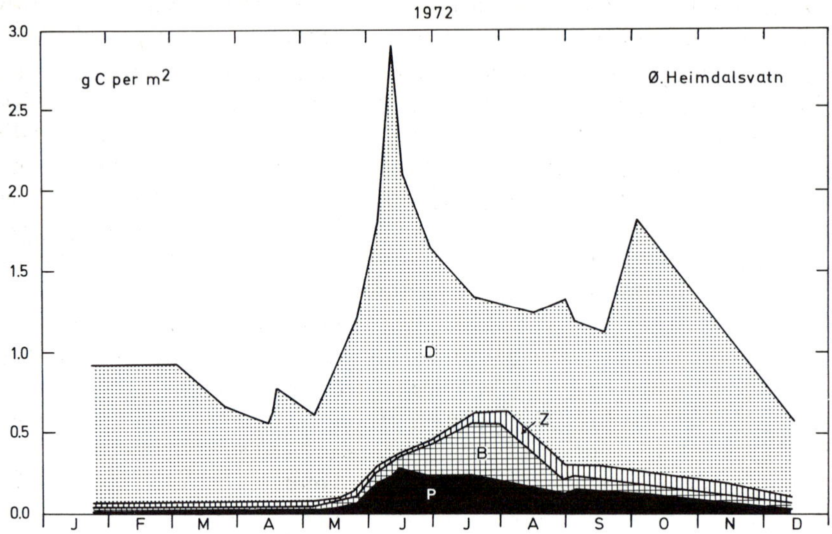

Fig. 3. Succession in the plankton community during 1972. *D* Non-living organic particles, *Z* Zooplankton, *B* Bacteria and *P* Phytoplankton

(BRETTUM, 1971), together with the macrophytes in the littoral zone, are estimated to give a net production not far from 3 gC m^{-2} yr^{-1}. Thus the sum of macrophytic production should be less than 8 gC m^{-2} yr^{-1}.

Production of epiphytic and benthic algae has been measured, and the values which have been found by the shore (MÜLLER, pers. comm.) are of the same magnitude as those found for an average water column of phytoplankton. Since production of benthic algae decreases with increasing depth, the benthic and epiphytic algal production is assumed to be less than that of the phytoplankton.

If the four fractions of primary produced organic material are compared, they are found to be of approximately the same order of magnitude. The most underestimated fraction is obviously allochthonous organic material, and thus it seems clear that this fraction must be the largest. The amount of dissolved organic material has not been measured in Øvre Heimdalsvatn, and in a total carbon budget it might change the picture in some respects (WETZEL et al., 1972). The dissolved organic carbon may, however, be mainly of terrestrial origin, so even if there are considerable amounts of this fraction it will only increase the relative significance of allochthonous organic carbon in the system.

The step from primary produced material to its utilization by animals, fungi and bacteria is not fully understood, neither in Øvre Heimdalsvatn nor in other lakes. Some results, however, indicate the links between the autotrophic and heterotropic organisms. Figure 3 shows the succession in the plankton community of Øvre Heimdalsvatn. Non-living POC predominates throughout the year and is probably mainly of terrestrial origin. Among the living particles, bacteria represent the highest biomass. The large amount of bacteria may indicate that they utilize allochthonous organic carbon to a large extent, but since they probably also live on dissolved organic carbon, there may not be a direct link between bacterial biomass and the non-living POC in the lake.

Among the zooplankton, the filter-feeding *Bosmina longispina* is estimated to give rise to more than half of the zooplankton production in Øvre Heimdalsvatn. Gut analyses of this species gave the impression that it mainly eats the non-living fraction of the POC. Another important zooplankton species in Øvre Heimdalsvatn is *Cyclops scutifer*, which was previously believed to be a strictly carnivorous form. However, studies both by ALIMOV et al. (1972) and PERSSON (1973) give rise to doubts about this, and our own experience indicates that this species at least sometimes lives on non-living organic particles.

Plankton production in lakes with considerable input of allochthonous material can be compared with those with minor input. Lake Latnajaure in Abisko, Sweden an Lake Char on Elsemere Island in Canada are two arctic lakes with little allochthonous input. The yearly phytoplankton production is about 5–8 gC m^{-2} yr^{-1} (NAUWERCK, 1967; BRYLINSKY and MANN, 1973), which is less than that found in Øvre Heimdalsvatn, but of the same order of magnitude. If zooplankton production is compared, there is a production of about 1 gC m^{-2} yr^{-1} in Øvre Heimdalsvatn (LARSSON, in prep.), which is 4 to 40 times more than that found in the two other lakes (NAUWERCK, pers. comm.; RIGLER et al., 1974).

There may be objections to such a comparison, since estimates of zooplankton production have low accuracy and there are differences between the lakes other than input of allochthonous material. However, the differences are so pronounced that allochthonous material probably account for a great part of it. Two Russian lakes near the arctic circle by the White Sea, Krivoe and Krugloe (ALIMOV et al., 1972), both have a primary and a secondary production in the plankton community of the same order of magnitude as Øvre Heimdalsvatn. These lakes are surrounded by trees and must have a considerable input of allochthonous material.

In the benthic community, *Gammarus lacustris* together with several species of chironomid larvae are the most abundant forms in Øvre Heimdalsvatn. Laboratory experiments carried out on *Gammarus lacustris* (BORGSTRØM, pers. comm.) indicate that this species prefers leaves in its diet, especially those of willows. Further, it was frequently found that the chironomids had gut contents which most likely had been dead organic particles prior to consumption. Only benthic diatoms occured as frequently (AAREFJORD, pers. comm.).

Field and laboratory studies (LILLEHAMMER, 1974) have been carried out to determine the importance of allochthonous organic material for the aquatic insect fauna. Attention has been given to both the streams in the catchment area and the lake itself. There was a decrease in the number of stonefly (Plecoptera) species, which consume allochthonous material on the way upstream from the subalpine zone to the low alpine and mid alpine zones. This reduction was paralleled by a reduction in the amount of *Salix* vegetation along the streams. Around the main inflow of Øvre Heimdalsvatn, the number of stoneflies per m^2 is about 30% higher than that around the outflow. Among the species in the vicinity of the inflow, 74% were dependant on allochthonous material, while the corresponding figure for the outflow was 60%. Laboratory feeding studies have shown that herbivorous stoneflies prefer large leaf fragments to smaller organic particles.

Some species discriminate between leaves from different plants, and *Salix* leaves are among those preferred.

The aquatic insects and *Gammarus* may play an important role in the mechanical breakdown of leaves and aquatic marcrophytes, making them available for other groups such as chironomids, zooplankton, fungi and bacteria.

Although most of the allochthonous material comes into the lake during the spring, and probably much of it is utilized together with the autochthonous-produced material during the summer, some still remains for the heterotrophic organisms as a winter supply. Prior to ice cover, the zone near land receives an additional supply in the form of wind-blown leaves. The organisms in this zone have therefore especially good food conditions for growth during the winter.

Winter growth is illustrated by two mayfly species forming about 70% of the total mayfly fauna in this zone, having over half of their growth during ice cover (BRITTAIN, pers. comm.).

The allochthonous plant material is transported into Øvre Heimdalsvatn by the inlet streams in amounts comparable to the autochtonous production, and in the diet of the heterotrophic organisms it seems to be an important component. This, together with a comparison between lakes with differing allochthonous input, leads one to suggest that in freshwater ecosystems at higher altitudes and latitudes, input of allochthonous material may produce an increase in secondary production.

Acknowledgements. We are grateful to our colleagues in the Norwegian IBP/PF group, who generously contributed unpublished data to this paper. These include cand. real A. E. KLOSTER—sediment data and parts of the macrophyte data, Cand. real PÅL BRETTUM—phytoplankton biomass data, fil. lic. C. MÜLLER—periphyton and macrophyte data from shallow water, cand. real. F.R. AAREFJORD—data on lake zoobenthos, cand. real. R. BORGSTRØM—data on *Gammarus* feeding, cand. real A. LILLEHAMMER and DR. J. E. BRITTAIN—insect data from the streams and the lake.

DR. BRITTAIN also helped with comments on the manuscript and in improving the English.

T. ANDERSEN, M. BASBERG, G. BLOCH, B. HOLTH-LARSEN, L. KLAVENESS, M. KRISTIANSEN, B. LARSEN, O. MÆHLUM, and J. NYDAL all gave valuable assistance both in field and in the laboratory. Special thanks go to Professor R. VIK, convener of the Norwegian PF-section, who stimulated us to write this paper and gave valuable advice and criticism regarding the manuscript. The production project on Øvre Heimdalsvatn has been financed as part of the Norwegian contribution to the International Biological Programme.

References

ALIMOV, A. F., BOULLION, V. V., FINOGENOVA, N. P., IVANOVA, M. B., KUZMITSKAYA, N. K., NIKULINA, V. N., OZERETSKOVSKAYA, N. G., ZHAROWA, T. V.: Biological productivity of lakes Krivoe and Kurglow. Productivity problems of freshwaters. In: KAJAK, Z., HILLBRICHT-ILKOVSKA, A. (Eds.): Proceedings of the IBP-UNESCO Symposium on Productivity problems of freshwaters, pp. 39–56. Warszawa-Kraków: Polish scientific publishers 1972.

BRETTUM, P.: Fordeling av biomasse av *Isoëtes lacustris L.* og mosen *Scorpidium scorpioides* (Hedw.) Limpr. i. Øvre Heimdalsvann, et høyfjellsvann i Sør-Norge. (English summary). Blyttia **29**, 1–11 (1971).

BRETTUM, P.: The Phytoplankton of lake Øvre Heimdalsvann, Central South Norway, in 1969–1970. Norw. J. Botany **19**, 79–90 (1972).

BRYLINSKY, M., MANN, K. H.: An analysis of factors governing productivity in lakes and reservoirs. Limnol. Oceanogr. **12**, 606–617 (1973).

FRANCISCO, D.: Glucose and acetate utilization by the natural microbial community in a stratified reservoir. Unpublished Ph. D. Thesis, Department of Environmental Sciences and Engineering, School of Public Health, University of North Carolina, Chapel Hill, N.C. 1970.

LARSSON, P.: Vertical distribution and estimation of standing crop of zooplankton in a mountain lake with fast renewal. Verh. Internat. Verein. Limnol. **18**, 334–342 (1972).

LARSSON, P.: The zooplankton in a subalpine lake (in prep.).

LILLEHAMMER, A.: Norwegian stoneflies, Part II. Distribution and its relationship to environmental factors. Norsk ent. Tidsskr. **21**, 195–250 (1974).

NAUWERCK, A.: Das Latnajaureprojekt. Untersuchung eines fischfreien Sees vor und nach dem Einsatz von Fisch. Rep. Inst. Freshwat. Res. Drottningholm **47**, 56–75 (1967).

NIELSEN, E. S.: The use of radioactive carbon (C^{14}) for measuring organic production in the sea. J. Cons. Int. Explor. Mer. **18**, 117–140 (1952).

PERSSON, G.: Zooplankton. Experiment med gjödsling av sjöar i Kuokkelområdet. Kuokkelprosjektets rapport 2. (mimeographed). Uppsala Universitetet, Limnologiska Institutionen 1973.

RICH, P. H., WETZEL, R. G., THUY, N.: Distribution, production and role of aquatic macrophytes in a southern Michigan marl lake. Freshwat. Biol. **1**, 3–21 (1971).

RIGLER, F. H., ROFF, J. C., MACCALLUM, M. E.: Production of the zooplankton in a polar lake. J. Fish. Res. Bd. Can. **31**, 637–646 (1974).

WETZEL, R. G., RICH, P. H., MILLER, M. C., ALLEN, H. L.: Metabolism of dissolved and particulated detrital carbon in a temperate hard-water lake. In: MELCHIORRI-SANTOLINI, U., HOPTON, J. W. (Eds.): Detritus and its role in aquatic ecosystems. Proceedings of an IBP-UNESCO symposium, pp. 485–523. Mem. Ist. Ital. Idrobiol. **29** suppl. (1972).

WINBERG, G. G.: Symbols, units and conversion factors in studies of freshwater productivity. International Biological Programme. Section PF (Productivity of Freshwaters). London: IBP Central Office, 1971.

Subject Index

Oecologia

In Cooperation with the International Association for Ecology (Intecol)